Pre-contract Studies

Pre-contract Studies

Development Economics, Tendering and Estimating

Third Edition

Allan Ashworth

Blackwell
Publishing

© 2008 by Blackwell Publishing Ltd

Blackwell Publishing editorial offices:
Blackwell Publishing Ltd, 9600 Garsington Road, Oxford OX4 2DQ, UK
    Tel: +44 (0)1865 776868
Blackwell Publishing Inc., 350 Main Street, Malden, MA 02148-5020, USA
    Tel: +1 781 388 8250
Blackwell Publishing Asia Pty Ltd, 550 Swanston Street, Carlton, Victoria 3053, Australia
    Tel: +61 (0)3 8359 1011

The right of the Author to be identified as the Author of this Work has been asserted in
accordance with the Copyright, Designs and Patents Act 1988.

Designations used by companies to distinguish their products are often claimed as trademarks.
All brand names and product names used in this book are trade names, service marks, trademarks or
registered trademarks of their respective owners. The Publisher is not associated with any product or
vendor mentioned in this book.

This publication is designed to provide accurate and authoritative information in regard to the
subject matter covered. It is sold on the understanding that the Publisher is not engaged in
rendering professional services. If professional advice or other expert assistance is required,
the services of a competent professional should be sought.

First edition published 1996 by Addison Wesley Longman Limited
Second edition published 2002 by Blackwell Science Ltd
Third edition published 2008 by Blackwell Publishing Ltd

ISBN: 978-1-4051-7700-9

Library of Congress Cataloging-in-Publication Data

Ashworth, A. (Allan)
    Pre-contract studies: development economics, tendering, and estimating / Allan Ashworth.
— 3rd ed.
            p. cm.
    Includes bibliographical references and index.
    ISBN-13: 978-1-4051-7700-9 (pbk. : alk. paper)
    ISBN-10: 1-4051-7700-4 (pbk. : alk. paper)   1. Construction industry—Finance.
2. Real estate development.   3. Building—Estimates—Great Britain.   I. Title.

    HD9715.A2A84 2008
    690.068'1—dc22

                                                                2007040594

A catalogue record for this title is available from the British Library

Set in 11/13pt Bembo by Graphicraft Limited, Hong Kong
Printed and bound in Singapore by Fabulous Printers Pte Ltd

The publisher's policy is to use permanent paper from mills that operate a sustainable forestry
policy, and which has been manufactured from pulp processed using acid-free and elementary
chlorine-free practices. Furthermore, the publisher ensures that the text paper and cover board
used have met acceptable environmental accreditation standards.

For further information on Blackwell Publishing, visit our website:
www.blackwellpublishing.com

Contents

Preface

This is a book that is primarily concerned with the aspects of the pre-contract phase of building development, excluding the actual design process. It aims to provide the kinds of information and solutions that are pertinent to the development of projects in the construction industry. The various chapters deal with, for example, the development process and the capital appraisal of projects prior to and during their design. It assumes broadly that projects in either the public or private sectors need to consider the financial implications very carefully and very early. This is to avoid abortive work and their associated wasted costs. Too many projects have become far too advanced in their planning process and only then abandoned, sometimes entirely or in other cases for more appropriate proposals. Whilst this is not always due to financial imbalances, there are many examples that can be cited where this has been the case.

This book focuses on the financial aspects of development, whether through capital investment, funding sources, the economics of development, the evaluation of financial data, whole life costing and tendering, estimating and cash flow considerations. Money and profits are keywords, almost above all else, in the development of construction projects whether in the UK or overseas. The final chapter outlines a number of different issues that would have been totally absent from our knowledge and experiences just over 25 years ago. This indicates a construction industry that is not standing still or relying only on traditional and well tried and tested methods and techniques. The book also includes a short bibliography and directions for further reading.

The third edition of this book is aimed at supporting students' learning and understanding on a large range of undergraduate programmes. In this edition, a large number of the charts and graphs have been revised, replaced or introduced to take account of changes that have affected the industry even over a short number of years. These paint a different picture of the industry to that even at the start of the twenty-first century. The industry and its contextual framework never stand still but continue to evolve in an ever developing world. Additional material has been added in various places to reinforce aspects of valuations including references to the recently re-branded REITs (Real Estate Investment Trusts) and the RICS Red Book that is used for valuation practices.

More information has been added about the major property companies and who owns the land in Britain as well as new sections on Conservation, the Private Finance Initiative and Value Management.

I would like to acknowledge the help and assistance of a great many individuals who have provided information and encouragement during the writing of this edition. These are, of course, appropriately referenced in the text. I am personally encouraged by those lecturers and students who have passed on their comments and suggestions to me.

I would like to express my thanks to my wife Margaret, who has helped me in numerous ways during the updating of this text.

Allan Ashworth

Acknowledgements

I am especially grateful to the following organisations who have allowed me permission to reproduce material from their own publications. An appropriate acknowledgement is also provided in the text.

Barclays Bank Review
Building Societies Association
DTZ Money into Property
Hargreaves Lansdown Insurance Brokers
Halifax Bank
Hong Kong and Shanghai Bank Economics Review
Jones Lang LaSalle
Scottish Widows Life
The WM (World Market) Company

I am also grateful for comments from colleagues and students, which I have received on the second edition, and which I hope I have managed to incorporate in the third edition.

Chapter 1
The Development Process

Introduction

The statutory definition of development is defined in the Town and Country Planning Act 1990 as 'the means of carrying out of building, engineering, mining or other operations in, on, over or under land, or, making of any material change in the use of any buildings or other land'. Development can therefore be classified broadly as undertaking construction works, such as building and engineering, or making a material change of use to the land or property.

The first stage in property and construction development is to identify the need for a project. This usually arises from a client or developer wishing to construct a project. They may need to see a return on investment funds or may need to address requirements for housing, health, education or other purposes. Such projects will induce a relevant infrastructure to ensure effective operation and function.

Those wishing to build must consider the trends and patterns in a society. Such factors include demographic trends, especially age trends, changes in lifestyles, the implications of new technology, the importance of fashion and the balance between urban and rural living. The selection of an appropriate site, the choice of design consultants and constructors, and finding sources of finance are three of the early decisions that the client or developer will need to make.

Demand for development

Throughout history there has been a demand for buildings for different uses and purposes, such as housing, industry, commerce, religion, entertainment, leisure, etc. Over a period of time existing buildings decay and require maintenance repairs, adaptation and modernisation or become obsolete (Raftery, 1991). The life cycle of buildings – from inception to construction, use, renewal and demolition – is diverse. The demand for buildings and their upkeep and adaptation is continuous, driven by changes in the

1

social, economic and technological aspects of society. The following developments have occurred to meet the needs and demands of society in achieving expectations:

- General rebuilding after damage from the First and Second World Wars
- Improvements in housing standards and quality
- Schools for an increased number of pupils
- Hospitals and health buildings for the National Health Service created after the Second World War
- Universities to satisfy the increased demand for higher education by young people
- Redevelopment of commercial and retail town and city centres
- Factories and warehouses required for increased automation and new technologies
- Office buildings for the service sector and to accommodate new technologies
- Roads and highways for the motor car
- Multi-storey car parking
- Power stations for the increased demands from all kinds of users
- Airports to meet increased demand for air travel
- Out-of-town shopping malls
- Change of use of existing premises, refurbishment and conservation
- Decay, obsolescence and redevelopment resulting in demolition
- Tourism and leisure developments.

Future emphasis is likely to be directed towards maintaining these assets and future developing trends focus upon changes in lifestyles (single occupancy dwellings, increased leisure opportunities (including travel), changes in employment patterns (shift towards home working, the demise of manufacturing and a corresponding increase in the service sector), demographic shifts (increase in the population age) and the long-term focus on health and education.

A recent report (DTZ, 2006) indicated that both insurance companies and pension funds had buying and selling priorities in terms of their investment needs. These priorities have a profound effect on the demand for development. Selling priorities included prime high street shops, provincial city offices and in-town shopping centres. Buying priorities included central London offices, offices in the area of the M25 motorway and retail parks.

Customer satisfaction

Table 1.1 represents a survey of customer satisfaction (Latham, 1994). The data is based upon attitudes of clients who frequently commission construction projects and have an interest in the ownership and use of property. It compares a number of recognised attributes of buildings with those of the automobile. Some may not recognise the results of such a comparison, but the points made cannot easily be ignored. The Building

Table 1.1   Comparison of construction projects with those of the motor car

	Modern buildings			Motor car
	Domestic	Commercial	Industrial	
Value for money	5	3	4	4
Pleasing to look at	4	3	3	4
Largely free from defects	3	1	2	5
Timely delivery	4	4	4	4
Fit for purpose	4	2	3	5
Guarantee	4	1	1	5
Reasonable running costs	4	2	3	4
Durability	3	2	2	4
Customer delight	3	2	2	5

(5 = good; 1 = poor)
(Source: Latham, 1994)

Research Establishment (BRE) regularly commissions surveys into defects in housing. Although defects are generally reducing in number after the dwellings have been handed over to the client, the number still remains excessively high. This single statistic compares poorly against the number of faults found in a new motor car.

When clients and developers are satisfied with their building projects, it is more likely that they will want to build again or to recommend the process to others. This is good news for the industry and all who are employed by it either directly or indirectly. This information should also help to inform the industry on the standards expected and how these are perceived in practice.

Further studies, such as Egan (1998), have gone one stage further by suggesting a rethinking of the way in which the industry is organised and managed and its cultural and trade practices, from the perspective of a major client or the construction industry. Since the publication of *Constructing the Team* (Latham, 1994), the construction industry has inched forwards largely assisted by the good work of the Construction Industry Board which was set up in response to this report. The Board concluded its work in July 2001, after establishing a number of forward-thinking committees and development work. But there remains much to do.

Site identification

The development of land and buildings is promoted by:

● Property developers who have identified the need for entrepreneurial projects
● Commercial and industrial corporations, developing for their own use
● Public and statutory authorities developing for their own functions

- House builders and housing associations
- Charitable organisations
- Government-sponsored development agencies
- Other private clients.

The most important factor that influences development is the developer's perception of market conditions. This knowledge is important in identifying the type of development, the region in the country in which the development is likely to take place and finding a suitable site that satisfies the developer's own criteria. The developer's ability and knowledge are important in identifying areas of potential growth for development, considering that it will take some time before such a project can become operational. Whilst there are always likely to be risks involved, the developer will use a range of market research techniques in order to identify levels of supply and demand, trends in expectations and measures to stay ahead of possible market competition. The different considerations of sites for different development purposes are described in Chapter 6.

The development process

There are several different ways of describing and presenting the property development process. In its simplest analysis it consists of *design, construction, use* and *disposal*. The process should be looked upon as a cycle of activities, where land, once brought into development use, undergoes these different stages. In prime city centre locations, where sites are more valuable, disposal and demolition are usually only a precursor to further property development. Returning land to its greenfield site status is not a common occurrence in a society that continues to expand its range of activities, with a consequent need to use more of this scarce and limited resource.

The development cycle for a construction project can be best classified into five different stages, and associate phases as shown in Table 1.2. These are not discrete activities and are based on the Royal Institute of British Architects' (RIBA) work stages. The different functions merge into each other as the project moves through its life cycle. Emphasis should always be on securing those developments that best satisfy the criteria identified by the developer or client at the beginning of the development. A brief will have been written that identifies the type and scale of the project, its standard of construction, funding availability, its costs and the date when the project should be available for handing over for occupation. During each of these activities, those involved with the development will have different tasks to perform, in respect of designing, costing, forecasting, planning, organising, motivating, controlling and coordinating. These are some of the roles of the professions involved in managing property and construction, whether it be new build, refurbishment or maintenance. To these activities should also be added research, innovation and improvement of quality and standards.

Table 1.2   RIBA work stages

Stage	Phase
Appraisal	Inception
Strategic briefing	Feasibility
	Viability
Outline proposals	Sketch design
Detailed proposals	Detailed design
Final proposals	
Production information	
Tender documentation	Contract documentation
	Procurement
Tender action	
Mobilisation	
Construction to practical completion	Project planning
	Installation
	Commissioning
After practical completion	
In use	Maintenance
	Repair
	Modification
Demolition	Replacement

(Source: RIBA)

Inception phase

Appraisal

This first stage in the process involves both identifying the client's requirements and the possible constraints on development. The development process is initiated principally by either a project looking for a suitable site, or an available construction site looking for a project. The available site may be an existing building awaiting demolition or redevelopment. The development project may cover the whole spectrum of different building types constructed for the public or private sector, including housing, commercial, industrial, recreational, social or activities as remote as forestry and agriculture. The initiative may come from a developer, the site owner or a client seeking a site for a proposed development. The planning authority too may make recommendations or designate revised land use patterns in an area. Many projects arise from long-term construction development programmes, where clients consider the scheme as a part of the overall objectives of their own organisation. At this stage, studies will be undertaken to enable a client to decide on whether to proceed and if so which procurement route should be selected, on the basis of the outcomes required.

Strategic briefing

The strategic briefing stage, which is done by or on behalf of the client, identifies the key requirements and constraints involved. It identifies the procedures, organisational structure and the type and range of consultants to be used. It is important, during this early part of the process, to consider a range of issues that are going to determine whether the project has any chance of coming to fruition. There is little point in expending large sums of money or time on a project that will never be constructed. The client or the developer will prepare an outline brief of the proposals and the issues that need to be settled. These will include an analysis of the market potential and the costs of the development in very broad terms. Broad alternatives will be considered, such as whether to purchase and adapt an existing building.

It may be necessary to consider whether, for example, any public funding may be available in respect of grants or loans for a project built by the private sector. During this stage it will also be necessary to determine whether the proposed site, if one has been located, is suitable for the project envisaged in terms of its location, topography, aspect, size and ground conditions. It is important to establish as soon as possible whether the project will receive planning consent from the local authority. Where this is not awarded, an appeal may be made, but it should be remembered that this can be an expensive process.

It is also prudent to establish matters of ownership, rights of way and other factors that might affect the whole process of development. When these are established and are acceptable, the legalities of site purchase can be initiated, but generally with the condition that outline planning permission for the project will be forthcoming.

The developer will need to arrange for land transfer and the finance for the development. Two types of finance are generally required. The first is known as short-term finance to cover the costs of the development until the project is disposed of to a client. The second, long-term finance, sometimes referred to as funding for development, covers the costs of owning the development. In the private sector the latter is frequently covered by a mortgage.

Feasibility and viability

During the feasibility and viability phases, the client's or developer's objectives for the project become more clearly established. Clients who are involved in one-off projects often come to their chosen consultants with a broad outline of their aspirations, a sum of money which is often insufficient and a time scale for occupation which is often impossible. Better informed clients, who are those involved in frequent capital development, usually have more realistic expectations of what can and cannot be achieved. The type of project will often determine whom the client or promoter appoints as designer. On building projects, this has traditionally been the architect. However, whilst traditions die hard, the developing role of the building surveyor is increasingly being appointed to oversee smaller works and schemes of refurbishment, on behalf of the client. With the growth of different combinations of procurement, such as design and build or management contracting, clients often now appoint the construction firm direct, choosing an alternative consultant as a main partner in the venture or appointing a project manager to be in overall charge of the project. Such consultants are retained for design and aesthetic considerations or for payment and cost control.

The feasibility phase seeks to determine whether the project is capable of execution in terms of its physical complexities, planning requirements and economics. For example, the available site may be too prohibitive in terms of its size or shape, or the ground conditions may make the proposed structure too costly. Planning authorities may refuse permission for the specific type of project or impose restrictions that limit its overall viability, in terms of, for example, the return on the capital invested. Schemes may be feasible but may not be viable. It is feasible to build a 10 000 bed hospital on a tiny island of 100 inhabitants, but probably not viable.

Design phase

Outline proposals

The various options of choosing a separate designer from the contractor, or using design and build in preference, are described in Chapter 11. Each has its own advantages and disadvantages. The fact that so many different procurement arrangements exist, suggests that none is suitable for all kinds of project.

In any event, it is first necessary to prepare schematic outline proposals for approval, prior to a detailed design. These proposals will need to be accepted by the client or developer in terms of the requirements which have been outlined in the brief. They will also need to be accepted by the relevant planning authorities for their permission and, in terms of funding through the preparation of an initial cost budget. An early estimate of the proposed costs and a developer's budget will be required, prior to the scheme becoming too advanced.

Detailed proposals

As the scheme evolves and receives its various approvals, a number of different specialist consultants will be employed. Some of these may be public relations consultants, particularly where a sensitive scheme such as a new road or public highway, building in the green belt or where a project which is out of character with the locality is being proposed. During the sketch design, the main decisions regarding the project's layout and form and the quality of materials and standards of construction will be agreed. A cost plan of the proposed project will be prepared by the quantity surveyor in order to guide the designers during the later stages of this process. A architect or planning consultant will be responsible for providing a well thought out scheme design in order to secure planning permission. It has been estimated that over 60% of all planning applications are now dealt with by professionals other than architects.

Final proposals

When the scheme has been agreed and approved by the client, further investigations will be undertaken in order to prepare the detailed design. Different solutions to spatial and other design problems will be considered and some of these will require revisions to other aspects of the project which have already been agreed. Each alternative solution must be costed to ensure that the cost plan remains on target. If the client's proposals or developer's budget will be significantly affected as a result of these costings, then proper authority should be sought before proceeding further with the design. It is still not that uncommon in the UK to find projects abandoned. It will also be necessary during this stage to consult firms who supply or install any specialist equipment that may be required.

In the United Kingdom, any change in use of a building requires the approval of the appropriate planning authority. This is normally the local authority. Where planning permission is rejected there is recourse to the Secretary of State at the Department for Transport, Local Government and the Regions (DTLR) who may initiate a planning enquiry. The acquisition of planning permission can be a highly complex and technical activity, needing a detailed knowledge of legislation and government policies as well as local knowledge relating to site of the proposed development. Obtaining planning permission may also involve the developer in additional planning agreements with the local authority, where additional conditions, described as planning gains, are sometimes required before planning permission is granted. These agreements inevitably increase the development costs for the proposed project. In some circumstances, such as alteration or demolition of a protected building, it will be necessary to obtain further approvals, such as listed building consent. It is proposed that a new planning framework will give Parliament more centralised control over major developments.

Clients and developers need increasing certainty about the costs of the proposal in order to input realistic and reliable information into their budgets. When it is known that planning permission will be forthcoming, the plans should then quickly achieve a level of detail in order for the quantity surveyor to provide a detailed estimate of the likely costs of construction. The cost plan will already have been prepared and this will be updated frequently to take into account modifications arising from planning and changes in design.

Production information

The production information is considered in two separate parts. The first part is concerned with providing adequate information of a sufficiency to obtain tenders. The second part includes the balance of information that will be required under the building contract, to complete the information for construction of the works on site.

Tender documentation

The documentation for tendering purposes will be prepared at the end of this process. This will also depend to some extent on the procurement method that has been selected. When the project is approaching the tender stage, the different firms that may be interested in constructing the project should be invited to tender. The long periods of time that elapse reflect the design and planning complexity required for solutions to bespoke designs for construction projects. However, it is also recognised that the process is too long and frequently still compares unfavourably with other countries of western Europe and in different parts of the world. During the latter decades of the twentieth century, considerable effort was made to reduce overall design and construction time periods as well as attempting to reduce the time spent on site by the contractor.

Tender action

Upon the receipt of the contract documentation, the contractors enter their estimating phase, since the awarding of the works of construction is still most frequently done through some form of price competition. Value for money has always been an important issue and added value and best value are now prominent in procurement practices (see Chapter 11). The contractors' bids are evaluated against price as well as other important considerations and a recommendation made to the client on the acceptance of the *overall* best offer.

Construction phase

Mobilisation

This is the award of the building contract to the successful firm and the formal appointment of the contractor. Additional production information is issued to the contractor and arrangements are made to hand over the site to the contractor, for the duration of the contract as defined by the contract period.

Construction to practical completion

This is the stage when the contractor commences the work on site. It is typically referred to as the post-contract period, since it commences once the contract for the construction of the project has been signed and work has started on site. Where the project is on a design and build arrangement or a system of fast track procurement, then this stage may start before the design is finalised, and then run concurrently. Contractors remain critical of the traditional arrangements. They are frequently required to price construction works, which although assumed to be fully designed, are often in reality not so. Throughout this stage, formal written instruction orders are given to the contractor for changes in the design, and valuations of the partially completed works are prepared and agreed for the purpose of interim payment certificates. Contractual disputes all too frequently arise, and all too often, due to misunderstandings or incorrect information being made available to the contractor. The contractor is also sometimes over-ambitious and enters into legal agreements that become impossible to fulfil. These create a right to liquidated damages on the part of the client. Project completion times can last from anything from a few months up to ten years or more. Upon completion, the formal signing over of the project to the responsibility of the client is made.

Occupation phase

After practical completion

One of the main tasks is to ensure that the project can be completed to the specified quality, the calculated costs and within the appropriate time scale. Commitments may have been made to future purchasers or occupiers, who will themselves have prepared their own plans for taking over the property. Anticipated problems need appropriate action to ensure that the project stays on target in respect of time and budget. Changing circumstances may mean that some variations to the scheme need to be instructed, in order to maximise the potential for the finished product. Some factors remain beyond the control of any of the parties involved, but the essence is how effectively and quickly these are resolved. The satisfaction of clients and developers with their completed project, centres around completion on time, at the agreed price and of a quality and standard that has been specified in their original brief. Satisfied clients are likely to recommend the industry to others, and thus offer a great marketing potential to designers and constructors.

In-use

This is the longest phase of the project's life cycle and one that the developers will keep at the forefront of their minds. The immediate aims of development are now satisfied, or so it is hoped, and the project can be used for the purpose of its design and construction. However, no development is complete and the success of others remains open, until occupiers or purchasers who are willing and able to pay the rents or purchase the property can be found. Forecasting the future demand for development projects is difficult owing to the long time lag between inception and completion. The collapse in the need for property, due to sudden changes in the economy can create financial disaster for a developer. The shrewd developer will attempt to make allowances for everything, even the unknown.

During this stage routine maintenance will be necessary. The correct design, selection of materials, proper methods of construction and the correct use of components will help to reduce maintenance problems and their associated costs. A sound understanding of potential problems, based upon feedback from project appraisals in practice, will help to reduce the possible future defects. Defects are often costly and inconvenient and minor problems sometimes require a large amount of remedial work to rectify; sometimes out of all proportion to the actual problem that has arisen. Many projects have only a limited life expectancy before some form of refurbishment or modernisation becomes necessary. The introduction of new technologies also makes previously worthwhile components obsolete. City centre retail outlets have a relatively short life expectancy (usually about 15 years) before some form of extensive refitting is required. Although the shell of buildings may have a relatively long life of up to 100 years or more, and some are able to last for centuries, their respective components wear out and need frequent replacement. Obsolescence is also a factor to consider in respect of component replacement.

Demolition phase

The final stage in a project's life is its eventual disposal, demolition, and a possible new beginning of the life cycle on the same site. Demolition becomes necessary through decay and obsolescence and when no further use can be made of the project (Ashworth, 1999). Some buildings are destroyed by fire, vandalism and explosion, or may become dangerous and require demolition as the only sensible course, perhaps years before the end of their expected lives. Other projects may need to be demolished because they are located in the middle of a redevelopment area. There are relatively few projects that last forever, and become historic monuments. Evidence from past centuries is all too apparent. Whilst life expectancy is attributable to decay, the style of living and the changing needs of space and buildings are constantly evolving to meet new challenges.

Some projects of notoriety become listed buildings. The Secretary of State at the DTLR has powers under the planning acts to compile lists of buildings of some special historic interest. It then becomes difficult to demolish, alter or extend these buildings in any way that would affect their character. Where non-listed buildings are thought to have special historic or architectural interest, then a planning authority may also serve a building preservation notice upon the owner. Although planning regulations are onerous, their aim is to allow development to take place in an orderly fashion. This must be the best policy for society in the long term. Projects that might have taken several years to plan and develop and have then been cherished for decades are finally removed from the urban landscape by demolition. In some cases this is swift, through modern demolition methods, where a lifetime's project can be reduced to a heap of rubble in a matter of minutes.

Environmental impact assessment

Environmental impact assessment was established in the USA as long ago as 1970. Such assessments are now worldwide and a powerful environmental safeguard in the project planning process. The original EC (now EU) Directive 85/337 was adopted in 1985 and since then the individual member states have implemented the Directive through their own regulations. In the UK, production of the resulting environmental impact statements has increased more than tenfold between the early 1980s and the early 1990s. As a result of the Directive, over 500 statements are now prepared annually in the UK. The required contents of a statement are given in Annex III of the Directive and are as follows:

- Description of the project: physical characteristics, production processes carried out, estimates of residues and emissions
- If appropriate, details of alternative sites and the possible effects of using them
- Description of aspects of the environment that are likely to be affected, such as population, fauna, flora, soil, water, air, climatic factors, material assets, architectural and archaeological heritage, landscape and their interrelationship
- Description of the likely effects on the environment of
  - existence of the project
  - use of natural resources
  - emission of pollutants
- Description of measures envisaged to prevent or reduce any adverse effects on the environment
- A non-technical summary of the above
- An indication of any difficulties encountered by the developer in compiling the above information.

In England and Wales the Directive is implemented in the Town and Country Planning (Assessment of Environmental Effects) Regulations 1988 as amended. These are corresponding regulations that apply to Scotland and Northern Ireland. Further guidance is included in *Environmental Assessment: A Guide to the Procedures* (DoE, 1989).

Sustainable development

Effective protection of the environment requires activity on many different fronts. There are four objectives of sustainable development that aim to:

(1) Limit global environment threats, such as global warming
(2) Improve the energy efficiency of buildings
(3) Combat fuel poverty through social action
(4) Provide economic growth through more efficient use of resources, such as reuse, recycling and recovery of waste.

Environmental protection is currently administered through DEFRA (Department for Environment, Food and Rural Affairs) and is concerned with, for example, air quality, contaminated land, noise and nuisance, pollution, radioactivity and waste and recycling.

The concept of sustainable development includes four main strands:

(1) Social progress which recognises the needs of everyone
(2) Effective protection of the environment
(3) Prudent use of natural resources
(4) Maintenance of high and stable levels of economic growth and employment.

The construction of buildings affects the environment in three main ways:

(1) The raw materials used for the manufacture of building materials are considerable. The quarrying of 250–300 million tonnes of material in the UK each year for aggregates, cement and bricks imposes significant environmental costs. Currently about 10–15% of the aggregate used in construction is from recycled or alternative sources. Efficient use of this material can save money, reduce waste for disposal, and reduce energy consumption and pollution from the supply process.
(2) Construction sites are often the cause of local nuisance such as noise, dust, vibration and the pollution of watercourses and groundwater. Good environmental protection is required to reduce costs and nuisance to neighbours and the immediate environment. For example, some 70 million tonnes of construction waste, including clay and subsoil are generated annually. Materials recovery will happen where this is economically viable.
(3) The construction industry uses about six tonnes of material per person per year in the UK. About 20% of this is for infrastructure and over 50% for repairs and maintenance. The use of buildings contributes significantly towards environmental problems, such as global warming.

The business benefits of adopting a more sustainable approach to construction of buildings are already being recognised by many far-sighted clients, designers and constructors.

Listed buildings

Listing began in Britain on 1 January 1950, under the austere post-war Labour government; a surprise to many who believed that conservation and conservatism went hand-in-hand. The UK was not the pioneer in the field, the French had been classifying historic buildings for the previous hundred years. Building preservation orders were introduced by the Town and Country Planning Act 1932.

Historic buildings are a precious and finite asset, and powerful reminders to us of the work and way of life of earlier generations. The richness of this country's architectural heritage plays an influential part in our sense of national and regional identity. The favourite views of England, street, village, town or city, almost certainly contain buildings protected by the process known as *listing*.

Buildings can be listed because of age, rarity, architectural merit, and method of construction. Occasionally English Heritage selects a building because the building has played a part in the life of a famous person, or as the scene for an important event. An interesting group of buildings, such as a model village or a square, may also be listed.

Under the Planning (Listed Buildings and Conservation Areas) Act 1990, the Secretary of State responsible has a statutory duty to compile lists of buildings or groups of buildings in England which are of special architectural or historic interest. Similar provisions exist in Northern Ireland, Scotland and Wales.

Listed buildings are grouped into three categories and classified as:

(1)   Grade I
(2)   Grade II★
(3)   Grade II.

Grade II offers minimal protection, but it is virtually impossible to alter or make any changes at all to a Grade I property. There are about 500 000 individual listed buildings in England, of which about 95% are Grade II listed. Almost all pre–1700 buildings are listed, as are most buildings between 1700 and 1840. English Heritage is carrying out thematic surveys of particular types of buildings with a view to making recommendations for listing. Members of the public may propose a building for consideration. The main purpose of listing is to ensure that adequate care is taken in deciding the future of a building. No changes which affect the architectural or historic character of a listed building can be made without listed building consent. This is in addition to the usual planning permission. English Heritage should always be consulted about proposals affecting Grade I and Grade II★ properties. It is a criminal offence to demolish a listed building, or alter it in such a way as to affect its character, without consent.

A Grade I listed building in a poor state can be an incredibly costly and painful experience to put right. English Heritage funds, which in theory are for anyone, tend to go to buildings that are of national importance and prestige, rather than individual homes. Such buildings in a good state of repair and condition are sought after and expensive.

The principles of selection for the lists are now carried out by the Historic Buildings and Monuments Commission (HBMC) and approved by the Secretary of State. They cover four groups:

(1)   All buildings built before 1700 which survive in anything like their original condition are listed.
(2)   Most buildings between 1700 and 1840 are listed, though selection is necessary.
(3)   Between 1840 and 1914 only buildings of definite quality and character are listed, and the selection is designed to include the principal works of the principal architects. Between 1914 and 1939, selected buildings of high quality are listed.
(4)   A few outstanding buildings erected after 1939.

In choosing buildings, particular attention is given to architectural or planning reasons or as illustrating social and economic history, for example, industrial buildings, railway stations, theatres, town halls, markets, almshouses, etc. Buildings that demonstrate technological innovation or virtuosity, for example, cast iron, prefabrication, or the early use of concrete or those associated with well-known characters or events are also listed.

In addition to compiling lists of buildings of special architectural or historic interest, the Secretary of State is also responsible for compiling a schedule of ancient monuments under the Ancient Monuments and Archaeological Areas Act 1979.

Anyone who wants to demolish a listed building, or to alter or extend one in any way that affects its character, must obtain *listed building consent* from the local planning authority. The procedure is similar to that for obtaining planning permission. It is an offence to demolish, alter or extend a listed building without listed building consent and the penalty can be a fine of unlimited amount or up to twelve months imprisonment, or both.

Some listed buildings enjoy a more favourable position on the payment of Value Added Tax on works than do unlisted buildings. Repairs and alterations to unlisted buildings are subject to VAT at the standard rate, but alterations to listed buildings that are designed as dwellings or used for qualifying residential or non-business charity purposes, together with those that are being converted to such use, are not subject to VAT as long as the work is done by a VAT-registered builder and with listed building consent.

Many churches are of special architectural or historic interest, and are listed as such. But so long as they are used for ecclesiastical purposes they remain generally outside the scope of the listed building controls.

Conservation areas

In 1967, the Civic Amenities Act provided for the identification and designation by local planning authorities of conservation areas. These were defined as *areas of special architectural or historic interest, the character or appearance of which it is desirable to preserve or enhance*. Conservation areas are often centred on listed buildings, but not always.

Home owners want to live in conservation areas because of their cachet, architecture and a perception that property in them is expensive. There are some 17 000 clusters of UK streets that are classified as conservation areas, and the number has increased considerably in recent years. They are designated by local councils because they contain buildings of special interest and historic streetscapes. Authorities as diverse as St Albans in Hertfordshire, with some of Britain's oldest streets, and Lambeth in south London, which has a high proportion of Victorian housing without infilling, have been amongst the most prolific creators of them.

Councils have only limited powers to preserve standards, and therefore effectively rely on the goodwill of residents. Planners issue guidelines (see Table 1.3) but have few enforcement powers if maverick residents ignore them. A small proportion of streets within these are areas designated by section four of the Town and Country Planning Act

Table 1.3   Some rules for conservation areas

Brickwork	No painting, rendering and absolutely no stone cladding.
Doors	Should be traditional, appropriate to the age of the house. No PVCu replacements are allowed.
Windows	Try to maintain the original design, shape and materials.
Extensions	Planning permission is always required. Try to ensure that the new addition matches the original house.
Conservatories	May be refused planning permission if they do not appear sympathetic to the house.
Satellite dishes	Try to locate these in lofts or on the house side. Camouflage them with colour wash.
Frontages	Concrete front gardens for car parking are frowned upon, as are inappropriate front walls.
Burglar alarms	Planners want them hidden under the eaves or painted to match the house.

(Source: St Albans Council)

(1990), which aims to retain or reinstate historically correct features. This gives local councils widespread powers to control types of railing, front doors and even paint colours. Property professionals support conservation areas, not least because they boost house prices. The lack of enforcement can bring disadvantages. Problems frequently occur because of the inconsistency of planning officers.

The relevant Acts of Parliament that are concerned with listed buildings and conservation areas are as follows:

- Town and Country Planning Act 1932
- Historic Buildings and Ancient Monuments Act 1953
- Local Authorities (Historic Buildings) Act 1962
- Civic Amenities Act 1967
- Town and Country Planning Act 1971
- Town and Country Planning (Amendment) Act 1972
- Town and Country Amenities Act 1974
- Ancient Monuments and Archaeological Areas Act 1979
- Local Government, Planning and Land Act 1980
- National Heritage Act 1983
- Finance Act 1984
- Local Government Act 1985
- Housing and Planning Act 1986.

Property investment companies

There are about 100 quoted property companies in Britain. The following FTSE 100 members include British Land, Hammerson, Land Securities, Liberty International and Slough Estates.

British Land is a property investment company based in London that invests in prime, modern properties. Their portfolio is valued at £10.6 billion: the majority is directly owned and managed. The balance is held in joint ventures and partnerships, of which British Land's share is valued at £1.2 billion. British Land's investment approach is to concentrate on the fundamentals of each asset. A key criterion is a property's enduring attraction to occupiers, because of its business suitability, location and efficiency.

The portfolio focuses on areas where the principles of supply and demand are strong over the long term. Some 41% is invested in out-of-town retail properties, including the Meadowhall Shopping Centre, which is one of only six regional centres in the UK, 88 supermarkets and 66 retail warehouses. A further 40% is invested in Central London offices, including Broadgate, the premier City office estate. Its high quality properties, balanced portfolio and long lease profile have secured total compound ungeared property returns of 10.8% per annum over the past 10 years.

Hammerson plc is a leading European property company, with operations in the UK, France and Germany. The group invests in and develops shopping centres, retail parks and prime offices. Its high quality portfolio of around 1.2 million m^2 of retail space and over 270 000 m^2 of prime offices, is valued at £6.3 billion. A hallmark of Hammerson's approach is its objective of working closely with major occupiers to create buildings that provide exciting, vibrant and functional working environments.

In 2006, 73% of its activities were in the UK with the remainder in France and Germany. Investment properties account for 86% and the remainder is invested in property development projects. It has a mixture of interests in shopping centres (57%), offices (29%) and retail parks (14%).

Land Securities has a £14.4 billion investment portfolio, substantial development programme and 1800 employees. It is able to provide commercial accommodation and property services to a wide range of occupiers. For over 60 years it has been the leader in property investment, management and development activity in the UK. Its market focus is retail, London offices and property outsourcing. Its vision is to be recognised as the UK's leading property company.

Land Securities own and manage approximately 6 million m^2 of commerical property and provide property services to more than 2500 private and public sector clients. It all began in 1944 when Harold Samuel, Land Securities' founder and chairman, bought Land Securities Investment Trust Limited, which at the time owned three houses in Kensington together with some government stock. Through the prudent purchase of London property, the acquisition of companies and post-war development activity, by 1969 Land Securities had established itself as the UK's leading property company, a position it still holds today.

Liberty International is a major UK property company, with property investments of over £7 billion, of which regional shopping centres amount to over 85%. The company is engaged in three principal activities. Capital Shopping Centres, the leading company in the UK regional shopping centre industry, owning interests in four major out-of-town centres; Lakeside, Thurrock; MetroCentre, Gateshead. It also has eight major in-town regional centres. It has a substantial development programme, involving both new centres, in Cardiff and Oxford, and extensions to existing centres. It also owns Capital and Counties, a successful retail and commercial property business, increasingly concentrated in Central London, the South-east of England and California, USA. Its investment activities looks to use the substantial capital resources at its disposal to access profitable real estate–related financial market opportunities.

Its aim is to produce outstanding long-term returns for shareholders, through capital and income growth, with a relatively full dividend distribution policy from recurring income. The group focuses on premier property assets, particularly shopping centres and other retail, which have high potential, scarcity value and require active management and creativity.

Slough Estates International is a property investment and development company focused on the provision of Flexible Business Space in Europe and North America. It is the leading provider of flexible business space in Europe. Flexible business space is space on industrial sites or business parks which can be put to multiple uses, such as manufacturing, light industrial, distribution (both small and big-box), research and development, offices and warehousing.

Its headquarters are in the UK but the company has operations in nine countries, serving a highly diversified customer base of over 1700 tenants operating in a wide range of sectors and representing both small and large businesses, from start-ups to global corporations. It has property assets of £5.6 billion, over 4 million m^2 of business space, an annual rent roll of £289 million and the weighted average unexpired lease length is 11.9 years.

In 2006, the quoted property companies were worth in total about £50 billion. During 2006, property prices in general were in the ascendant, although some commentators suggested that their values were not sustainable over the longer period. A recession or a fall in stock market prices can trigger a fall in capital values. These will also reduce when property prices generally begin to fall, as occurred in the early to mid-1990s. The decline in the property asset values of industrial companies and financial institutions can also lead to a decline in investment and lending. The following observations can be made about the property industry:

- Commercial property in the United Kingdom is estimated to be worth about £1000 billion.
- Property is essential to the activities of manufacturing, services and agriculture.
- The total value of commercial property is more than double that of outstanding government gilt-edged stoke, and is worth about one half of the value of the total UK equity market.
- Economic activity directly concerned with commercial property and construction contributes just under 8% of total gross domestic product (GDP). A broader contribution of the sector results from its role as an essential factor of production. This is estimated to be around 10% of GDP.

- The direct contribution of commercial property and construction to GDP is the same size as the UK energy sector, about a quarter of the contribution of the manufacturing sector and about one third of the banking and financial sector.
- Commercial property accounts for one third of total investment in physical assets in the economy which is about the same as investment in plant and machinery.
- Property represents an important asset in the balance sheets of companies, representing an average of 150% of net assets, 30–40% of total assets and 100% of capital.
- Commercial property accounts for about 20% of the total assets of financial institutions.
- Banks and institutional lending to industrial companies is based partly on profit projections but partly on collateral provided by the assets of the firm.

According to *Money into Property* (DTZ, 2006) there was a vast improvement in the financial resources of many property companies in the mid-1990s. This provided a clear indication that economic recovery had started to take place in the property sector. The property share index fell from over 1500 points at the height of the boom, to below 500 at the depths of the recession. By 1993, this had recovered by almost 90% to around 1350 points. By the beginning of 2007 the index was more than double this value. During 2004 property companies outperformed the All Share Index for all companies. This rapid turn around in fortunes was reported to be based upon expectation rather than firm evidence, since many of these companies were still burdened with large debts.

The quoted property companies are currently outperforming the wider equities market. Share prices have benefited from intense corporate activity, the disenchantment with the new economy (technology) stocks and the good performance of the underlying asset class. Property fund managers expect UK real estate to outperform other asset classes at the present time. The total return of property is expected to be higher than either equities and government bonds (DTZ, 2006).

Developers

Property development involves a range of different activities, and by their nature will involve developers with different objectives. All of them are concerned with future projections and expectations. Commercial development is undertaken, for example, often on the basis that prospective clients will continue to have an overwhelming preference to rent floor space rather than as occurs in some European countries to purchase for owner occupation. Property development is a business that works on margins between cost and sale price, sometimes based upon cost and sometimes upon sale price. In the absence of this margin there would be no financial incentive for development.

There are many different arrangements that can be provided for development. However, essentially developers can be broadly classified in two categories. Each may undertake work speculatively based upon market intelligence.

- *The investor developer*: the intention with this kind of development is to retain ownership of the project by the developer. Short-term bridging finance may be required for construction and then a long-term loan, often from one of the institutional sources.
- *The merchant developer*: in this scenario, the project is developed with the use of short-term funding and, when complete, disposed of to an owner or occupier. This sort of development has taxation advantages and these are examined in detail later.

The aims and objectives of property development are wide and diverse. The different types of developer involved each have their own particular needs and desires regarding the project.

Occupiers

Occupiers require buildings that suit their particular needs of occupation. The prime objectives are to provide a building that best serves these personal needs, with benefits achieved from occupation and with less concern for its market valuation. An industrialist, for example, will require premises that allow the production process to be carried out in the most effective, efficient and economic manner. The profits from such a business far outweigh any changes in the market value of the property.

A developer who also intends to become the occupier is able to specify a building that meets personal requirements as closely as possible. In some cases the development may be so inflexible in its design that it cannot be easily utilised by others and its relative market value may therefore be small. Buildings such as schools and hospitals are essentially designed and constructed to suit the functions undertaken, with only limited concern for site values and possible resale opportunities. Where such projects are adapted for other uses, extensive conversion work is often required. It may also be necessary to build such projects on sites that might otherwise have only limited development potential. Occupiers are generally more concerned about spatial arrangements and function rather than the possible long-term investment.

Property companies

The objective of property companies that are concerned with development is almost solely related to the profit potential of the development. The project represents the company's production and in this sense the developer is no different from any other company manufacturing a product for sale. The process is frequently long and the development may be constructed speculatively or for a specific client. The property company identifies types of development that are required in particular locations, finds a site, appoints designers and constructors and often sells to the customer through its own network without the need for an independent agent. In some cases the company may also be the designer and constructor offering an all-in service.

Some property companies may choose to specialise in terms of location, whilst other companies may offer only certain types of property, such as offices, industrial premises or housing. In other cases, the company may provide a package deal arrangement to supply and construct factory made (industrialised) units that have the advantage of being available very quickly. Some will choose to concentrate on a particular process, such as conservation or refurbishment, in order to develop a niche market for their services. The advantages of specialisation enable the company to gain an above average knowledge and expertise.

The financial objectives of property companies remain of paramount importance, and these may be identified in several different ways and achieved through a combination of rental, leases and the sale of the completed development.

Investors

The investor's view of property is similar to that of the property company, that is, for financial gain. However, investors tend to take a longer-term view, expecting both the capital and income to increase over the invested life, which may be several years. The acquisition and disposal of property investments can be costly, and it is therefore necessary to have allowed an investment to gain at least some maturity before converting it into other assets. When investors become involved in the development itself, they expect higher returns. The risks involved in project development are greater than those of a completed building.

Investors generally are cautious, disliking unconventional investments that because of their nature may be unpredictable and difficult to dispose of at some future date. Property that may have an unusual design, is constructed with new methods or materials, that has unconventional lease terms or involves substantial management capability will not be favoured as a potential investment.

Investment companies generally manage a portfolio of investments to spread risk across a number of geographical as well as different industry markets and commodities.

Builders and contractors

A building company may seek to enlarge its range of activities by carrying out development work of a speculative nature. This is often done for taxation and legal reasons through a separate company charged solely with this task. In this case, a company will become involved with the additional risks associated with land purchase, finance acquisition and sales or lettings. In this respect the building firm is largely acting in the same way as a development company, with the added bonus of being able to profit from the building construction operations. When the firm acts only as a building company, then the profits accrued largely result from the activities relating to the construction work being performed. There are additional benefits of combining both development and construction work. This allows the resources of the building contractor – the workforce and expertise – to be retained through, for example, undertaking more development work when contracting work is not available or where prices for it are too competitive. The building developer is therefore the reverse of the property company, which employs a building contractor for a proposed development.

The public sector

The public sector has had a growing involvement with development projects usually as a client. In the 1970s, for example, over 50% of all new building projects originated from the public sector. These projects included, housing, hospitals, education, roads, public utilities, etc. By the early 1990s, this percentage had fallen by a half to about 26% of all new projects. This was partially due to the privatisation of many parts of the public sector. However, the decline was more marked than the figures suggest, since there was an overall decline in new orders for construction work during the recession of the mid-1990s. The introduction of the Private Finance Initiative (PFI) has, in part, helped to rejuvenate the flagging public sector.

Public sector policies are influenced by political ideals and the characteristics of the government in power and control, both nationally and locally. There is also the need for accountability in terms of raising and spending finance and the social needs relating to the community at large. Many public sector development projects would not be undertaken by the private sector, since they are unable to show financial profitability. However, their provision can often demonstrate other non-monetary benefits that accrue from the expended costs, and benefits that are of tangible worth to the country and the community. It is important that a public authority can demonstrate that it has acted lawfully and that all of its dealings are free from even a hint of suspicion of corruption. Public accountability often includes some element of public participation.

Most public works building projects, undertaken either through central or local government or other government controlled agencies, are directly related to the particular interests of the authority or government department. Traditionally much of this work would be commissioned by, for example, a county or borough council. The design might have been undertaken by its own staff or private consultants and for projects other than the smallest, a contractor would be appointed. However, since the early 1980s some local authorities have undertaken development projects, such as the provision of industrial units or business parks in order to attract commerce and industry to their locality. These organisations are, however, different to the private developers in several different ways. They can only undertake work within their powers, otherwise they will be acting *ultra vires*.

Parties involved

The development of construction projects involves a range of different interested parties to the process. These include the following.

Landowners

These include the traditional landowners such as the Crown Estates, the Church Commissioners and the landed aristocracy. Whilst these types of landowner are partly motivated by economics, such as obtaining a return on the capital invested, they are also concerned with the political and social issues concerned with land. Another category of land owners comprises industrial corporations, who use and own land because it is incidental to their production processes. This group also includes farming and agriculture, retailers, etc. The former nationalised industries also fit comfortably into this group. Their principal motives for land ownership are with the use of land, rather than for any financial investment that the land might otherwise provide. A further category of landowners are the financial institutions, where land is seen solely as a means of investment and this remains their prime, if only, reason for owning land. They are the most informed group regarding land and property values.

Landowners can have a major influence over the type of development that might be undertaken. Only the state has a bigger influence through its planning procedures. It can encourage and discourage development and has the powers to prohibit development that does not fit in with the plans produced by the local planning authorities.

Families who own London

The heads of four London families, together with the Crown Estate, control some of London's finest addresses. The Crown Estate alone is worth an estimated £6bn and includes Regent Street, Regent's Park and Kensington Palace Gardens. The Queen does not benefit directly since the profits go to the Exchequer and she and members of her family receive payments from the Civil List in return.

The estates shown in Table 1.4 were established centuries ago, largely through wise purchases and canny marriages. They have endured thanks to Britain's tradition of

Table 1.4   Families who own London

Family	Creation	Worth	London Portfolio
Grosvenor Group	1677	£2.3bn	100 acres in Mayfair; 200 acres in Belgravia
Cadogan Group	1712	£1.9bn	90 acres between Knightsbridge and Albert Bridge
Howard de Walden Estate	1086	£1.5bn	90 acres central Marylebone
Portman Estate	1532	£1.1bn	110 acres south of Marylebone around Portman Square
Crown Estate	1066	£6.0bn	Regent Street, Regent's Park Kensington Palace Gardens

primogeniture, under which the eldest son has traditionally succeeded to the entire estate and not been obliged to share it with siblings. The pattern that grew up in London is unique. No other cities look like London in that they are owned by single large families. Fortunes have fluctuated over time. A previous Duke of Westminster (Grosvenor Estates) was obliged to sell off his Pimlico Estate in the 1950s to pay death duties. However, business is now booming. Pre-tax profits run into the tens of millions each year for each of these groups.

The estates and their supporters claim that London benefits from this collective ownership of some of its most expensive districts. Contrast, for example, the smart appearance of Mayfair and Belgravia, still retained by Grosvenor, with the comparative shabbiness of Pimlico, where there are many shareholders.

The Crown Estates

The Crown Estates is one of the most important landed estates in the United Kingdom. It includes substantial urban, rural and marine interests. It is part of the hereditary possessions of the Sovereign in right of the Crown that is managed by the Crown Estates Commissioners. The net surplus is paid to the Exchequer.

The roots of the Estate go back to the reign of Edward the Confessor but its true origins lie in the Norman conquest in 1066. It has over 300 000 acres of agricultural land in Great Britain, making it the largest agricultural landlord in the United Kingdom. Until the time of George III, who came to the throne in 1760, the reigning sovereign received its rent and profits. In the early 1990s, the Estate achieved a revenue surplus of almost £80 m and its property values were worth £1947 m. Since then these assets have more than doubled. It has a wide ranging and quality property portfolio. In its ownership are more than 1000 listed buildings, with 750 of these located in London. Almost 50% are Grade I listed, compared with the national average of 2%. It has property located in central London and the West End and is one of the capital's largest landowners, with more than 8 million square feet (800 000 m^2) of office space, 2.5 million square feet (250 000 m^2) of retail space and over 1 m square feet (100 000 m^2) of miscellaneous property, including hotels, clubs and residential accommodation.

The Estate does not have borrowing powers, which create both a constraint and a discipline on its activities. It resisted the temptation to invest in Docklands, and in 1990 the start of the decline in property prices introduced a moratorium on development, which was lifted towards the end of the decade.

Statutory bodies

The system of statutory planning control in the United Kingdom is one of the most sophisticated and complex in the world. It is necessary to obtain planning approval for virtually all kinds of development, within the definition described on p. 1. Whilst local authorities are responsible for interpreting the various Acts and planning regulations, the ultimate authority in these matters is vested in the Secretary of State (SoS) at the DTLR. The SoS must approve all the county structure plans, and these then will influence the outcome of any planning application or possible appeal against permission that might be lodged with the authority. The main purpose of planning is either to prevent undesirable development or encourage development. Planning permission discourages the former, whereas perhaps one of the best ways of ensuring that required development will take place is to support the project through some form of financial incentives that are made available particularly in those locations or regions that require the development of industry, commerce or housing. A criticism expressed by those wanting to carry out development is the slowness of the process. Whilst successive Acts have introduced more legislation, including that to comply with wider European union requirements, there has been a desire to speed up the process. This has tended to be achieved through increasing flexibility, but at a cost of increasing the uncertainty of the outcome.

Professional advisers

Because the development process is so complex, and because most developers do not employ the range of skills and expertise that are required, it is necessary to employ a range of different professional advisers to advise on funding, design, costs, construction, letting, etc. These advisers will vary depending upon the type, nature and size of the project being envisaged, and might include some or all of the following:

Architects

Traditionally the architect was the first point of contact with a client who was contemplating the construction of a building. However, this position has been eroded as clients have opted for different procurement relationships to suit their own particular needs, such as design and build, and other similar single point responsibility arrangements. These have, since the end of the 1970s, resulted in clients entering into single arrangements with contractors, and this has shifted the focal point of the project away from the designers.

The architect's function is to determine a proper arrangement of space within the building, its shape, form, type of construction and materials to be used, environmental requirements and aesthetic considerations. This is all now done within the concept of a whole project life approach. Architects prepare the design, obtain planning permission and building regulation approval, prepare detailed drawings and specifications and advise on the appointment of a constructor. They may be involved in the appointment of other consultants. During the construction on site, the architect's duties are largely one of inspection of the works to ensure that they comply with the contract. They do not supervise the contractor on site. The amount of inspection which is required will vary with the type of project. Refurbishment schemes, for example, are likely to require more frequent inspection than new buildings. During the post-contract phase the architect issues instructions and certificates to the contractor.

Surveyors

There are a number of different types of surveyor. The *quantity surveyor* has developed from the role and function of a measurer to that of cost and value consultant. The emphasis of their work has moved from being solely concerned with accounting functions to those involving the financial forecasting of construction projects. The quantity surveyor's role is threefold. First, as a cost consultant at the strategic and conceptual phase of pre-design, both on an initial cost and whole lifetime basis. Second, in preparing tendering and contractual documentation for use by general and specialist contractors. Third, in an accounting role during the construction phase where reports are made for interim payments, financial progress and the control, the adjudication of contractual claims and the preparation and agreement of the final project expenditure. The quantity surveyor may be employed on behalf of the client or the constructor. The role of the quantity surveyor has shifted from one focusing on cost to one now mainly concerned with added value.

Traditionally the *building surveyor's* role was concerned with the maintenance and repair of buildings and in the preparation of survey reports for prospective purchasers and property users. The relative importance now attached to the need for adequate maintenance and repair of building stock has provided building surveyors with many opportunities. Commercial owners now realise that their buildings are a major investment. Building surveyors are also involved with the conversion and renovation of existing buildings and the design of small building projects as well as contributing their expertise to other aspects of building work.

General practice surveyors are employed in four main areas of work: agency, valuations, management and investment. Knowledge and understanding of local property markets and land and property values are the particular attributes of this part of the profession. Valuation is one of the main skill bases, as it is vital to investment work. Valuations may be required for a variety of purposes such as sale, lease, insurance, investment or loans and for a range of different clients, such as developers, purchasers and property owners. General practice surveyors may be involved at the outset of a new development project and are sometimes the client's first point of contact on a proposed development. They also advise the financial institutions on investment in order to yield the best result for their shareholders or members.

Engineers

A wide range of different types of engineer are employed both within and outside the construction industry. On civil engineering projects the design and supervision of the works is undertaken by the *civil engineer*. *Structural engineers* really work in a branch of civil engineering concerned with the analysis of structural capability of buildings and engineering structures.

Building services engineers claim that they are the people who bring a building to life. Services in buildings today include heating, lighting, plumbing, energy supply, telephones, computer systems, fire, security protection, etc. Without these services the building is a shell providing little more than the traditional function of shelter. In modern buildings, 30% of the costs can often be allocated to services and in extreme examples this can rise to above 60%.

Builders and contractors

Builders are engaged in the administrative, commercial, managerial, scientific and technical aspects of building. They are responsible for managing the construction process on site including planning and programming the works, budgeting and costing and ensuring that the standards and quality meet the expectations set out in the contract. Their skills ensure that the project meets all the requirements of the developer or client. The growth of design and build arrangements over recent years is an indicator of the level of satisfaction that they have been able to provide in the industry.

Planners

Planners are employed in the planning departments of local authorities, by many other statutory undertakings and in consultancy. Planners recognise that land is a limited and scarce resource and that the increasing demands placed upon it require an adequate system of allocation of use. Planners need to forecast changes in use, but the pace of change in the twenty-first century makes it difficult to plan for all eventualities. The effects of development are wide-ranging. Consideration needs to be given to the protection of the countryside, the landscape, archaeological features, historic buildings, mineral reserves and water gathering grounds. The planner needs to consider aspects of whether or not a proposed development will cause pollution or increased road traffic, what effects it will have on neighbours and whether it will create even greater demands for further development. *Planning consultants* will advise on how best to obtain planning permission where this is difficult and prepare applications to the planning authority on the developer's behalf.

Others

It may also be necessary to employ a range of other specialists such as tax advisers, accountants, economists, etc. At the preliminary stages of some developments and in some locations it may be necessary to employ archaeologists or soil mechanics firms to establish the nature of the ground conditions of the proposed site. Those with expertise in marketing will need to be employed to supplement the work of estate agents, and when the project is in use some of the above professions will be involved in the management of projects such as shopping mall developments.

Planning and control

Government has wide powers of control over the development of construction works. It seeks to resolve the conflicting demands of industry, commerce, housing, transport, agriculture and recreation by means of a comprehensive statutory system of land use planning and development control. Government's aim is for the maximum use of urban land for new developments, with the intention of protecting the countryside and assisting urban regeneration.

The system of land use planning in Britain involves a centralised structure under the Secretary of State at the DTLR. Strategic planning is primarily the responsibility of the county councils, while the district councils are responsible for local plans and development control, the main housing function and environmental health. The development plan system involves the structure and local plans. The structure plans are prepared by the county planning authorities and require ministerial approval. They set out the broad policies for land use and ways of improving the physical environment. Local plans provide detailed guidance, usually covering about a ten-year period. These are prepared by district planning authorities but must conform with the overall structure plan. The two-tier planning system has been simplified in the non-metropolitan areas by replacing it with a single tier of district development plans.

Before a building can be constructed, application for planning permission must first have been obtained. This is made to the local authority in the required form. In the first instance, outline approval would be sought to avoid the expense of a detailed design which could fail to secure approval. Full planning permission must still be obtained. If a scheme fails to obtain approval then a planning appeal can be made to the Secretary of State. The Building Act was introduced onto the statute book in 1984. The 1985 Building Regulations are framed within this Act and allow two administrative systems to be applied; one through the local authority building control department and the other via certification.

Such controls are necessary to do the following:

- Secure improved standards of design and construction
- Ensure the safety and health of the occupants
- Provide for the proper location of buildings and industry
- Make the best use of the land which is available
- Provide for the safety, health and welfare of those engaged in the construction process and those affected by it.

Marketing

Where the development project is to be disposed of after completion, then its promotion and marketing become an integral part of the development process.

A clear understanding of what is being provided and a knowledge and understanding of the potential market for the project are essential. The time lag between inception and completion is considerable. Many projects have been quickly started on site in a time of relative prosperity and could have been disposed of many times. But sometimes by the time of completion, perhaps five years later, the market has deteriorated or the needs have been switched to other types of property. It is then that those responsible for its disposal may have to work very hard to secure potential purchasers. A clear view, supported by data analysis of the past and projections of the future are important at the outset and need to be monitored during development. Where the project allows, it may be possible to switch a design for say offices to flats, where demand for the former declines and for the latter increases. It is generally not easy to do this unless the design is flexible, perhaps just a shell for fitting out later, and planning permission for a change of use would be granted.

What to build

The first stage that a developer will need to consider is just what type of project should be envisaged. Where a scheme can be designed that might usefully serve a range of different client types, then this clearly has some advantages, if planning permission can be obtained. The developer will need to consider the other types of property in the location where the development is envisaged, particularly those that are in direct competition to the proposal. The decision involves an assessment of costs and income. It will be necessary to establish through market research if there is an unsatisfied demand in an area for a particular type of project. The relative shortage of types of property can generally be identified by agents.

Where to build

Developers need to construct projects in areas where there is the expected demand. Some developers work on a national basis. Others confine themselves to a close geographical area with which they are familiar. They understand the regional planning framework and have built up close working relationships with designers and contractors. As described earlier, the where to build scenario depends upon sites looking for projects and projects looking for sites. A developer will have an interest in both of these options.

How to build

In essence, the principle today must be in adaptability and durability resulting in estab-lishment of the principle of *long life, loose fit* and *low energy*, sometimes referred to as the *3 Ls* concept. A building's life expectancy may be up to 100 years. During that time, there are likely to be many changes in ownership or tenancy and even more so in technological developments. The property may also undergo a material change of use during its lifetime. The managing agents for a development project will be able to offer advice on letting, supervision of repairs, rent reviews, etc. and how these factors are likely to have an impact upon the designing of the development. There is currently an emphasis upon low costs in use; certainly where this approach is adopted as a design principle, it will reduce the occupants' ongoing costs. This factor will then have the likely effect of enhancing the rental values of the property. Generally higher value locations attract a kind of occupier who will expect commensurately high standards, with consequently higher prices. For example, new office projects constructed around London are now unlikely to find purchasers unless they are equipped with air condi-tioning systems, computing and telecommunications networks.

When to build

It is desirable to build when construction costs are at their lowest. This implies a general scarcity of construction work and consequently a better availability of skilled craftsmen. In times of recession, there is also a better chance of obtaining a better quality project at a less expensive price. Taking into account that projects may take at least two to three years to bring them to fruition and, assuming that a recession will not last forever, then this seems to be a good time to develop. During a recession, finance charges may also be lower. All development, however, relies upon an analysis of forecasted trends in activ-ities and it is essential that projects are available to meet these trends otherwise the opportunity is lost. In some countries booms and slumps in construction activity are kept to a minimum by the public sector only building when prices are lower in times of a recession.

One of the secrets of effective property development is anticipating the building needs of tomorrow and the future. The developer must be able to anticipate when pro-jects will be required for occupation and be able to assess long-term trends and then backdate these to the development process. Towards the end of the 1980s, there seemed to be an insatiable appetite for high quality office accommodation in and around London, with needs mirrored elsewhere throughout the United Kingdom. There seemed no end to the possible demand for such types of property, with site values and rents rising accordingly. The square mile of the City, with the Bank of England at the centre was then bursting at the seams. Many offices were old-fashioned, too small and most important did not have the space that was required to install the maze of cables for computer terminals. These were now the tools of the financial services industry. However, by the beginning of the 1990s, the country and world markets headed into one of the worst recessions that had been seen that century and which was to last for more years than many analysts predicted. Later, forecasters predicted that any upturn in the economy would only result in moderate levels of economic activity for the foreseeable future. Consequently new development proposals were curtailed, but in their wake several well-known development companies ceased to trade, most notably Olympia and York, the developers of Canary Wharf in London's Docklands. This firm of developers were accompanied by others; building contractors and even professional practices either merged or also ceased to trade. This emphasises the importance of taking the longer term view and accepting that slump follows boom as night follows day.

By the end of the twentieth century, a new government with new ideas had come into power in Britain and during its first period of office more stable and sustained growth occurred. This suited the construction and property industries, with high levels of employment, new ideas identified by Latham (1994) and Egan (1998), increasing property prices and good fundamental returns to those involved in property and construction. These were more marked in the south than in the north, although there were pockets of rapidly rising prices throughout the country.

Chapter 2
Property Investment Economics

Introduction

Property investment economics is a branch of the subject and study of general economics. It uses the various techniques and expertise that have been developed to study the subject of economics associated with property. The basic problem facing all types of economics is that there is an insufficient supply of resources such as land, capital, etc. to satisfy all individual and collective needs. The scarcity of these resources requires individuals to make choices. Policy makers face the difficult choices of deciding to what degree economic efficiency in one area of activity should be sacrificed or expanded at the expense of another. In order to do this it is necessary to investigate the inequalities that are present in any system of activity. These choices exist in the construction and property industries. For example, a site for development may have several different competing uses, only one of which can be satisfied. Investors have a limited amount of capital for investment and therefore need to choose against their own criteria which projects should be supported. In times of expansion, there may be insufficient capital and other resources available to undertake all projects. In times of depression, it may be necessary to entice the use of these resources from the economic activities in other markets.

Money may also be used for purposes other than investment. It may be used to purchase articles for use, for consumption or for pleasure. This chapter is not concerned with these aspects, but only with aspects of investment.

Construction and property industries

Whilst construction and property, as industries, are intrinsically linked together, they are nevertheless separate activities. They are both concerned with adding to the quantity and quality of building stock of residential, commercial, industrial and other types of buildings. The skills of construction are chiefly concerned with the design and erection of buildings. These may be new build or refurbishment. The people involved include architects, surveyors, engineers and builders. The skills required for the property industry are more in the province of the general practice surveyor and investment and financial analysts. Whilst both industries are geared to profit making, one achieves this primarily through the construction of buildings and other structures and the other through the development and letting of property assets.

Principles of investment

The principal objective of any investment is to maximise returns whilst minimising possible risk. In practice, however, investments that incur low risks frequently produce low returns. The building society account is typical of this. Comparison with any investment over time will indicate that money invested in any building society account offers high security but a low financial return. Alternatively, money that is invested in some unknown and highly speculative venture will be high in terms of the risks involved but with a wide variation in the possible return, varying from nothing to almost infinite. In these circumstances there is also the very real risk of the loss of the capital invested. In practice, a majority of investors are looking for some compromise situation. Some investors will be prepared to accept some risk for a higher possible return than might be usual. This same principle applies to investments in property. High-risk investments may result in high gains or losses. Conversely, a low risk will never produce high returns for the investor.

The basic aim of the investment analyst is to asses the value of investments to enable the best decisions to be made. Since investments depend upon the prediction of risk and return, these are the principal problems associated with investment analysis.

Investment is the giving up of a capital sum now in exchange for benefits to be received at some time in the future. These benefits may be represented by income received from the investment, an increase on the capital amount invested or a combination of both. Investment today may also enable expenditure in the future to be reduced. Since property is only one of the possible options for investment, it is necessary to consider not just the various types of property that are available, but also the possibility of investing in other commodities that may offer better returns. Property is often more correctly described as landed property and relates to freehold or leasehold interests in land and buildings. The different types of possible investments competing with property include, for example, the following:

- Stocks and shares
- Debentures, i.e. loans to companies and local authorities
- Insurance policies and bonds
- Unit trusts and investment trusts
- Works of art, antiques, etc.

- Banks and building societies
- Gilt-edged securities, i.e. government loans and bonds.

Table 2.1 indicates that the arguments for investing in the UK stock markets are powerful ones. Today's investment conditions of low inflation, sustainable economic growth and financial stability are the foundations for long-term financial growth. Table 2.1 draws startling comparisons between the performance of savings accounts and inflation with that of equities. Over the longer term there is really no comparison. Even at the end of 2001, after two years of poor performance of equities, fund managers were suggesting that investors should sit tight and not panic. Table 2.2 identifies a number of landmark events which may have had some effect on investment performance.

Table 2.1   Investment returns (£) 1945–2000

	Index 1945	Dividends reinvested	Index 2000
Equity index	100	yes	97 023
Equity index	100	no	7727
Gilt index	100	yes	3296
Retail price index	100	yes	2348
Building society index	100	yes	1471

Note: None of the above are straight line calculations. They each to a varying degree accept the ups and down of the market place. Some of the above investments can fluctuate widely and different investments perform at different rates at different times. (Source: Scottish Widows and Barclay's Bank)

Table 2.2   Impact of military attacks or threats to the USA on the Dow Jones index

	Percentage movement			
	Initial reaction	One month later	Three months later	Six months later
Pearl Harbor	−6.5	3.8	2.9	−9.6
Korean War	−12.0	9.1	15.3	19.2
Cuban Missile crisis	−9.4	15.1	21.3	28.7
US bombing of Cambodia	−14.4	9.9	20.3	20.7
US bombing of Libya	−2.6	−4.3	−4.1	−1.0
Gulf War	−4.3	17.0	19.8	18.7

(Source: Ned Davis Research)

At the time of writing (2007), the current volatility of world events makes investing a white-knuckle ride. However, history has demonstrated the benefits of long-term investment. The only caution to this is in purchasing investments at the top of an economic cycle. Reactions to the attack on the World Trade Center atrocity, for example, caused stocks and shares to fall dramatically. The world stock markets' performance have very often recovered rapidly in the aftermath of previous events as shown in Table 2.2. The initial reaction to the events listed, in some cases, lasted only days or weeks. The economic effects of 9/11 have, however, been much longer lasting. These have in part been affected by concerns from other world events, such as war in Iraq and Afghanistan.

Functions and features of the investment market

One of the most important functions of an investment market is to bridge the gap between income and expenditure. There are individuals and institutions that have a surplus of capital or income above their rates of expenditure. For others, their expenditure patterns are above their current income and capital capability. The function of the investment institutions is to collect the surplus funds that are available, by way of offering investors a return above the capital loaned. These funds are then made available to others in need of loans, at rates that are in excess of the investor's returns. The investment market functions to bring the supply of existing and new financial claims into some sort of equilibrium to meet the demands that are required.

Where the supply of an investment at a particular time is greater than the demand for its quoted price, then its price will fall and the yield will rise as a consequence. This may in turn stimulate an increased demand, because of its increased yield. It may also restrict the availability of such investments since this may deter existing investors from selling. The process continues until the supply and demand reach a state of equilibrium. Where there is economic uncertainty, then there may be a tendency to retain surplus capital rather than to risk investing it in a market that may be spiralling downwards. The effects of this result in depressed prices and enhanced interest rates. The level of interest rates is also influenced by the decisions made by the Bank of England.

The investor's choice of possible investments should be those with which the investor feels comfortable. The different sorts of investment will differ widely and it is necessary to consider their different attributes. These can be summarised as follows:

- Security of capital
- Interest
- Capital growth
- Access.

Security of capital

This is one of the most important features and distinguishes investments generally from gambling. In the latter, there is no capital security, in fact, just the opposite. The majority of investors will want to place their capital where there is the strong possibility that they will be able to recover at least the amount invested should all other things fail. The greater this possibility, the greater will be the willingness to invest. However, there are many types of investment where the capital may also be at risk, albeit a calculated risk. Stocks and shares are examples of this.

Interest

It needs to be remembered that investors are giving up the use of their capital today and that capital will be worth less tomorrow. In order to persuade investors to do this it is necessary to reward them with a higher amount than that which is originally invested. The amount of interest payable on the capital will depend upon current rates of borrowing, the length of time that the capital is invested and the amount of capital concerned. Larger amounts generate higher rates of interest as do longer periods of investment. Some consideration may also be given to the payment of interest. This may be monthly, annually or when the investment reaches its maturity date.

Capital growth

In addition to interest received, the investor will also want the capital sum invested to grow in value. Property, like many other investments, is able to show capital growth over long periods of time, in addition to any rents received. During periods of general economic prosperity with boom conditions in the construction and property industries, capital growth in property has often far outstretched that of other investments. There have been few examples of the principle of negative equity as occurred in the mid-1990s throughout the UK (see Fig. 3.3). However, since that time, virtually all property prices have recovered and in many cases have surged ahead of competing investments.

Access

Investments that are easy to purchase and sell are more attractive to investors. Those that are cumbersome, take time or cost money are less attractive and are required to show higher returns if they are to attract potential investors. Where capital has, because of the nature of the investment, to be invested for a number of years to make it worthwhile, then the consideration of this aspect will also require higher financial returns to be offered. Where immediate access to the capital is required, then the options available to the potential investor are more limited. Where there are high costs involved in converting the investment back into liquid capital then this will also require further monetary consideration to make it worthwhile. Property, as an investment, is not an easily accessible commodity.

Most investments should also be considered as long-term ventures. Quick, short-term investing which is highly speculative and a risky business is not recommended.

Investment in property

Investment in property has traditionally, and since the 1970s, been seen as a good hedge against inflation, and therefore a good, sound investment. It may not, of course, represent the best investment that is available. However, like all investments, performance in the past is not necessarily a guide of its possible performance in the future values. Investment in property is also much less volatile than investments in, for example, stocks and shares, where prices rise and fall daily. However, there are some disadvantages associated with property investments as already outlined. The overall risk attached to investment in property will be a reflection of these attributes, and these should be compared with the other possible options that are available for investment. Generally, the more risk that is associated with an investment the higher the yield and return that might be expected. The risk, of course, might become a reality and outweigh any possible returns. Different investors will assess the risks involved in different ways.

Good property investments have security, produce a return and compare favourably with other types of investments. Good quality property, in desirable locations, has a high demand, with little likelihood of possible risks. This is often the case even in times of a recession within an economy. Where for example, leases are not renewed and the property becomes vacant, then this type of property is quickly able to find new owners or occupants. However, property is not easy to dispose of at short notice. In some cases it may take several months or even years to find a potential purchaser who is able to proceed before the owner can realise the capital asset.

Institutional investors are able to identify those factors that return good yields and in many cases will specify the standards of property that they are willing to purchase from developers. While owning the building they will have little interest in its operation, other than to keep it in a good state of repair. In this way they are seeking to protect their investment. This is also facilitated by the fully repairing lease, whereby the tenant agrees to repair and maintain the building throughout the life of the lease. Institutional investors' aims and objectives are shown in Table 2.3.

Investment in property may be described as *direct*, in which the property is purchased by the owner for occupation and use. Investment in property of this type requires a careful consideration, since a wrong decision in respect of design or location, for example, cannot easily be rectified at some later date. Direct property investment is really a spin-off from ownership and use. Any investment that accrues is therefore of a secondary interest. Alternatively, property may be described as *indirect*, in which case

Table 2.3   Institutional investors' aims and objectives

- High yield
- High retained value of the property
- Low management costs
- Secure tenants
- Blue chip tenant companies
- Happy tenants
- Long leases
- Frequent upward only rent reviews

the interest in the property is solely in respect of the potential financial gain associated with such an interest.

The property sector can be separately analysed into a number of different subgroups according to function, type, size, location, etc. A direct investor would include, for example, an industrial manufacturer. Such a firm would need to define as closely as possible its needs in terms of its property ownership. However, it considers its interest in property largely as a means of production instead of as an investment. Other groups, such as pension and insurance fund managers will be largely indifferent to many of the property characteristics associated with their use, being concerned only with the investment potential of the property. There are also other circumstances such as sale and leaseback, where an owner occupier may wish to raise capital on a project but still retain its use. In these circumstances, an investor will purchase the property and then immediately lease it back to the original owner on agreed terms. This type of arrangement may be preferable to raising the required capital in some other way, often at much higher interest rates. The disadvantages are the loss of ownership and the periodic rising of the rental in the lease terms. An alternative to sale and leaseback is to mortgage the property, the mortgagee having an indirect interest in the property. The purchase of property bonds or purchasing shares in property companies are other ways of having an indirect interest in property.

The amount of yield or rate of interest received from an investment will vary according to some of the reasons described below. Inflation and taxation will also have some influence on interest rates. Changes in taxation structures, especially capital gains tax and the availability of grants or loans will influence possible investors in property. Over a period, like all other forms of investment, prime property will show comparative yields to other commodities in the market. At different times, the returns on prime property investments will considerably outstrip investments of other types. Prime property enjoys those characteristics that are frequently most sought after and are in high demand, even in times of a general recession. The valuation of prime high street property, for example, is typically 20 times the annual rental income. For example, if the Gateshead Metro shopping centre on Tyneside was valued on this basis, although it would not be classed as a prime site, then its annual rental of £20 million could make it worth as much as £400 million (February 1995). In 1993, it was actually valued at £250 million, since when of course its valuation would have increased to in excess of £500 million at the turn of the century. Today it could well be worth £1 bn.

Value

Value is a subjective term. It is interpreted by different people in different ways. It is, like beauty, in the eye of the beholder. Expert valuers should assess value on a similar basis to each other, although their conclusions are unlikely to be identical, since it is to some degree a subjective assessment, even within a carefully designed set of rules. In the economic theory of value, an object must be scarce relative to demand to have a value. Where there is an abundance of a particular object and only a limited demand for it, then using economic criteria, it may have little or no value attributes. Value therefore is a measure of supply and demand. In order to increase the value of an object or commodity then either its demand must be increased or its supply reduced. Where a small change in price creates a large change in demand, then demand is said to be *elastic*. Where a large change in price leaves the demand virtually unchanged, then the demand is *inelastic*. The elasticity of demand is influenced by the availability of alternative substitutes. There are also the short-term and long-term requirements between supply and demand to be considered.

The market value of a property is defined as the money obtainable from some other person, who is both willing and able to purchase the property if it was offered for sale. A property that is to be put up for sale will have its market value assessed by a valuer. This is only an assessment or estimate of its value. Its real value represents the actual selling price of the property that is paid by a purchaser. The supply of land is relatively fixed, in that no more can be created. The supply of property is also fixed in the short term, due to development lead-in times, but as a whole and over a long period of time may be variable, as new buildings are constructed, existing ones altered and changes in use are made from one classification to another. The market value of a specific property is influenced by the amount of similar property coming onto the market, together with the demand for that property. These considerations are of more importance than the entirety of a type of property that might be available. The demand for any property is influenced by the sorts of changes shown in Table 2.4.

Table 2.4   Reasons for the demand for property

• economic prosperity	• interest rates and rental values
• population mobility	• technology and innovation
• local amenities	• trends in population growth and demography
• type of society	• transport and communications
• standards of living	• planning regulations and restrictions
• money supply	• fashion in respect of property and its location

Where a whole range of market prices of any product can be studied, then a pattern of consumer choice can be observed. In terms of the housing market, the fact that similar houses in different locations are priced differently, indicates that occupiers are prepared to pay different amounts to live in different locations. They also place different values on property according to a range of different attributes. The examination of range of prices paid for houses allows comparisons and predictions of values to be made by identifying the attributes concerned. Huge amounts of transactions data are available to guide both buyers and sellers.

Major factors affecting property values

There are many factors that affect the value of property and are therefore considered by those wishing to invest in property. These are shown in Table 2.5.

Table 2.5   Major factors affecting property values

• International issues
• National scenarios
• Government policies
• Local economy
• Geography
• Fashion and demand

International issues

The effects of world events have an effect on the UK economy, UK investments and hence UK property prices. Investment is thus now a global concept. World events might include wars or rumours of wars, famine or changes in trading arrangements, such as the effects of oil pricing as was experienced in the 1970s. The shifting of world trade from one country to another will also affect property prices. For example, the demise of the textile industry in the UK in the 1950s created a glut of cotton mills. Their values before this shift in the location of manufacture and hence their decline, were several times the amounts of their actual selling prices in the 1960s. In many cases, their final values represented little more than their site value. In other circumstances, new uses were found for the buildings, in an attempt to at least recoup some of their lost values. The City of London, at the present time, is one of the world's major financial centres and this influences the rents that can be charged on property and hence their investment capability. Forecasters are predicting that by 2025, the USA and Europe will no longer be the epicentre of the developed world, being replaced by those countries surrounding the Pacific Rim and the influences of China. If this prediction runs true, then this will have a disparate effect on property prices in the various countries concerned. Like all long-term forecasts, only an element of the truth will be realised.

National scenarios

All countries have an economic or business cycle. In some countries these are more pronounced than others. This cycle is discussed more fully later in this chapter. Where economic indicators show positive trends then investors become more confident. Trends in manufacturing output, unemployment statistics, bank lending, interest rates and inflation are all indicators that are used to measure a country's performance. House sales and house prices have in the past provided a prime indicator of future economic patterns in the UK. The different indicators must be considered together, since in isolation they may represent only a limited forecast. Within the overall economic analysis it is also important to consider separately the forecasts that are of particular relevance to property investment. An examination of these economic indicators is considered in Chapter 7.

Government policies

The policies outlined by government are of special importance to the investor. A government that is sympathetic to the needs of investors will encourage investment to take place. In respect of property, government may also provide different kinds of financial incentives and awards in an attempt to ensure that its policies are achieved, particularly where circumstances would otherwise discourage development from taking place. These incentives are examined further in Chapter 8. Perhaps one of the biggest discouragements to a developer is the uncertainty of government policies which then creates a lack of confidence. This becomes particularly acute at election time, where the possibility of a change of government and hence policies may result in a paradigm change of policy direction for a developer in either the construction or property industries. The stop–go nature of the British economy in the latter part of the twentieth and early years of the twenty-first centuries characterised a boom followed by slump. This has had a detrimental effect on the fortunes of both the construction and property industries.

Local economy

Whilst the performance of the country generally is important to developers, the individual regions around the country also have an important effect upon development projects. These might include market led projects such as the commercial developments of the late 1980s. They will also include socially engineered projects, such as the rejuvenation of industrial areas that have fallen into decline because industrial processes are no longer required in their present form. This decline continues to occur as a shift between the manufacturing and service sector employment continues to take place. Developers will be particularly concerned with the return on their investments under these circumstances. In some regions and locations, projects will be successful, whereas in other areas similar projects will remain unsuccessful.

Geography

The various factors associated with development sites are described in Chapter 6. In addition, it will also be necessary to consider the location of the site in respect of its position in the country, and whether the development site is in the countryside or in an urban area. The existence of rivers and other waterways that might cause flooding have been a cause for concern since the major floods that occurred at the start of the twenty-first century. Other matters that might have a positive or negative effect upon the land and its buildings and hence their value will also need to be carefully investigated. It will be necessary to consider aspect and other climatic conditions, particularly where the project demands certain specified requirements. For example, a holiday village set in an area that experiences poor climatic conditions is unlikely to be successful in attracting potential visitors who are seeking the sun. Access to the site through one of the different transport networks will also be important. One of the reasons given why Canary Wharf did not succeed in its original form was due to the limited concept of what constituted an adequate transport infrastructure. The initial financial collapse of Canary Wharf is further discussed in Chapter 7.

Fashion and demand

These are difficult issues to properly address as evidenced by speculative developments, such as offices and houses that can remain unoccupied for several years after their completion, even though at their inception they would have been considered to be viable projects. In some cases assessing the need may have been done correctly, but external influences, beyond normal forecasting capabilities, attributed more risk than was originally anticipated. Also if all risks were fully expected, then few developments would ever materialise in practice. Developments that lag behind rapidly changing fashion and technology will inevitably be more difficult to let or sell. This can easily be explained by the changes that have occurred in the housing market since the mid-1950s. Amenities taken for granted today, such as a garage, central heating and double glazing were always considered as extras during that period.

Where the design or the project's aesthetics become unfashionable, then these may have a major influence on the possible disposal of a property and hence its price. High-rise local authority housing developments that were constructed widely throughout the 1960s, have in some cases had to be demolished even though debt finance still exists. This was due in part to changes in fashion as well as the high costs associated with their upkeep. Most commentators now accept that these high-rise dwellings were a quick fix to housing problems, rather than soundly and correctly thought out solutions to a long-term problem.

New developments must always attempt to take into account the latest in terms of fashion and technology, but to avoid gimmicks that will soon become outdated. The individual features of some property may require extensive refurbishment to bring them up to modern equivalent standards. In some cases this may not be possible. This is especially true in the case of warehouses that lack a sufficient headroom for modern plant and machinery or offices where modern communications technologies cannot easily be installed. Other factors to consider are the availability of services, the state of repair, the potential for its improvement or the capability for vertical or horizontal extension.

The nature of property investments

Property has its own sector in the United Kingdom stock market and is thus classified separately in the Financial Times All Share Index (FTSE). Some of the major UK property companies have been described in Chapter 1. The property sector can generally easily be categorised unlike other investment sectors where debates occur as to where particular quoted companies should be classified, especially because of changes in their activities and practice over time. The multinational conglomerates, for example, are difficult to clearly gather under a single heading. Some of the large building contractors also have subsidiary companies, concerned primarily with property development. Some of these subsidiary companies were formed on the basis of providing work for the main contracting business, during times of a lull in the market, in addition to their making a profit in their own right. Some of the building contractors also have financial interests in activities beyond contracting or materials manufacture. The largest contractor in Europe, Bouygues, earns a large slice of its revenues from its associated media companies.

The property sector owned property valued at almost £40 billion in 2000 and at least double this amount by 2007. This compares with a valuation of £25 billion at the peak of activities in the late 1980s. It owes its creditors, principally banks and other lenders, about £5 billion more than its combined cash, debtors and other assets.

In line with the remainder of the United Kingdom stock market, about 70% of the shares in the property sector are owned by the investment institutions. Over 90% of the total valuation of the property companies are located in Great Britain, with the remainder in the major cities of North America and Europe. There is a rapidly growing interest and activity in activity in the former countries of Eastern Europe, and at the present time a continued effort in identifying the ownership rights of property in these post-communist societies. Countries with high population densities limit site availability but generate high site values, as in Britain. The property stock market is thus much more UK-orientated, and especially England-biased owing to the higher densities there, than the average stock market company, who will typically earn about one-third of its trading profits in a range of overseas companies.

The portfolios of most property companies can be subdivided into the three main classifications of *office, retail* and *industrial*. Residential investment, in new buildings, now typically represents about 15% of the construction industry's output but has been declining since its peak in the 1970s. In 2001, this sector was, however, very buoyant with rapidly rising prices and easy access to mortgages. Typically houses remained for sale for a matter of weeks, although there was some variability across the country as a whole. However, as far as property investment by developers or institutions is concerned, it is negligible, since most of this kind of development is either for owner occupation or for the declining public sector, post local authority housing sales. An analysis of the quoted property sector's portfolios is shown in Table 2.6 and Fig. 2.1, varying over time with a shift towards retail investment, away from offices. There has been little change in the proportion of industrial assets over the past 25 years.

Table 2.6   Institutions' property assets (percentage over time)

	1980	1990	2000
Office	56	53	38
Retail	26	35	45
Industrial	14	10	14
Other	4	2	3
	100	100	100

(Source: DTZ)

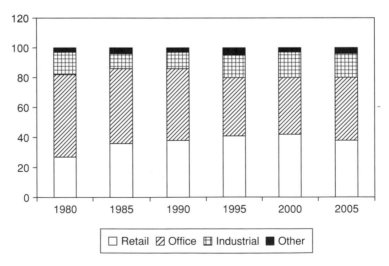

Figure 2.1   Institutions' property (percentage) assets by sector (Source: DTZ)

The industrial classification may be misleading since this will include the newly developing business parks that straddle the traditional classifications of office and industrial categories. The office data implies that in physical space terms, the quoted property sector owns about 40 million square feet of offices, where perhaps half a million people may work. The retail sector includes about 15 million square feet of shopping space, equal to about 15 000 small high street shops. The industrial space covers approximately 120 million square feet, of which about 20% is in decline and will eventually be subject to demolition. Rental growth indices indicate that retail property has consistently provided the best returns since the early 1980s. Table 2.6 confirms this, as the big financial institutions have switched their funds into this sector since then.

Whilst a large proportion of the property investments are dominated by the institutional investors the reverse is not the same. Only about 10% of the £800 billion UK investment portfolios are invested in property (see Fig. 8.2).

Most institutions will have a spread of investments in order to reduce the possible risk of an entire sector performing badly. The institutions are supported by financial industry analysts, who retain a close watch on the performance of share dealings, expected announcements on performance and projections of future performance, based on a combination of company expectations and the general state of the UK and world economies. In addition, there are of course, a very large number of owner occupiers, such as private investors, public limited companies, central government and local authorities, who retain their own personal financial interest in the property that they own and occupy.

Distinguishing features of property

Property has several distinguishing characteristics that separate it from other forms of investment. In economics there is the principle of perfect competition. The property market is far away from this description. Table 2.7 lists some of the reasons that distinguish property from other forms of investment.

Table 2.7   Distinguishing features of property

• Uniqueness	• Supply of land and buildings
• Durability	• Transfer of property ownership
• Proof of ownership	• Legislation
• External conditions	• Indivisibility
• Sources of finance	• Imperfect knowledge
• Changes in fashion	• Income generation
• Decentralised market	• Government intervention

Uniqueness

Probably the main feature of any property is that it is heterogeneous, i.e. no two properties are alike. Hence they will each therefore each have an independent value. An individual plot of land or the building that stands on it is unique. A row of speculatively built semidetached houses may appear to be the same, but each will have its own individual characteristics. Some will have larger gardens, others will be at the more desirable end of the street and others may have more elaborate interiors. The valuation of any land or property is affected by size, design, condition, etc. Along a street of essentially identical houses, the differences in value may be attributed to their closeness or distance from different amenities and other types of property. In some circumstances these differences will be of a minor nature, in others the distinction will be sufficient to make considerable differences in market value. Similar considerations apply to other forms of property.

Location, location and location are said by some to be the three most important characteristics to consider. This attribute is vulnerable and easily affected by other development works.

Supply of land and buildings

The supply of land is virtually fixed. It was Mark Twain (1835–1910) who said, 'Buy land, I understand that they are not making anymore of it!' Even its uses are relatively inelastic due to the planning constraints and the zoning of land for different designated uses. Where there is a considerable demand for products or goods in general, then increased manufacturing can often be undertaken to satisfy that demand. This is generally not so with land, if one ignores the small amount of land reclamation. Owners of land or property are, for example, restricted in the way that they may use their possession. The time taken to develop land through financing, planning, design and construction is slow to respond to any increase in the demand for the different types of property. It is a little less difficult to postpone a development that is under construction, where, for example, housing developers may vary their rate of building to match those of their sales. It is more difficult to accelerate building to meet a current demand that might otherwise not exist when a development is completed and ready for sale.

Land that is susceptible to mine workings, subsidence or flooding may in extreme circumstances have little or no value in terms of building development potential. This became all too apparent after the floods in 2000–1. Land that requires extensive and expensive preparatory work prior to construction operations will only be used where more desirable sites are not available for the purpose desired. Contaminated land and its reuse for building purposes has now become a thriving concern of some companies and organisations. The need to make such land safe on which to build has a serious effect on the price that can be charged.

Durability

Whilst many goods wear out and deteriorate quite rapidly, for example, the automobile, property, on the other hand can last for many years, particularly where it is well maintained and has thus a relatively slow rate of deterioration. Even when the property has become obsolete, in terms of its useful life, the site on which the property stands will probably have increased in value which may at least compensate for this depreciation. In this respect, property is quite different from many other investments in that its durability offers good security over a long period. It is one of the few consumable products that increase in value over time.

The physical life of a building is difficult to estimate being dependent upon a combination of the physical attributes of the building, the function for which it was designed and economics associated with its maintenance and repair. Decay and obsolescence each play their part.

Transfer of property ownership

Compared with other sorts of investments the transfer of property from one owner to another is very high in terms of time and money. Other investments can be completed, by comparison, very quickly, some almost just for the length of a telephone call. There are time and costs involved in property transfer because of the many different individuals, such as agents, lawyers, surveyors and valuers acting for the different interested parties and the need to exchange all sorts of different documentation relating to the transaction. The process has, even with the advent of electronic communication, not become more efficient or economic.

Proof of ownership

In the majority of cases the ownership of property presents no problems, with the legal rights properly established at the outset of purchase. In some cases ownership of an existing property is not clear and can be challenged by individuals with different legal and financial interests. The legal position may not always be clear and straightforward and title deeds that may go back for centuries may be incomplete and even imprecise. Where such problems arise, then it can be both expensive in time and money to resolve. With other sorts of investments it is usually possible to produce receipts showing when and where the investment was purchased. Where property is concerned, possession may not always be an indication of ownership.

Legislation

Unlike other types of investments, land and property are subject to a wide range of legislative powers (see Chapter 6) that are likely to have an effect upon its price. Prime agricultural land, without planning permission to develop, will suddenly see its value escalate if planning permission can be obtained. Other legislation, aimed particularly at curbing the attentions of speculators have been introduced by successive governments. The type of legislation, the rates and taxes applicable change to meet the needs of government in respect of revenues and political expediency.

External conditions

The value of land and buildings may be directly influenced by external activities that are part of the site or adjacent to it. Severe subsidence problems may make a property and even its site almost worthless. The provision of new highways or the designation of industrial estates or business parks will affect the values of property within their vicinity. The development of new out-of-town shopping malls has affected the values of local shops within the catchment area. The construction of projects that may create noise, pollution, vandalism or increased traffic will tend to have a downward effect upon the values of surrounding properties.

Indivisibility

Another feature of property that distinguishes it from most other investments is the relatively high cost of each unit. These units cannot be separated easily into smaller manageable parts. Even a moderate interest in property will require a considerable amount of capital for its purchase. Investors therefore have to either purchase the whole, often for occupation, or acquire shares in property companies, property bonds and other indirect means.

Sources of finance

Unlike many other types of investment, property is often purchased with a loan or mortgage. This frequent use of credit in purchasing property results in property being highly susceptible to both the availability of the credit supply and the rates of interest charged on such loans or mortgages.

Imperfect knowledge

The transactions of the majority property interests are generally imperfect. The details of the transaction are often kept secret and some are even unknown to the seller. There is a movement to provide much greater information and access to this information about all types of property. Imperfect knowledge also arises because potential sellers will attempt to increase prices, i.e. the value of the property, whereas potential buyers will be seeking reasons to reduce the price below some theoretical market valuation. In other circumstances an anxious buyer or seller will distort the price agreed between the two parties still further.

Changes in fashion

Building techniques and standards of accommodation are constantly changing and improving and these will affect the value of the property. It is necessary to anticipate the longer-term demand for certain types of property in particular locations. Fashionable areas in towns and cities are known to shift over time. The various different locations for residential property around London are a good example of this. It is also important to be aware of the changes that may occur throughout the different sectors of the property market. The prediction that many office workers might be working from home at the end of the last century resulting in a glut of office accommodation and hence a reduction in its value failed to materialise. However, property that no longer satisfies a demand in its present form may need expensive building alterations to be made, and even these may still result in reduction in the yield that was expected. The decline in traditional church attendance, and the demise of some city centre shopping arcades have resulted in the need to find alternative uses for such buildings. Although cinema-going is enjoying something of a revival, this is taking place in new, purpose-built auditoria in new locations since former cinemas were often unsuitable for modern projection technology. In some cases, even the ubiquitous bingo halls have been short-lived, with new uses having to be sought.

Income generation

Investment in property is attractive, since unlike many other investments, it is able to generate an income, when in use. A painting by an old master, of course, can be purchased and then loaned to a user for a rental sum, but this is much less common. Other forms of investment are much more limited in this respect. Investment in property allows the owner to both receive income from rents whilst at the same time seeing the capital value of the property increase. In this context it meets the basic requirements of an investment, i.e. an increase in the capital sum and a regular return.

Decentralised market

Unlike stocks and shares, there is no central agency for dealing with property transactions. In individual town locations there is no central clearing agency for even domestic property. Property is normally purchased through estate and property agents. It is also not possible to obtain details of the availability of, for example, restaurants that might be for sale other than to gather the information from all those agents who might have or know of the likely availability of such property. The exceptions to these are the national agency chains, who operate throughout the country or firms who work constantly for major clients such as the different institutions with property interests. Even with the advent of computerised property databases, many property sales remain excluded.

Government intervention

Government has a keen interest in property and is able through a wide range of legislative powers to control the types of development and to discourage those that it considers undesirable. These powers have been enacted through various Acts of Parliament and are concerned with the implementation of planning regulations, rent control and security of tenure, compulsory purchase, the encouragement of development through loans and grants, the application of taxation to property such as mortgage relief and the control of credit through interest regulation. Each of these in turn will influence the investment, either directly or indirectly that may be made in property. Such action is able to control the supply and demand of property, its condition and its price. The details of such measures are regularly reviewed.

Institutional investment in property

The most important criterion concerned with any property for investment purposes is its location. This will determine not only its present and future rental values but also its capital growth and the ease whereby it can be disposed of when it is no longer required. The concept of prime property has as much to do with its location, as any of its other individual attributes. Other factors to consider are that the building's design meets modern uses and expectations, and its tenure and the tenancy agreement.

Offices

Location and context

It has been stated that perhaps as many as 7 million, or 25% of the working population of the United Kingdom are now employed in offices of some kind. It is also estimated that about half of these work in the south-east of England, even after many relocations of firms both in the public and private sectors to provincial cities throughout the country. However, many companies continue to desire a presence in and around London, particularly those firms that are concerned with city activities such as stock broking, banking, insurance and the other financial markets. Some organisations, especially government departments, have a desire to be close to Parliament and the workings of the legislature. The office revolution in respect of computing and information technology with their predicted knock-on effects of working from home, has not really materialised to any significant extent, but this may change in the future. Recent surveys continue to show that there is a trend amongst office employees and so-called hot desking is an increasing feature of some organisations. The only factor militating against this happening on a huge scale is the necessity for human interaction with one another.

The highest value properties are likely to be found in the business areas of towns and cities. Efficient transport communications are a key to their success and these are frequently lacking at the present time. Roads are often choked with traffic, especially at rush hours, and rail travel has become singularly unreliable. Air travel is expanding at a considerable rate, especially between the large cities. Another feature that affects the rental values is the ability of a development to attract high profile companies to the project. Like any sort of building development the attraction of different types of tenant will have a marked effect upon the long-term value of the property. Coupled with road transport is the need for accessible and secure car parking. This may be more important for provincial cities than in London, where access by public transport is much more common and easier because of its underground rail systems. A continued lack of investment is, however, threatening this very real asset.

Design and specification

Generally there is no ideal size for investment properties, although small buildings are not generally favoured due to the disproportionate management and associated costs involved. Investors may be interested in an office building, but not usually the individual units themselves. The external appearance, whilst a matter of opinion and subject to changes in fashion, is nevertheless an important aspect to consider, since it is the first impression that is received. It is important that the building is in a good state of repair or can be easily and inexpensively renovated. Although layout is important, the key criterion is the building's flexibility, to allow for open-plan, individual offices or even multiple clients.

Structure

The following parts should be borne in mind:

- Best use of good and expensive site
- Minimum six storeys (taller in prime areas)
- Floor to ceiling height sufficient for the provision of service pipes and cables
- Suspended access floors for the installation of cabling for information and multi-media technologies
- Avoidance of possible early obsolescence

- Overall floor to ceiling heights at least 3 m
- Floor design for superimposed loadings (increased for corridors, store rooms and plant rooms)
- Double-glazed windows desirable but not essential although their inclusion has been well rehearsed, in terms of heating, sound and other comfort criteria
- Fitting-out costs will include demountable partitioning by the individual tenants.

Services

Basic services should be considered in the following areas:

- A lift is essential in buildings with more than one storey (size of building and number of storeys will determine the number of lifts)
- Lighting to cover a wide range of different occupations (specific task lighting provided to supplement where required)
- Low pressure hot water systems
- Ducted warm air systems are common
- Air conditioning not now viewed as a luxury in the UK
- Avoidance of complaints from sick building syndrome, etc.
- Toilet facilities for male and female provided on each floor to aid possible subletting.

Finishings

Internal finishings typically account for between 10% and 14% of the overall costs.

- Carpeting required to maximise rental value
- Walls are typically painted, other than entrance foyer area that would be more elaborate
- Tenants have the opportunity of applying wall coverings if required
- Acoustic suspended ceilings of different grades and quality are standard.

Industrial

Location and context

Industrial property includes both warehouse and buildings for manufacturing industry. In terms of investment potential, industrial property sited in locations in conjunction with government grants or other financial assistance should be avoided. The fact that such developments occur in this way indicates a general lack of demand for property of that type in that location. Unless the location is likely to change because of such development, then it is likely to be difficult to dispose of when it is no longer required.

Industrial buildings are classified in accordance with the Town and Country Planning (Use Classes) Order 1987 (SI 1987/764). This includes premises designated as light industrial and general industrial. General industrial buildings that may have been designed to suit a particular manufacturing process are generally unsuitable for investment purposes. The importance of being close to motorway networks and rail systems as well as being easily accessible to a potential workforce are considerations that are likely to impinge upon rental values.

The high technology buildings that have replaced many of the older traditional industries, whilst offering a wider scope in terms of their location cannot be as flexibly located as might be imagined. The larger types of these operations tend to be located in those areas that provide for the most government financial assistance. Hence, they are often in those areas where traditional industry has declined, where there is a large potential workforce and a workforce that has developed certain skills and competencies. The manufacturing processes that are used are relatively sophisticated in terms of manufacture but often require less skill capability in respect of the employees. The Nissan car plant in Sunderland and the Toyota car plant in Derby are good examples of this.

Many regions have sought to develop high technology science parks, often in collaboration with universities. Such manufacturing companies are often at the forefront of science and innovation and attract academic staff from the universities as well as a largely graduate workforce.

Design and specification

There are no absolute specifications, although the key criteria are to provide working or storage environments relatively inexpensively. There is no ideal site cover, i.e. the area of the site covered by buildings. This will depend to some extent upon the nature of the industry and the type of vehicles that are used to transport the raw materials and the finished goods to and from the factory in the case of manufacturing. For convenience, single-storey, rectangular shapes suit most situations. There should be few internal obstructions, such as steel columns that might limit the building's use. The standard width is about 20 m with a height to eaves, to allow fork lift trucks easy access, of about 7 m. About 10% of the space is allocated to toilets and offices, sometimes as an extension to the main industrial area, where the same storey heights are not then required.

Structure

As with office buildings, there are structural points to consider:

- Typically of steel or concrete portal frames covered with a corrugated sheeting, of which 10% is made of translucent sheets
- Thermal insulation to satisfy the Building Regulations requirements

- External walls brickwork up to about 2 m and then PVC-coated metal cladding composite with insulating panels
- Brickwork is provided at the lower levels for resilience to possible damage from vehicles or vandals
- Floor loadings should be adequate to support heavy vehicles
- Loading doors should be capable of allowing a fully laden lorry to enter the building.

Services

Services to an industrial building differ from the considerations that apply to offices:

- Heating depending upon building use, i.e. storage or manufacture and the type of products
- Artificial lighting will help supplement the natural lighting provided through the roof lights
- Necessity for heavy duty service cables if manufacturing is anticipated.

Finishings

Finishings of an industrial building involve the following considerations:

- External appearance designed to require minimum maintenance
- Internally self finished as a part composite wall and roof panels
- Internal finishings account for 4–6% of the overall cost.

Retail

Location and context

The different types of retail shopping outlets are described in Chapter 6. The locations of shops include those in local communities, town and city centres and more recently, since the early 1980s, out-of-town shopping centres, such as hypermarkets. In the larger cities, many of the shopping centres have a regional dimension including the major stores that could not be supported in small towns alone. Large towns of about 100 000 people are able to support an extensive range and variety of shops, but much less so than the regional cities that will have upwards of 500 000 people.

Shopping has undergone major changes since the middle of the twentieth century. Many specialised shops, such as ironmongers and milliners, have disappeared or diversified. The local shops have been unable to compete with the supermarkets and have simply ceased to trade. In some areas they have been converted into ethnic restaurants.

Supermarkets have in turn given way to the hypermarkets and out-of-town shopping malls, that offer a comprehensive range of shopping experiences.

Design and specification

Since there are a wide range of different retail shops, it is difficult to make more than general comments in the space provided. Shop units with a frontage onto a street or in a shopping mall should be about 7 m wide by about 20 m deep. Floor ceiling heights are typically 3.7 m, but will vary with the overall plan size of the shop. The preferred shape is rectangular and with an absence of piers. Staircases should be placed at the rear of the building where the sales space is cheaper. Access should be direct without any step up or down, to assist the old, the disabled and parents with prams and pushchairs. The trend towards self-service outlets, that began in the 1960s, encouraged a high turnover per square metre. Storage is normally provided at first floor level, since stairs discourage would be shoppers. Escalators to upper or lower floors generate sales, but are only cost effective for large areas. Older shops with three or four floors, and no means of vertical transportation are now obsolete in the context of modern shopping policies. Shop units are often provided as a shell for subsequent fitting out by the tenants. The shop fronts, whilst provided by the tenants, often become the landlord's fixture. The lighting, heating and internal finishings are all about creating an image in the shopper's mind. Individual units have a life expectancy of about 15 years at the most. When one shop begins this renovation work, others then need to follow. Many of the chains of shops around the country have developed standard frontages and interiors to allow for easy recognition by the public. Intruder alarm systems and closed circuit television (CCTV) monitors have become common place, if not yet universal.

Relationship between income and value

The relationship between income received at the present time and value will give the yield. This can be expressed in the following way:

$$\text{Value} = \frac{100 \times \text{Present income}}{\text{Present yield}}$$

$$£50\,000 = \frac{100 \times £2500}{5\%}$$

A knowledge of any two of these three factors will enable the third to be calculated. The present yield will be a reflection of current investment rates that might be expected. In respect of property investments it will be necessary to allow for probable outgoings that are associated with the ownership of property, such as repairs, community charges, business rates, general upkeep etc. Through convention, such outgoings are generally deducted from the gross income that might be expected, in order to find the net income that may be due. Many of these outgoings will need to be estimated amounts based upon historical data and future expectations.

The yield of an investment is a particularly important concept as it indicates to an investor the level of earnings of that investment, or the speed at which it will earn money. All things being equal, a higher yield will be more attractive as an investment. The higher yield will also to some extent reflect the risk that is involved. An investor may therefore be prepared to accept a higher level of risk in anticipation of a greater return on the investment.

Property investment and equities

Insurance companies and pension funds have traditionally invested in government securities, since these have been able to provide the investor with a known rate of return on which they could depend. These companies need such a guaranteed return to enable them to be able to meet all possible claims. In times of rising inflation, the institutions found that returns above the rate of inflation could also easily be achieved by investing in equities. In stock market and property jargon, equity means the interest in a company or a property that bears the full risk and receives the full rewards with no upper limit on potential earnings. It is effectively the entrepreneur's or owner's money, as opposed to that which is borrowed from elsewhere. Thus equities on the stock market are simply ordinary shares. Preference shares are not strictly speaking equities, since they have preferential rights and fixed dividends.

The major institutional investors are always seeking better ways of investment, and industry analysts are able to advise on those areas, industries or companies that provide the best possible return. The institutional investors still invest large amounts where risk is minimal and returns are almost guaranteed. They also take advice in attempting to predict those opportunities where return will be higher and the investment will still largely be secure. Property offers a good location for funds, at times considerably in excess of other types of investment. The advantages of direct property investment include the following:

- The investor has more control over the property purchased.
- Fewer acquisitions need to be made where large sums are involved.
- Property values are generally less volatile than stocks and shares, whose prices change daily.
- Property companies are often highly geared, making their return less secure than other shares or direct property investment.
- When the stock markets are depressed, it is still possible to sell property at a reasonable price.

- The tenants of property often pay their rent in advance, usually quarterly, whereas the dividend on equities is in most cases paid half yearly in arrears.
- Rent continues to be paid by tenants, even if the company concerned is making a loss.
- Rental income is a more secure form of investment since it is paid ahead of bank interest.
- Even if a tenant enters into liquidation, the investor still retains the property asset.
- Following liquidation, a company's equities will have minimal value, if at all.
- Most modern property leases include provision for rent reviews, enabling the investor to charge current rates irrespective of the tenant's profitability.

Direct investment in property also has a number of disadvantages set out below:

- The time and costs involved in purchase or selling. The volatility in the prices of, for example, stocks and shares are partly offset by their rapid trading capability.
- Property values do not always match those of inflation.
- Of critical importance with property is deciding which is the most advantageous property to purchase.
- Changes in technology, design or working and living practices will all affect the property's value. In some cases where these factors have not been properly considered, they may have the effect of making the project obsolete before their decay.
- Changes in communication networks will have a positive or negative effect.
- Town planning decisions may also have a similar effect both upon the valuation and the demand for a property.

Real Estate Investment Trust (REIT)

A Real Estate Investment Trust or REIT is a tax designation for a corporation investing in real estate that reduces or eliminates corporate income taxes. In return, REITs are required to distribute a large majority of their income, which may be taxable in the hands of the investors. The REIT structure was designed to provide a similar structure for investment in real estate as mutual funds provide for investment in stocks.

Like other corporations, REITs can be publicly or privately held. Public REITs may be listed on public stock exchanges like shares of common stock in other firms. REITs can be classified as equity, mortgage or hybrid. The key statistics to look at in REIT are its net asset value, and its adjusted funds from operations and its cash at disposal.

The legislation laying out the rules for REITs in the United Kingdom was enacted in the Finance Act 2006 and came into effect in January 2007 when many major British property companies converted from plc status to REIT status, including the five that were FTSE 100 members at that time: British Land, Hammerson, Land Securities, Liberty International and Slough Estates. British REITS have to distribute 95% of their income. They must be a close-ended investment trust and be UK resident and publicly listed on a stock exchange recognised by the Financial Services Authority.

To support the introduction of REITs in the UK, the REITs and Quoted Property Group was created by several commercial property and financial services companies. Other key bodies involved are the London Stock Exchange (LSE) and the British Property Federation (BPF). The Reita campaign was launched in 2006 by the REITs and Quoted Property Group, in order to provide a source of information on REITs, quoted property and related investments funds. Reita's aim is to raise awareness and understanding of REITs and investment in quoted property companies. It does this primarily through its portal www.reita.org, providing knowledge, education and tools for financial advisers and investors.

These trusts are already popular in other countries, such as the USA, France, Australia and Japan. However, since the introduction of the tax-efficient REIT regime was introduced at the beginning of 2007 all but one of the newly created companies have failed to convince the market that they are worth it. Hammerson is the sole company to trade at a significant premium. It was reported that capital values of prime shopping centres would fall by 8% over the next two years and that City of London offices would drop by 3% in 2008. The underlying performance of the property market is likely to affect the performance of the fledgling British REIT market. Three years of nearly 20% returns have left investors fearing a slowdown at best and at worst a reversal in profits.

Strategy of property investment

Investment institutions such as insurance and pension funds managers will seek to reduce their risks by spreading their investments across a wide range of companies and industries. Fund managers are constantly transferring funds, they hope into those areas or sectors that offer the best possible returns. The level of investment reflects the level of performance. The information in Tables 2.8 and 2.9 is based upon a survey undertaken by the World Markets Company plc. This company measures over 2000 pension funds,

Table 2.8   Asset allocation of investments; property is highlighted

	1977	1980	1985	1990	1995	2000	2003	2004	2005
UK equities	49	45	47	54	53	49	39	37	33
Overseas equities	5	8	20	18	27	24	28	29	32
UK bonds	23	22	18	9	6	9	12	13	12
Overseas bonds	0	0	0	3	4	4	3	2	3
Index Linked	0	0	2	3	4	5	9	9	9
Cash/Other	3	3	3	4	3	4	3	3	4
Property	20	22	10	9	3	5	6	7	7
	100	100	100	100	100	100	100	100	100

(Source: The WM Company plc)

Table 2.9 Annual return across a number of investments

	Annual percentage returns										
	1977	1992	1994	1996	1998	2000	2001	2002	2003	2004	2005
UK equities	**14.2**	**20.8**	−5.6	**16.8**	12.0	−4.1	**−12.6**	−22.5	**21.1**	**12.9**	**21.8**
Overseas equities	11.3	**19.7**	**−3.7**	2.6	**17.2**	−6.7	−15.1	−24.2	**24.7**	10.3	28.8
UK bonds	12.3	**19.1**	−8.4	7.6	**20.6**	9.4	**4.2**	**9.6**	4.0	7.4	9.4
Overseas bonds	9.3	**29.1**	−5.3	−5.8	11.6	**14.6**	**4.9**	**9.3**	4.3	4.8	11.6
Index Linked	8.0	17.8	−8.1	6.5	**20.3**	**4.3**	**−0.1**	**8.7**	6.8	8.5	9.3
Cash/Other	9.6	13.2	**5.4**	6.1	7.3	**11.5**	**2.3**	**2.6**	3.8	6.9	11.4
Property	**17.1**	−1.1	**12.3**	8.6	12.8	**10.1**	**7.6**	**9.1**	11.0	**18.9**	**20.4**
Totals	**13.0**	**18.6**	**−3.9**	**10.7**	**14.0**	**−1.3**	**−8.9**	**−13.9**	**17.0**	**11.2**	**20.1**
Property ranking	1	7	1	2	4	3	1	3	3	1	3
Retail price index	4.7	2.6	2.9	2.5	2.8	2.9	0.7	2.9	2.8	3.5	2.2
Average earnings	8.9	5.0	4.4	4.9	4.5	4.5	2.1	4.1	3.6	4.7	4.1

Note: Percentages above the total assets performance are shown in bold.
(Source: The WM Company plc)

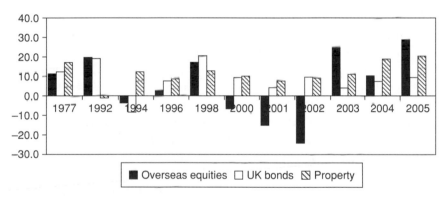

Figure 2.2 Annual returns of investments 1977–2005 (Source: The WM Company plc)

of which 1336 provided data that contributed to the data shown in Tables 2.8, 2.9 and
Fig. 2.2.

Table 2.8 illustrates a typical asset mix allocation based upon funds between 1977 and
2005. Whilst the predominant investment of these funds is in equities, this has fluctuated over the period between 1997 and 2005 with a definite shift or mix towards
overseas equities. This emphasises the importance of investing in equities for the best
expected return over a period of time (see Table 2.1). Institutional investment in property has shown a considerable decline since 1977, although by the turn of the twentieth
century, returns were ahead of all other forms of investment.

The analysis in Fig. 2.2 is obviously an historical perspective whereas investors are
inevitably concerned about the future performance of the investment. All funds state that
the past results are no guarantee of future performance. Even with the careful analysis of

data there is also always an element of chance involved and there are losers as well as those who gain considerable amounts through investment.

Whilst the performance of property investments compared with other forms of investment is not as attractive as some other forms of investment, there are other factors that must also be borne in mind. The distinctive features of property have already been summarised. These especially include the individual nature of each item of property, the costs involved in the transfer of ownership and the fact that it takes time to complete property transactions. Equities are liquid allowing for their easy sale and purchase.

Table 2.9 and Fig. 2.2 compare the annual rates of return on different kinds of investment between 1977 and 2005. During the late 1980s, there was a boom in property generally but this was followed by the rapid collapse of the property market that occurred in the early 1990s. Figure 3.4 indicates this clearly where many purchasers of property experienced a loss in values, known as negative equity. The downturn began in 1991 and did not recover, especially in the south of England, until the end of the decade. Compared with other forms of investments such as equities, for example, property investment consistently underperformed, other than in a few individual years. However, between 2000 and 2005 it exceeded the average of all investments in five of the six years. DTZ (*Money into Property*) reported a significant inflow of funds into property by institutional investors in 2004–5. By comparison, equities outperformed the average of all investments in only the last three of these six years.

At the turn of the twenty-first century, property was enjoying a comeback, albeit probably brief. Equities had been poorly performing against a background of a predicted world recession, fuelled by uncertainties in world events. Economic forecasters and the Royal Institution of Chartered Surveyors were all suggesting a slowdown in UK property prices that would occur during the latter part of 2007. In some areas of the country, property prices were expected to fall marginally.

Overseas property has only ever represented a very small proportion of total investments and is less than half of one per cent of total assets. Overseas property investments have poorly performed, being one of the few types of investment to yield returns below the retail price index.

Using the retail price index (RPI) as a measure of comparison, property has generally performed very well throughout the period of time under examination. Over the period 2000–5, property returns considerably exceeded the percentages of the retail price index and average earnings in all years. However, measured over a longer period, investment in UK property has overall performed at about only half the rate of United Kingdom equities.

Investment in property is also now much more of an international venture, with Japan and the USA having considerable investments in European property, and the UK investing abroad in the USA and Australia. Recent surveys have indicated that London, amongst European cities, remains the most popular and sought-after location for commercial premises. It remains ahead of Paris, Frankfurt and Brussels and in recent years has been able to consolidate its premier position. The analysis is based upon a survey of businesses using eleven indicators such as access to markets, costs and availability of staff, communications and availability of accommodation. The cities of southern Europe, such as Barcelona, Milan and Madrid have all become more important in recent years, whereas some of the provincial cities of the United Kingdom have declined, relatively, in their importance.

International investors also have to consider currency issues in addition to the other common problems that face investors. For example, during the early 1980s the UK investor could have gained over 18% per annum through currency alone. Over the subsequent period, most of these gains were reversed and for the three years up to 1987, the loss on the US dollar reached almost 15% per annum. Until 1988, the Japanese yen showed consistent strength against sterling, never falling below 8.5% per annum. In the early 1980s the German mark displayed consistent gains against sterling, but since 1992 it has displayed a sharp rise in variability. The reluctance of successive UK governments to join the common European currency (the Euro) has produced criticism from overseas investors. Clearly in terms of manufacturing, this has presented difficulties, especially bearing in mind the value of the strong UK pound.

The percentage change in investments in overseas equities is shown for comparison in Fig. 2.3. Investment in UK property has now to be set within a global scale, considering all possible opportunities.

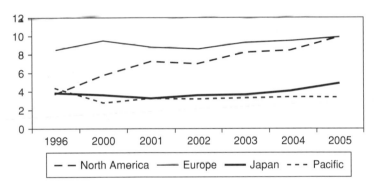

Figure 2.3   Overseas equities performance (Source: The WM Company plc)

According to the WM Company (UK Pension Funds, 2005), property remains the only alternative asset to have a significant weighting in the average pension fund. By the end of 2005, the average fund held just under 7% of assets in this area. Whilst this is up from the average of 5% ten years ago, this increase has been driven up by the strong returns achieved rather than by any significant cash flow. Despite this high average weighting, it is interesting that 40% of the funds of the world do not hold property. Investment in property is thus seen as steady, following the principle, *as safe as bricks and mortar*. Equities offer, especially in the longer term, much more substantial returns but are also much more volatile with shares rising and falling daily.

The property cycle

The property industry, like the rest of British industry, tends to develop a pattern known as the business cycle. Alternating periods of boom and slump in property development, investment and performance coincide with movements of trade generally. However, like the construction industry these often occur earlier and often last longer. As such they are often much more severe than experienced in other industries. In the nineteenth century, booms or peaks in activity tended to appear every seven to ten years. At these times, the activities of construction and property were stretched in terms of personnel, and the availability of other resources such as finance and building materials. In the early part of the boom years of the 1960s, for example, a provisional order for materials such as facing bricks needed to be placed during the design phase in order to be sure of obtaining these at the appropriate stage during construction.

Immediately after the end of the First World War the pattern or cycle altered and the country faced almost two decades of high unemployment with the slump reaching its trough in the Depression of the mid-1930s. From the end of the Second World War until the early 1970s, Britain and other world economies were running at relatively high levels, with little unemployment other than a blip in the late 1960s. Unemployment was never higher than 2.5% and was often as low as 1%. Concurrently with the stable employment levels, the economy experienced a much lower inflation rate than the country was to face in the 1970s. These saw a return of the slump–boom pattern, but characterised by ever deepening troughs and slumps and more modest recovery in the boom years. The time span had also fallen from a previous 7–10 year period to one of only 3–4 years.

The early 1970s were recession years followed by a bubble in 1974 that was to burst by 1976. By 1978, the economy had started to pick up slightly but this was followed by a recession with over 3 million unemployed; many of whom should have been at work in property and construction. This slump was worldwide with few countries escaping the misery of high inflation. Inflation peaked in the mid-1970s in excess of 25%. Since 1985 the economy rose to peak in 1989, but by 1990 was already in decline. It was predicted to bottom out by 1992/93, but even by the mid 1990s there was only scant evidence that the economy was improving. It was not until the late 1990s that real prosperity returned coinciding with the election of the new Labour government. Unemployment levels, for example, fell to their lowest recorded levels by the end of 2001. However, the economic problems and their knock-on effect were to return, exacerbated by the recession in the USA and in mainland Europe, where its effects were much more severe than in the UK.

Both the construction and property industries have in the past been useful indicators for economic commentators to follow. Historically, a decline or increase in construction activity has been a sign that recession or prosperity is to follow. It is an indicator for the rest of the economy. The changes in the costs and sales of domestic property also signal changes in general economic performance. In 1994, for example, whilst private housing was beginning to show higher prices and increased numbers of transactions, this was very shaky. By the autumn of that year both of these indicators were showing a reversal in trends measured for the year as a whole. The housing market took much longer to recover, but by the end of the 1990s, negative equity had disappeared and house prices were showing increases that were well above the levels of inflation.

The use of construction and property as an economic regulator

Government is a major client of both the property and the construction industries. It is therefore tempting to suggest that these industries may be used as regulators to control the economy. Whilst there are no doubts that these industries are damaged by the stop–go nature of economic activities, there is only scant evidence to suggest that government effectively turns the tap on or off, in order to regulate economic perform-ance. The tap of activity is turned on and off, but often as a result of government policy rather than through intention. Government may defer or cancel construction projects or property purchases for other reasons, such as to reduce the public sector borrowing requirement or to meet other commitments in expenditure. This in turn often creates a situation that severely affects growth in other areas of the economy and the demand for property of its different kinds. Cuts in public expenditure may also sometimes include a high construction consequence, but these are often accompanied by other measures, rather than targeting this industry.

However, government is able to intervene in the construction and property markets, chiefly in three ways; through *finance, legislation* or *regulation and provision*. The state can intervene in the market through finance by grants, benefits, subsidies and taxation. It can choose to offer grants for the construction or purchase of industrial or commercial premises in order to entice a potential employer to an area of high unemployment. It can also offer incentives for the development of certain types of project such as private housing. It has done this through mortgage tax relief, but the uncertainty of maintain-ing this policy in the future causes future potential buyers to be cautious about a possible commitment. In order to encourage all kinds of property to be properly maintained it could introduce a form of taxation relief for the annual maintenance of building projects. There are many sorts of combinations of financial measures which can be used to stimulate development activity, if this is so desired. Changes in regulations, such as in town and country planning, can also create opportunities for development. For example, the allowance of a wider range of projects to be constructed in certain designated areas may be just the sort of stimulus a developer is looking for. Third, since government is a major client, it has considerable scope to influence development activity through new build, refurbishment or the repair and maintenance of projects.

In conclusion, the property cycle feeds directly off the economic cycle at various key points. Fluctuations in output and employment drive the occupier markets through the demand for space; changes in the financial climate impact upon property yields and investment allocations. Property cycles are marked off from economic cycles by additional sources of potential instability inside the property industry. Development is volatile not because it is tossed about by building manias or waves of purely speculative finance. Building booms and slumps can be explained simply by developers' responses to current market signals, and the development lag.

The investment markets are potentially another source of instability. But, in practice, since the 1970s, any tendencies for yields or investment allocations to exaggerate the cycle have been dampened by two factors. First, yields have mostly anticipated the major swings in the rental cycle a year or so ahead. Second, there are offsetting and balancing forces: yields, for example, are pushed up by the major building booms, and the rapid capital growth in upswings itself curtails investors' appetite for property.

A report by HSBC (2001) stated that by the end of 2001, growth in the economy would slow down. Manufacturing has already set this trend. It is increasingly likely that the USA will head into recession, even discounting the events of September 2001. The UK will also be affected. The old adage, 'When America sneezes, the whole world catches a cold', remains largely true. However, despite the forecast of such a recession, a soft landing is forecast for the UK. This means that levels of insolvencies, unemployment and debt will be less than experienced in the recession of the late 1980s and early 1990s. The UK does not have the debt levels or current account deficits that are a feature of the US economy. These characteristics were present in the early 1980s and it is these imbalances that determine the severity of the economic slowdown.

By 2007, increases in property prices came to an abrupt halt. These had been forecasted for sometime by economists at banks such as ABN AMRO and publicised on web sites such as the aptly named www.housepricecrash.co.uk. The upward spiral of prices had been sent into reverse largely because of the sub-prime market lending in the USA. The reason why it affected property prices in the UK was due to the fact that many banks, and especially Northern Rock, had over-exposed themselves to lending in the USA market. Sub-prime mortgage lenders provide higher-priced loans to consumers with poor credit ratings. Due to the cooling world economies thee have been increases in the defaults and amongst sub-prime borrowers. Consequently a number of lenders have shut down or scaled back their operations.

Chapter 3
Capital Investment

Capital budgeting

Introduction

Capital budgeting is the process involved that establishes criteria for investing in the resources of long-term projects. These projects include plant, equipment and machinery and buildings. The acquisition or development, maintenance and subsequent disposal or demolition of such capital assets are essential to the majority of organisations. The importance of these assets, especially buildings, is due to the following:

- They represent a large item of expenditure
- This sum is often the largest single item of expenditure
- Investment is often committed for a long period of time
- Once the capital project has been started it is difficult to curtail
- The need for the investment is forecasted today but it will not come into use until some time in the future, when conditions and circumstances may have changed
- Capital investment decisions can have an impact upon whether a firm achieves its financial objectives
- There is a relationship between working capital and the size and use of fixed assets.

Basic principles

The primary function of capital budgeting is to invest in an asset that over the long term produces a positive cash flow outcome.

There is a preference for current rather than future cash flows. Investors must therefore be compensated for postponing the recovery of their investments and their returns on investment. The benefits of capital item acquisition are received over the long term and this factor lies at the centre of any capital budgeting. In the shorter term, therefore, investment decisions will produce negative cash flows.

In addition, the longevity of capital assets and the substantial outlays required for their acquisition suggest that some form of capital discounted cash flow analysis is desirable. This will anticipate the outflows in terms of capital expenditure on the development in the case of buildings and inflows that might represent sale, lease or rental arrangements over the ensuing years. The principles associated with discounting are considered in Chapter 5.

The analysis of cash flows may differ from accounting income since the two aspects measure different attributes of the project. For example, project depreciation or appreciation considerations (see Chapter 4) may represent different amounts in financial reports from those allowed in a cash flow analysis. A project that appears worthy of investment in the longer term may show negative effects early in its life. This issue may create difficulties for shareholders and investors, bearing in mind that all developments based upon some future expectations include some aspects of speculation within their analysis. Even after making some attempt to analyse the possible risks involved (see Chapter 4), there can be no guarantee of the certainty of future events since any prediction can only be made within the context of a range of possible outcomes. Prevailing conditions are often very fragile and susceptible to events that occur worldwide. Under possible, but extreme conditions, the majority of capital projects are likely to be able to show that they might not be worthwhile investments. Techniques are employed to help justify a particular course of action.

Every capital project has to be financed, and there are no free sources of capital. Grants can sometimes be obtained from charitable organisations, but these usually represent only a small proportion of the total financial demands. The majority of companies will wish to maintain a capital structure of a mixture of debt and equity finance that will minimise the overall financial requirements for the project. The different sources of finance are discussed in Chapter 8.

Capital budgeting always involves allocating resources among a range of possible competing proposals and investment opportunities. The constraints are a mixture of financial and managerial resources. Both of these commodities are a scarce resource.

There are also other aspects of capital budgeting to be considered. These emanate from the microeconomic environment and include the fluctuations within the business cycle that require careful analysis. An example is the interest rates that might be applicable and the amounts of inflation that can be expected. A proposed investment in a period of prosperity may be a very different consideration in a period of recession. It is often desirable to build during a recession, when costs are likely to be lower, with slogans from the construction industry such as 'There has never been a better time to build'. There is the difficulty, however, of forecasting the likely end of a recession and just what the sort of demand expected for property will be. The property cycle does have periods of growth and recession and also more extensive times of boom and slump.

There is a general pattern, but the interval years vary (see Chapter 2). The recession in both property and construction during the 1990s was severe and lasted longer than other periods of recession.

Interest rates will also have an effect on the timing of the capital proposals. Similarly, periods of high inflation complicate the process of forecasting cash flows, undermine the utility of financial statements and distort the financial ratios that are derived from such statements. In the public sector a more sensible approach is probably to undertake capital works projects in times of recession, when costs and prices will be lower. The adoption of such a philosophy is also helpful to the construction industry, in helping to smooth out peaks and troughs in activity and thereby encouraging greater overall efficiency within the industry. However, the immediate demand for construction projects mitigates against such a philosophy.

Leasing

Leasing is a process whereby the owner (*lessor*) of an asset (e.g. a building) enters into an agreement (*lease*) with the user (*lessee*) to allow the latter to use the asset for a specified period of time, usually in return for payment. Real estate still constitutes the largest single category of leased items. Many items of plant and equipment are also now often leased in preference to making the full payment by way of purchase.

There are many advantages of leasing rather than purchasing, although a primary consideration is often the cost differentials that are involved. Part of the choice is also influenced by taxation issues that may differ between different firms. For example, a firm that makes only small profits may be unable to enjoy the full benefits of taxation relief available. However, a leasing company might be able to take full advantage of the taxation benefits and pass these on to the lessee in the form of reduced payments. Where equipment is concerned, the firm may have an insufficient capacity to allow it to make the full use of an item and leasing may therefore be the preferred option. The fundamental decision between lease, purchase or hire is financial, since the product and its attributes usually remain the same. Most of the issues involved can be dealt with in this way. The following are some of the key variables to be considered:

- Leasing generally provides for a 100% finance agreement since no initial payments are usually required
- Lease agreements are normally fixed for a term of years in the case of property, so removing some of the issues associated with risk
- Leasing provides an alternative source of obtaining facilities or premises for companies that may have limited capital budgets
- Leases are often available over much longer periods of time than conventional financing agreements

- Leases allow owner's capital to be used for other purposes
- Leasing may conserve a firm's existing sources of credit for other uses and often does not restrict a firm's borrowing requirements
- Leasing can be completed very quickly and does not incur the same legal charges that are required with purchase
- Leasing is much more flexible than capital purchase, since it allows the firm to relocate much more easily
- Capital purchase normally involves substantial front-end costs that, if required, can be built into the lease agreement payments
- The costs associated with leasing are often 100% tax deductible
- Leasing avoids the need for underwriting and the raising of share capital
- Leasing may avoid the possibility of obsolescence, particularly where technology and fashion are changing rapidly.

Strategic management

Capital budgeting decisions function within an overall plan and expected outcomes of the organisation. In its broadest sense, firms strive for a set of conditions that are harmonious with the survival of the organisation. The responsibility of senior management is to define the aims and objectives that are capable of achieving this primary objective. Management is also charged with the tasks of managing change in order to optimise its course of development. It has therefore to develop a strategy that will translate the stated aims and objectives or mission of the firm into specific policies.

Company finance

Finance, it has been said on many occasions, is the lifeblood of a company. The flow of income into a company's accounts through receipts from work done and the expenditure on payments for materials, labour, plant and equipment result in either a profit or loss for a particular project or accounting period. Figure 3.1 illustrates the flow of funds in and out of a firm, such as a building contractor. This model can be adapted for other types of firms or organisations. The majority of companies start up in business trading with the owner's capital and loans obtained from a bank. To this may be added investments from shareholders. Other income is then received from the profits made on work done. Expenditure is required for the purchase or rental of offices and workshops, for the purchase of materials and the hire or purchase of mechanical plant. One of the largest items of expenditure is for wages and salaries and the associated costs of employing labour. Finance will need to be set aside for the payments of taxes and any dividends paid to the shareholders of the company. Some aspects of trading are likely to risk non-payment for work completed, by way of bad debts. Over a period of time, loans and the

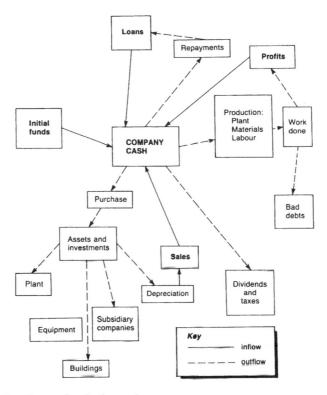

Figure 3.1   The flow of cash through a company

interest on loans will need to be repaid and income may be generated from the sale of unwanted or under utilised assets that have previously been purchased by the company.

Turnover

An impression of the size and importance of a company can be gained from the financial *turnover*, that is the total amount of income received from all of the different sources. All companies above a certain size have to prepare audited accounts and large organisations produce annual reports showing many other indications of the company's profile. British Petroleum is the parent company of one of the largest international oil and energy groups. It was the largest company in the UK. Its turnover in 2001 was expected to be about £45 billion. By comparison, the turnover of the British construction industry for the same period was £60 billion (currently £90 billion in 2007). The largest building and civil engineering contractor was BICC plc with a turnover of in excess of £4 billion, although even this figure disguises the wider activities of this firm beyond that of general contracting or the construction industry alone. Measuring the turnover of an individual public limited company (PLC) is relatively straightforward since the information is quoted in their annual reports. Measuring the turnover of an industry, such as construction, is more complex, since many firms are involved in activities outside of the industry. Also is it appropriate to include materials and component manufacture, or non-contract work such as housing and the large and growing DIY market?

In the public sector, the term turnover is not used because its services are not sold in the market in the same way as goods and services are in the private sector. In local authorities, the term *budget* is used as an indicator of size and the services that might be provided. A large county council serving a population of one million inhabitants might be operating on a budget of about £2 billion.

There is a relatively high concentration of private ownership in the UK. Of the ten largest firms, for example, none are now in the public sector. Some formerly public sector corporations were privatised during the 1980s such as BT and British Airways.

Turnover in the construction industry is represented by new construction work which is typically worth about 60% and the remainder is derived from repairs and maintenance projects. The diminishing public sector accounted for less than a third of this turnover. This sector continues to decline, in real terms, which is due in part to the effects of privatisation of previously nationalised companies, such as those engaged in the electricity, gas and water industries. Civil engineering now typically accounts for about 20% of the total output. New PFI projects have helped to bolster the public sector's share of the industry.

Profit

Profit is the amount of cash left in a company after the payment of all expenses. It is the difference between income and expenditure. It is also the amount that is available for distribution to the shareholders, although this is at the discretion of the board of directors of the company. It is unusual for all profits to be paid to a company's shareholders; some will always be retained for future investment or to be offset against past or future bad debts. Bad debts are an unfortunate feature of the construction industry because of its high risk activities in manufacturing predominantly bespoke products. Insolvency in the construction industry is high when compared with other industries. The construction industry does not normally build prototypes and this in itself adds further to the risk involved in construction and development. Profits often do not bear out the risks involved, and comparison with a number of indicators shows that the industry scores badly when compared with other industries and activities. Whilst profitability is relatively high in times of boom, such profits may be negative in times of depression. Whilst individual building and civil engineering contractors' profits can vary considerably, they typically as a group account for less than 5% of turnover over a period of time. By comparison, in the middle of 2001, BP was reported to be making record profits of more than £1.3 million per hour.

Return

It has already been noted that the capital invested in a company can come from several different sources. In the case of sole traders and partnerships, the finance usually comes from the owners themselves, supplemented by bank overdrafts and loans. In a limited company, the capital consists of share capital provided by the shareholders plus the company's own reserves. Wherever the initial capital comes from, all persons providing it have the similar objectives of seeking a return on the capital that they have invested in the business. This is assessed by determining the *return on the capital employed* (ROCE) which is explained more fully in Chapter 4. This is the measure which relates profits earned to investment or capital employed. Building and civil engineering contractors will aim to utilise the capital employed many times per year. This may be in a ratio as high as 10:1; although it can be much lower. The construction industry is by nature, stop–go and this discourages long-term investment in and by the industry. The amount of investment will also vary depending upon the capital intensity of the firm, and whether for instance the firm has a large land bank or uses property as an additional source of income.

Cash flow

A *cash flow forecast* is an attempt to show the anticipated inflow and outflow of money from a business over a period of time, such as a financial year. The importance of cash flow cannot be underestimated, since although a company may be trading profitably, without access to cash to pay its employees and suppliers, it may be unable to continue to trade. Cash flow is the life blood of any industry. There are examples of many companies that have gone into receivership, not because they failed to make a profit, but because they did not have access to cash or finance by which to trade. Cash flow forecasting is further considered in Chapter 10. Chapter 4 examines some of the ways in which a firm can supplement its short-term financial needs. A longer-term strategy might include:

- The avoidance of work in non-profitable sectors
- The abandonment of unprofitable companies within a group
- Changing the pattern of working or financing
- The purchase of new companies for
  - diversification
  - elimination of competitors
  - ownership of raw materials
  - market for the company's own products
- Investing in further mechanisation
- Developing a conglomerate portfolio.

Market capitalisation

Market capitalisation is the value that is placed on companies by shareholders and investors. This value changes daily, sometimes on the basis of rumour rather than on the basis of actual performance. It is the expressed worth of a company (see also Chapter 4). The market capitalisation of a company is the total number of shares the company has in issue multiplied by the share price. This is constantly changing. The market capitalisation values (2007) of the largest ten companies quoted on the London stock exchange are shown in Table 3.1. The size of these represents in many cases mergers and acquisitions, which has been a common feature in the past twenty years. This feature is also becoming more evident in construction and property industries. Some of the companies listed in Table 3.1 have their own property holdings and undertake design, surveying and construction using their own in-house staff. Of these ten largest companies, eight were on the list five years ago. So whilst there is some shift and jockeying for position the giants of British industry remain at the top of the London stock exchange.

Table 3.1   Capitalisation values of the ten largest companies on the London Stock Exchange

Company	Market capitalisation (£m)
HSBC	106 584
BP	104 484
Glaxo Smith Kline	83 384
Vodafone	78 817
Royal Bank of Scotland	65 074
Royal Dutch Shell A	62 881
Barclays	50 527
Royal Dutch Shell B	46 991
Astra Zeneca	44 777
HBOS	42 887

Investment

Investment is necessary for wealth creation. Wealth creation requires investment. Where the latter is small, then the distribution of the former is largely academic, so said Lord Scanlon. Some of the products of the construction and property industries, i.e. buildings and other structures have been seen as a good source of long-term investment. In periods of rapidly rising property prices and when the demand for accommodation exceeds supply, the return to investors has often been high and much higher than other forms of investment. Sometimes the return has been artificially high and only the informed investor has been able to see that on occasions this has been short lived. Whilst the majority of other manufactured goods and products depreciate after purchase, buildings and other structures have in the past tended to appreciate in their values in the short and long term.

The phenomena of the early 1990s experience of declining property and rental values were unprecedented in recent times, resulting in mortgage lending on some properties being unsecured, with the investment being higher than the actual value of the asset. The term *negative equity* came into our vocabulary to describe the downward trend of property prices. However, pension funds, insurance companies and private investors continue to see prime property as a good hedge against inflation. This has, in some cases, reaped considerable financial rewards over a matter of a few years. Over the longer term, investment in property has shown a real mark up above many other commodities. However, the future can never be certain. We also live in a world that requires gains today rather than promises tomorrow. Past performance is also no guarantee of the future predictions.

The cost of capital

An important element in the use of resources is the cost of those resources. This is true of the cost of finance just as it is true of any other resource. The cost of finance is generally termed the cost of capital. The individual components of capital may have quite different costs, being obtained from different markets.

Ordinary share capital

The dividend paid on ordinary shares depends upon the availability of profits earned by the company. The amount is determined by the directors of the company, who may choose to retain profits for investment elsewhere in the company. Whilst there is no obligation to pay a dividend, companies would soon find themselves short of shareholders if they did this too often. The cost of equity capital is therefore taken to be the rate that has to be effectively paid in order to maintain the value of the equity. Equity has two parts: growth in *capital* and in *yield*. For example, a company has the following profile:

- 1 000 000 £1 shares
- Annual profits £100 000 (none retained)
- Current market share value £2.

The cost of equity capital is the same as the yield on the equity to the investors:

$$\text{Cost of capital} = \frac{\text{Dividends per share} \times 100\%}{\text{market value of share}}$$

$$= \frac{£100\ 000/1\ 000\ 000 \times 100}{200\text{p}}$$

$$= 5\%$$

In practice, the full amount is rarely paid, but a proportion retained and invested and thus expected to produce a future growth in earnings and dividend, and hence capital value.

$$\text{Cost of capital} = \frac{\text{Dividend per share} \times 100\%}{\text{market value of share}} + \text{expected growth due to retained profits}$$

$$= \frac{£100\ 000/(1\ 000\ 000/2) \times 100\%}{200\text{p}} + 5\%$$

$$= 7.5\%$$

Cost of retained earnings

It cannot be assumed that the cost of retained earnings is nil. If a company retains some of its earnings, rather than distributing them, then this is equivalent to the issue of fresh equity to existing shareholders. Collectively they are required to make a further investment in their company. The cost of existing retained earnings is therefore the same as the cost of existing equity. The cost of new retained earnings is the same as that of a new equity issue. There are also taxation considerations to consider.

Mergers and takeovers

A merger occurs where two or more firms that were previously autonomous come under common control. The process is normally termed a takeover if the party initiating it is much larger and or more powerful than the other party. There are frequently commercial advantages relating to the size of a firm. These include:

- *Economies of scale*: large-scale production techniques can be employed to lower the firm's unit costs
- *Monopoly power*: larger firms have a greater degree of market influence than smaller firms
- *Risk reduction*: size gives the opportunity for rational diversification to reduce risk
- *Cost of capital*: the larger well-known firm is perceived by individual and bank lenders as a safer investment and is thereby able to borrow funds at a cheaper rate.

The growth in the size of a company can of course be achieved by policy and strategy within an existing firm, and this in turn will lead to some of the advantages listed above. However, it is also commonly achieved through a merger or acquisition. The former is referred to as *internal* growth and the latter as *external* growth. The latter is often so called for the following reasons:

- *Financing*: Growth requires funds for expansion. These may be generated internally from profits or externally through debt or equity finance. Where a merger occurs, the existing finance of both companies can be used to a much greater effect. In many ways this is a cheap way of securing finance for development purposes.

- *Speed*: Growth is achieved much more rapidly by means of a merger than by other means and the advantages of size are thus achieved more quickly.
- *Risk*: The actual process of growth, like many other activities, includes an element of risk. Companies that expand too quickly can overstretch their resources and capabilities resulting in operating difficulties and sometimes liquidation. Expansion through the acquisition of an established business involves much less risk.
- *Asset acquisition*: A firm may seek a takeover or merger in order to gain the under-utilised assets of another company; for example, its land bank, that may have a potential for development that the existing owner is unable to develop owing to a lack of financial capability.
- *Competition*: Regardless of the other possible benefits, a merger between two firms may take place solely to remove competition from the market place or to resist the possible takeover by another company.

There are several different ways that a merger can take place. These include the winding up of one of the companies, retaining a legal identity of both firms where they each have a well known name or forming a new company under a new name. A basic requirement for any merger is that it must provide advantages for both parties. In the case of a takeover, the smaller firm may be unable to resist the advances of a larger company making an offer that cannot be refused.

The complexity of calculating the value of another firm and the making of an offer is considerable. Where rival firms are seeking to make bids then the original offer made will be subject to counter bids resulting in revised offers until either one is accepted or the other party withdraws from the process. The amount of the offer will differ depending upon which data it is based upon. For example, consider the following two companies, where Company A is seeking to take over Company B.

Example

	[A]	[B]
Number of ordinary (£1) shares	1 000 000	500 000
Asset value of company	1 500 000	600 000
Current earnings per share	25p	30p
Current market value	300p	400p

(1) *Net asset value*
This is calculated as

$$\frac{\text{Value of assets}}{\text{Number of shares}}$$

$$A = \frac{1\ 500\ 000}{1\ 000\ 000} = £1.50 \qquad B = \frac{600\ 000}{500\ 000} = £1.20$$

For the net asset value to be preserved so that the shareholders in each company feel that a fair deal is being made, A will need to issue 120 of its own shares in exchange for 150 of those of B. If it makes a more generous offer, then the net asset value will decline. If A wants to secure all of B's capital then it will need to issue the following number of new shares.

$$\frac{500\ 000 \times 120}{150} = 400\ 000 \text{ new } £1 \text{ shares}$$

(2) *Earnings per share*
On this basis, new 30 shares in A for 25 existing shares in B. For the full amount of B's shares this leads to an issue of:

$$\frac{500\ 000 \times 30}{25} = 600\ 000 \text{ new } £1 \text{ shares}$$

(3) *Market value*
In order to maintain the market value worth of the two shares the number of shares issued on this basis can be calculated as:

$$\frac{500\ 000 \times 400}{300} = 666\ 666 \text{ new } £1 \text{ shares}$$

It would appear that the value of a takeover using the above data is almost anybody's guess on whether the basis of calculation is fair. These calculations attempt to value the proposed company in different ways and act as a guide only. In many cases a bid submitted will be rejected due to it being too low, but other issues will also need to be considered. The initial takeover bid is frequently rejected and this is then followed by an improved offer from the acquiring company. This is particularly the case where other firms have also become interested in the takeover. In the real world it is also expected that the new firm formed from the two separate companies will be greater than the sum of the parts. This is referred to as *synergy*, where extra value is created that was not present in the individual firms. In the final analysis, therefore, the process is more one of bargaining rather than a strict application of the financial rules alone. These will act as a guide, but there will be a range and limits on both parties on what the actual takeover price is.

The case for investment in property

The property investment market during the early 1990s saw one of its worst recessions for many years. Yields and capital values were less than half those of the boom years of the 1980s. Property has also performed badly measured against the long-term yields in gilts and equities. However, this should not hide the fact that during the early 1990s all kinds of investment markets were depressed. A similar pattern existed at the turn of the century as the country formally entered into a short recession during the later part of 2001.

During the immediate period following the Second World War, investments tended to be dominated by gilts, debentures and preference shares. However, during the 1950s the UK financial institutions recognised the need to broaden their asset base beyond those of equities and government securities alone.

It is generally accepted that the property cycle is out of synchronisation with other forms of investments as shown in Figures 3.2 and 3.3. Also, while property investments have peaked four times since the early 1970s, this compares with six peaks for equities, but at different times. Property is viewed by some as a late cycle following on behind the revival in the remainder of the economy. The principle is that as business confidence

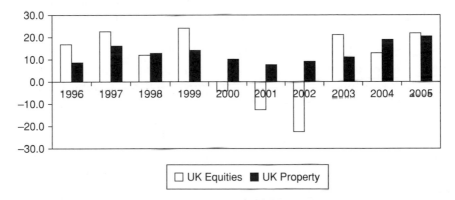

Figure 3.2 Comparison of percentage returns on UK property and UK equities (Source: The WM Company plc)

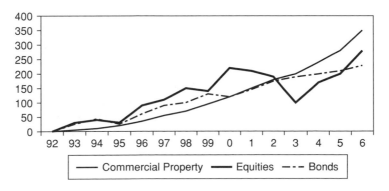

Figure 3.3 Commercial properties v equities v bonds over the past 15 years (Source: New Star)

increases, this creates an expansion, resulting in a demand for property space, whether in industrial, commercial or retailing. Fluctuation in the sales of domestic property has tended to signal a change, other recovery or depression, in the economy. This measure is believed to be one of the best economic indicators.

The performance of the different kinds of investment can be broadly described as follows, although there are exceptions and amidst these different divisions different investments will always buck the trend.

- *Gilts*: tend to perform the best in times of sharply declining inflation; conversely they tend to be unprofitable investments in times of rising inflation
- *Equities*: are the most successful in times of rising demand, prosperity and economic expansion
- *Property*: is best during periods of economic stagnation coupled with the rising costs of inflation.

This highlights the case for investing in a range of different assets and at different times of the economic cycle. There is also the realisation that the growth in property investments has at times outperformed both gilts and equities. The crash in the value of equities in 1987 and sterling's subsequent exit from the European Exchange Rate Mechanism created conditions that encouraged some investors to return into property. Similarly, the events that followed the 11 September 2001 terrorist attack on the World Trade Center buildings in New York had a partially similar effect. There is also the added difficulty of attempting to classify all gilts, equities and properties in a similar manner. Whilst they may generally behave in a similar manner, the different asset components will each respond in different ways. For example, retail property in the centre of London may follow a very different cycle from industrial premises in the north of England.

The security of income in any type of investment is of paramount importance to all kinds of investors. During the 1960s, investments that reduced the risks from the ravages of inflation were seen as advantageous. Property has always traditionally been seen as a good hedge against inflation. Figure 3.4 shows the trend in the rise of house prices between 1983 and 2007. The data are based on a large database managed by HBOS (formerly the Halifax Bank) and demonstrate that house prices were increasing faster than average earnings or retail prices, especially during the late 1980s. However, a recession terminated much of these gains with falling house prices up to the mid-1990s. This fall in house prices, which began to level off in 1992, was coupled with steady increases in average earnings and retail prices. However, there seemed to be little merit in acquiring assets, such as property, that required a large management input.

It was the cost push inflation of the late 1960s, and the inadequacies of equity shares that provided the real case for investing in property. There were also the advantages to be gained from the spreading of risks through the greater diversification both across markets and across other asset classifications. The addition of property investments to gilts and equities has helped to spread this investment risk admirably. Since the advent of

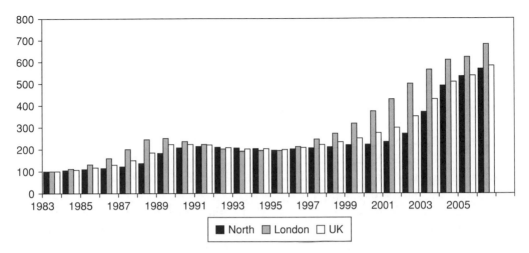

Figure 3.4 Regional house price indices, 1983–2006, £000 (Source: Halifax Bank of Scotland)

the Single European Market this has also included pan-European investments. The UK market has benefited from the outward investments from many countries in the EU. As with most commodities, there remains a strong relationship between its supply and demand. In this respect the general level of activity in the economy has a strong bearing on the demand for property. The effects of the recession in the closing years of the twentieth century and the early years of the twenty-first have inevitably focused investment managers on overall performance, increasing portfolio diversification and the consequent spreading of risk. This has benefited investments in property.

House prices

The value of property in the United Kingdom is estimated to be worth about £3 trillion, according to Britain's biggest mortgage lender. This is triple what it was worth ten years ago. All areas of the UK have seen the value of their residential housing increase considerably in value over this period. Unsurprisingly Greater London saw the biggest increase with Northern Ireland showing the second largest gain. This was from a relatively low base and represents a good response to the ending of the troubles seen in that area of the UK for too many years. Scotland was the only region not to record triple digit growth. Nearly 40% of the value of the UK's housing is concentrated in London and the South-east, an increase from around a third ten years ago. However, over time this fluctuates with the narrowing and widening of the North–South divide.

The value of property has grown much faster than the rate of retail prices. Between 1994 and 2004 the retail price index rose by only 28% compared with a 170% jump in the valuation of the nation's housing stock. The ongoing growth in house prices, along with home construction and renovation activity, has seen the value of the residential housing stock increase considerably.

The total value of private housing was 3.2 times higher than household debt in 2003. This is compared with being 2.7 times higher both five years and ten years ago. The rise in the value of housing assets has also had a significant positive impact on the health of household balance sheets. More than half of household wealth is derived from the value of residential property.

Table 3.2 shows average regional house prices across England and Wales, comparing prices in 2000 and 2006. It makes a distinction between house types under the four broad headings mentioned above. It should be understood that wide differences of price exist within these ten broad regions.

The highest percentage change was recorded in the North for terraced houses. This showed an increase between 2001 and 2006 of 161%. This compares with an increase of 64% for flats and maisonettes within Greater London. Wales recorded the largest percentage change in average house prices for all types of 139%. This compares with an increase in Greater London for all domestic dwellings of 72%. Yorkshire and Humberside recorded the largest percentage increase for both detached (136%) and semi-detached (153%) properties. These percentages may suggest that the North–South divide in house prices is closing. In percentage terms there may be some truth in this statement, although fluctuations in prices and ratios change over time and between property types and regions. Also whilst the average price of a detached house in Yorkshire and Humberside increased by £146 362, in Greater London the difference between 2000 and 2006 prices for this kind of property increased by £286 448, almost double.

Will house prices fall?

The question of course on everybody's mind is whether house prices will keep on rising. Some informed commentators have suggested that these might fall in the next five years. There are various reasons to suggest that house prices are overvalued (see p. 64) and likely to fall:

- The ratio of house prices to incomes has risen to an all-time high. In the UK, the ratio of house prices to incomes is 50% higher than the long-term average.
- The effect of this is that house prices are becoming unaffordable. It is increasingly difficult for people to be able to buy a house. This leaves the housing market in a vulnerable state.
- It is becoming increasingly difficult for first time buyers to get on the property ladder. This is mainly due to the rise in house price to earnings ratio. Some banks are willing to lend five times a borrower's salary. Banks are also considering mortgages over a longer period.

Table 3.2 Average prices of domestic dwellings by UK regions

	Detached		Semi-detached		Terraced		Flats/maisonettes		Average prices		Average difference 2006–2000	Average percentage change
	2000	2006	2000	2006	2000	2006	2000	2006	2000	2006		
North	109 019	245 367	57 537	141 200	41 459	108 398	44 920	114 485	61 417	143 297	81 880	133
North-west	123 971	280 306	65 197	155 102	41 081	105 056	61 603	135 006	66 944	152 491	85 547	128
Yorks/Humberside	107 324	253 686	57 499	145 569	43 532	112 838	56 379	128 820	64 983	153 560	88 577	136
Wales	98 151	229 049	59 121	143 046	44 659	114 363	54 937	128 151	64 767	154 741	89 974	139
West Midlands	134 490	286 549	69 862	156 344	54 761	128 147	55 913	120 758	80 300	173 778	93 478	116
East Midlands	109 026	237 331	58 627	139 854	46 748	116 227	48 096	117 366	72 132	163 075	90 943	126
East Anglia	126 740	263 527	74 482	168 704	61 754	146 673	52 808	132 893	87 346	194 543	107 197	123
South-west	157 575	315 926	91 541	197 074	75 037	171 480	69 703	154 320	102 535	217 222	114 687	112
South-east	225 674	404 899	122 214	230 705	96 318	191 116	77 654	157 860	131 107	251 008	119 901	91
Greater London	392 467	678 915	208 989	371 791	204 101	349 478	164 962	270 964	193 004	331 300	138 296	72
Average prices	158 444	319 556	86 507	184 939	70 945	154 378	68 698	146 062	92 454	193 502		

- For those who believe house prices can never fall it is worth remembering the case study of Japan. In the 1980s, there was a similar boom in house prices in Japan. But since the peak of 1991, house prices in Japan have fallen for 14 consecutive years.
- Traditionally the view of the housing market is that it is not just an asset but a place to live. Therefore, unlike the stock market, house prices won't rise and fall due to speculation. However, a lot of demand for UK housing is coming from buy-to-let speculators since private rents are buoyant.

The UK housing market suffers from severe supply constraints, because as a percentage of the total housing stock, the number of new houses built each year is very small. Therefore any change in demand magnifies any change in price. It only takes a small rise in demand to increase prices and vice versa. There are record levels of consumer borrowing in the UK. This is a combination of mortgage borrowing and personal debt like credit cards. The total level of debt is £1.168 trillion. Even a modest rise in interest rates could have a very adverse effect on consumer confidence and spending. Therefore the housing market is particularly vulnerable to any rise in interest rates that may occur. Many economists predict significant house price falls.

The alternative view suggests that house prices will keep rising because:

- There is a chronic shortage of building land and new homes. There is an estimated shortfall of 37 000 homes being built every year
- There is a population explosion due to immigration largely from EU countries
- There is a dire housing shortage in the South-east of England
- There is an increasing popularity in both buy-to-let and second homes
- There are huge social life-style changes from an ageing and wealthy population and many more adults choosing to live on their own.

HM Land Registry

HM Land Registry produces a quarterly residential property price report, based on the factual selling prices of all domestic property. It analyses property by towns and regions on the basis of four categories of *detached*, *semi-detached*, *terraced* and *flats and maisonettes*. The report provides an authoritative insight into what is actually happening to average prices and sales volumes in the residential market for England and Wales. The report is intended to complement information that is available from other sources. Any comparison with other data should have regard to the differences in volume, timeliness and coverage of contributing transactions. A breakdown of the average sale prices of old and new properties by property type is also incorporated. No weighting or adjustment is applied to the information collected to reflect any seasonal or other factors. The price data can be said to be actual unadjusted averages, drawn from the great majority of residential sales completed during the period (quarter) analysed.

New build or refurbish?

The decision whether to refurbish, rebuild on the same site, or to seek alternative premises or a new site for development depends upon a wide range of different issues and the answers obtained to the questions below. While the solution is heavily dependent upon the financial implications of the proposal, other factors must also be considered.

- What are the expected costs involved with the different projects?
- Must the existing building be retained owing to its listed status?
- Can the existing premises be refurbished?
- How much disruption is likely to be caused during the refurbishment?
- What new efficiencies would a new building offer?
- What disruptions will be caused to either the temporary or permanent relocation of the building's activities and its staff?
- How important is the time involved and is this sufficient for what is required?
- What are the nature, condition and quality of the existing premises?
- How suitable is the existing site location and the existing building to meet future needs?
- What are the costs and value relationships of the different proposals?
- If the existing site and buildings are not required, will they yield adequate capital in sufficient time?
- Does relocation offer advantages?

The decision to refurbish an existing building will also be influenced by the expected life of the existing structure and the lives of its various components. The structural elements will be the most important, since the costs of their refurbishment may be prohibitive. The life of the expectancy of the property will be influenced by some of the factors listed in Table 3.3. Life expectancy is a combination of decay or deterioration in the fabric and obsolescence.

Adapting to change

Development needs change with improvements in technology, fashion and lifestyle. The latter is particularly affected by the state of the country's and world economies. Where these are floundering then as far as buildings are concerned, it may be necessary to sit tight until improved times are not only forecasted but arrive. The pattern of commerce and industry also changes with the development of new products and improvements through innovation and invention. Changes in technology, especially information technology such as intelligent buildings may require a new project rather than the refurbishment of an existing structure. Changes in the supply price of energy sources will also influence the building form, with traditional heavy structures tending

Table 3.3   Building life and obsolescence

Condition	Definition	Examples
Deterioration Physical:	Deterioration beyond normal repair	Structural decay of building components
Obsolescence Technological:	Advances in science and engineering result in outdated building	Office buildings unable to accommodate modern information and communications technology
Functional:	Original designed use of the building is no longer required	Cotton mills converted to shopping units Chapels converted to warehouses
Economic:	Cost objectives are able to be achieved in a better way	Site value is more than the value of the current activities
Social:	Changes in the needs of society result in the lack of use for certain types of buildings	Multi-storey flats unsuitable for family accommodation in Britain
Legal:	Legislation resulting in the prohibitive use of buildings unless major changes are introduced	Asbestos materials, Fire Regulations
Aesthetic:	Style of architecture is no longer	Office building designs of the fashionable 1960s

(Source: adapted from RICS, 1986)

to offer better insulated solutions, making some of the lighter structures built during the 1970s and 1980s less useful. The traditional structures also tend to have lower failure rates and cost less to maintain than some of the newer forms of buildings.

As levels of affluence rise, life styles change with the demand for greater comfort, space and ease of maintenance. Fashions also change. Together, these aspects tend to increase the capacity and contents of buildings in terms of finishes and service installations. There is also a greater tendency to make changes more frequently, with some of these changes prebuilt into the lives of the components that are used. This leads to both the desire for new buildings and the refurbishment of existing structures.

Historic buildings

In the context of historic buildings, the position is more complicated owing to other pieces of legislation. The building may have been listed because of its architectural or historic interest under the Town and Country Planning Acts. In the context of these buildings the accent will always be on possible repair, renovation and conservation. It may also be necessary to find new uses for buildings that no longer fulfil their original purpose. Due to the extension and protection afforded to old buildings, there is likely to be an increasing number of applications amongst historic buildings for their change in use, alteration, extension or subdivision. Whilst the controls were originally designed for buildings with specific designated uses, major difficulties occur when they are applied to historic buildings that are to be altered, extended or where a change in use is proposed. This is because such buildings vary enormously in construction, materials and internal and external layout, resulting in a bewildering range of individual problems for which there are few guidelines. In some circumstances the problems have been solved by waiving standards or in some cases exempting buildings from normal requirements.

It is therefore necessary to reconcile a variety of objectives. These include provision for modern use and comfort, acceptable standards of safety, health and neighbourliness, retention of the historic context and the building's visual characteristics. The value of such reconciliation aims to combine modern use and preservation. This will be more guaranteed if the building is made structurally sound, hygienic, pleasant and convenient by adhering as far as possible to desirable standards of day-lighting, ventilation, damp proofing, insulation, fire safety, etc.

Legal considerations

All buildings need to conform to the various legislation such as:

- *The Public Health Acts*: these are concerned with dangerous structures and the means of escape in case of fire.
- *The Building Regulations*: these are concerned with the fitness of materials, exclusion of damp and water penetration, structural stability, fire resistance, design of stairs, etc.
- *The London Building Acts and constructional by laws*: these laws are often more rigorous than the Building Regulations alone due to the increased congestion in the capital.
- *The Fire Precautions Acts*: these have links with other legislation and are concerned with buildings that require the issue of a fire certificate.
- *The Housing Acts*: These are concerned with fitness for purpose.

Chapter 4
Evaluation of Financial Data

Accounting standards

The external users of accounts need to be sure that reliance can be placed by them on the methods used in business, in calculating its profits and losses and their balance sheet values. In 1975, the six main accountancy bodies in the United Kingdom set up a joint committee known as the Accounting Standards Committee (ASC). The standards issued by the ASC were called Statements of Standard Accounting Practice (SSAPs). In 1990, the Accounting Standards Board (ASB) took over the work and functions of the ASC. The ASB is more independent of the various accounting bodies and can issue its own recommendations, known as Financial Reporting Standards (FRSs) without the approval of any other body. The role of the ASB is to issue accounting standards. It is recognised for that purpose under the Companies Act 2006.

There is, however, no general law compelling people to observe the standards. The main method of ensuring their compliance is through the professional bodies using their own disciplinary procedures on their members. The ASB has, however, set up a review panel that has power to prosecute companies under civil law where their accounts contain a major breach of the standards.

The International Accounting Standards Committee (IASC) was established in 1971, from representative countries around the world, including the United Kingdom. The need for international standards of comparability and understanding has been due mainly to:

- Considerable growth in international investment
- Growth of multinational firms
- The need to harmonise standards across the world
- Needs of poorer countries who may be unable to have their own ASC.

Since 1971, 25 SSAPs have been issued and cover matters such as:

- Disclosure of accounting policies
- Earnings per share
- Accounting for government grants
- Accounting for value added tax
- Extraordinary items and prior year adjustments
- Accounting for depreciation
- Accounting for investment properties.

The true and fair principle

Accounts should be drawn up on the principle of the *true and fair* representation of a company. Accurate figures should be used as far as possible and reasonable estimates where these are not available although the use of the latter must be disclosed. The accounts should be free from wilful distortion, manipulation or the concealment of material facts.

There is a requirement in the Companies Act (section 226) that all financial statements should give a true and fair view. However, creative accounting techniques which appear to be used extensively and increasingly during recent years, seem to provide just the opposite results. We are all now too familiar with the names of well-known companies that were supposedly trading profitably, only to quickly find themselves in liquidation. Companies where the accounts had been audited by professional accountants. The book *Accounting for Growth* (Terry Smith) highlights many of the unwholesome creative practices that have been used by businesses. The author argues that the growth seen in company profits during the 1980s was in many instances not due to the improved management and efficiency of British industry, but to the manipulation of profits through creative accounting practices. Several leading industrialists found themselves imprisoned due to falsifying business accounts.

The true and fair principle indicates that a company is being managed to ethical standards. To an external user of the accounts or a potential investor this would provide a clear presentation of the financial state of the company rather than a distorted picture. It is not only those companies that are in financial difficulty that fake their accounts, in some cases it is companies that are otherwise sound in principle and practice. Greed still remains a huge temptation.

Accounting for depreciation

SSAP 12, *Accounting for Depreciation* lays down the principle that all assets with a finite life, i.e. a life that does not go on forever, should be depreciated by allocating costs less residual (or revalued amount less residual value) over the useful economic lives. The standard recognises that there are various different methods of providing for depreciation (see Fig. 4.1). However, it does not recommend or insist on which method should be used. The one selected will be the one that is the most appropriate to the asset concerned. The depreciation method selected should be used consistently and should not be changed unless it becomes unsuitable to use. Asset lives should be realistic estimates and not artificially shortened or lengthened. The figures should be revised on a quinquennial basis and, where these result in material changes, they should be treated as exceptional items under SSAP 6.

As freehold land normally lasts forever there is no need to allow for any depreciation, except in those cases of a reduction in value from factors such as land erosion, mineral extraction, contaminated land, dumping of toxic waste, etc. Since the material life of buildings is expected to be very long there is no need to charge depreciation. Where the estimated residual value is at least equal to the net valuation, then depreciation can be ignored. Accounts should clearly identify the method of depreciation that has been used, the economic lives of these items, the total depreciation for the period and the gross amount and accumulated depreciation.

Accounting for investment properties

SSAP 19 covers accounting for investment properties, where appreciation rather than depreciation exists. This statement requires investment properties to be included in the balance sheet at open market value. It is not necessary for the valuation to be undertaken by a qualified valuer. However, if investment properties represent a substantial proportion of the total assets of the company, then the valuation should be done by a professionally qualified valuer and at least every five years by someone external to the company.

Accounting conventions

Accounting, in its recording of the inputs and outputs of the organisation employs a number of basic conventions or concepts. There are seven conventions, as shown in Table 4.1, which provide a framework within which the function of traditional accounting takes place. These underlie all traditional accounting practices in both

Table 4.1   The seven accounting conventions

Money measurement convention
Entity convention
Going concern convention
Accrual convention
Prudence convention
Consistency convention
Periodicity convention

commercial and non-commercial organisations. However, this does not mean that all accountants will apply them or will necessarily apply them uniformly. There is conflict between the different conventions and this means that accountants are able to prepare accounts quite legitimately on widely differing bases.

The different conventions can be explained as follows.

- *Money measurement convention*: this recognises and records only those flows into and out of the organisation that can be expressed in monetary terms.
- *Entity convention*: accountants see the world as a series of interacting systems or *entities* and account for each of these systems individually. Frequently it may be necessary to consolidate the accounts of the smaller systems. The principal significance of the entity convention is that the organisation and the participants of that organisation are treated as separate accounting entities.
- *Going concern convention*: accounting in an organisation is on the assumption that it will continue in operation, in substantially its present form and for the foreseeable future. While financial statements record the cost transactions of the business, the statements also suggest some concept of value. The balance sheet implicitly suggests a value for the organisation, whilst the profit and loss account implicitly indicates the annual increase or decrease in this value.
- *Accrual convention*: this term refers to the process of ensuring that the financial attributes of an event fall in a particular accounting period and are then recorded in that period.
- *Prudence convention*. This refers to accountants' tendency to be conservative or cautious, almost to the point of pessimism. This caution shows itself in a number of unwritten rules such as:
    - don't record income until it has been realised
    - anticipate all losses and record these as soon as possible
    - when faced with a choice of asset or revenue valuations, always choose the lowest
    - when faced with different estimates of cost, always choose the highest.
- *Consistency convention*: this attempts to ensure, as far as possible, that the financial statements are prepared on the same basis every year and are thus comparable throughout time.
- *Periodicity convention*: this suggests that financial accounts are drawn up for a specific period, usually once per year.

Interpreting company accounts

The owners of companies want to know the answers to three basic questions:

- How much profit has the business made?
- How much does the business owe?
- How much is owed to the business?

However, it must be understood that an organisation's performance should only be measured against its own objectives, since different organisations have different object-ives or outcomes. For example, a trading company seeking to maximise its profits will have very different objectives to a government department serving the needs of a local community. However, there is the need in all organisations to provide a satisfactory return on investment and to be profitable and to use their resources as efficiently and effectively as possible. Otherwise they are likely to be loss making and go into liquida-tion or require some form of government or external subsidy.

It is always desirable to look at a company's performance over a number of years, rather than a single year which may have been an exceptionally good or bad year. In the case of a public liability company (plc) this can be done by obtaining copies of the company's annual report and accounts. As a general rule, at least three years should be examined, otherwise it is quite possible to select a freak year. Once the reports have been obtained, various trends and statistics in the company's performance can then be measured. Four main techniques of analysis can be used for this purpose.

Methods of analysis

Horizontal analysis

This technique involves making a line-by-line comparison of the company's accounts for each accounting period chosen for investigation.

Trend analysis

This is similar to horizontal analysis, except that the first set of accounts to be used in the analysis is given a value of 100. The subsequent accounts are then related to this base of 100 and the various items in the account compared. This adjustment enables people to see what changes have taken place more easily than by inspecting the actual accounts alone.

Vertical analysis

This technique requires all of the profit and loss account, and all of the balance sheet items to be expressed as percentages of their respective totals. The reason for showing the information in this way is that it helps to highlight the significance of the figures more easily.

Ratio analysis

The techniques of ratio analysis are discussed more fully later in this chapter.

Analysis of investments

One of the first figures that experienced readers of accounts look for is the *net asset value* of the business. This can be found in the capital and reserves, or the shareholders' funds, section of the balance sheet. It is useful to make a comparison between the net asset value, the book value of the company's assets, less debts, with the market capitalisation figure which appears in the newspaper share price listings. The market capitalisation is the total value of all the company's shares in issue.

The gap between book value and its market capitalisation can vary enormously. The share price may stand at a large premium to the net asset value in large successful companies, particularly after large takeovers. Where the company's shares are very unpopular, the price may fall to a discount to the asset value, creating a potential opportunity for a bargain against the market.

Debt

This is the key to the financial stability of a business. Interest charges, as a performance indicator, should be less than one-third of the operating profits to keep modest safety margins. Some would argue that these should be no larger than 20%. Another rule of thumb states that the cash generated from the company's operations year should not be less than half of the outstanding financial borrowings. This information can be found in the notes to the accounts that show borrowings with the first line of the cash report.

Profits and profitability

The outlook and potential for improvement in the business should be assessed. A strategy may be to invest in those companies with proven success or to seek out the high yielding bargains. The operating and financial reviews and the segmental analyses should identify where the profits and losses occur.

Liquidity and cash flow

The cash flow report should be examined to see to what extent a company is paying for its investment from its own cash flow and not generating new debt. For example, does the company suffer from liquidity problem or is it a profitable, low-debt company, choosing to use short-term funds?

Working capital investment

The main issue of concern here includes the way that the company is being managed and possible warning signs of trading problems. Hints of poor trading at the end of a financial year may be suggested by an accumulation of stocks or a sharp reduction in debtors.

Fixed assets

These consist of *tangible fixed* assets, *intangible* assets and *strategic* (fixed asset) investments. Key issues include ailing subsidiaries, their possible failure and the effects on the main company through cross guarantees. An accounting note on contingent liabilities will show whether guarantees have been given up by the company. The same note may show other contingencies such as potential product liabilities.

Tangible assets are those with a physical existence: buildings, machinery, vehicles, etc. Intangible assets are the assets of the business, some recognised, mostly not, which do not have a simple physical presence. The accounts often do not adequately reflect these items. Important issues include the extent of the investment of the company in its support of leading brands and in acquiring goodwill by takeovers, the extent to which brand names are recognised and the impact on various popular accounting techniques.

Financial ratios

The Companies Act 1985 requires all companies to produce an annual balance sheet that shows the capital invested and its classification and how the capital is being employed. The annual accounts of all public limited companies (plcs) are filed with the Registrar of Companies. In order to ensure that they represent a fair and true view of the company's trading position, they are subject to an annual audit by an independent firm of professional accountants. However, as noted above (Smith, 1992) even audited accounts may fail to reveal the true and fair state of a company. Also in practice, the accounts take some time to prepare and to be audited and hence by the time of their publication they will represent the state of the company at best a few months ago.

In an attempt to make some sense from company accounts it is necessary to measure the data that they contain. The technique used is referred to as *ratio analysis*. These are relative measures and have a meaning only in relation to either changes over time or to other ratios at given points in time. In themselves, the ratios may provide little more than indicators that need to be explored or explained. They need to be examined in the context of the firm and its history and by comparison with other similar firms elsewhere. The following are some of the more common ratios and these have been grouped under four headings to identify key areas of information.

Liquidity

The first concern of a manager is to ensure at least the immediate survival of a company. In order to assess whether the company can meet its short-term obligations:

$$\text{Current ratio} = \frac{\text{Current assets}}{\text{Current liabilities}}$$

Most commentators prefer to see a company with more current assets than liabilities. In the retail sector, for example, most companies work on cash sales (i.e. no debtors) which provides for healthy liquidity.

In some manufacturing companies stocks may be too high and contain obsolete and unsaleable items. The current ratio might therefore at first sight be misleading. To provide a more rigorous test of the company's ability to meet short-term obligations, this item, i.e. stocks, is removed from the calculation.

$$\text{Quick ratio} = \frac{\text{Current assets} - \text{stocks}}{\text{Current liabilities}}$$

Capital structure

The net assets of a company can be financed by a mixture of owner's equity and long-term debt. *Gearing ratios* analyse this mixture by measuring the contributions of shareholders against the funds provided by the lenders of the loan capital.

$$\text{Capital structure} = \frac{\text{Long term debt}}{\text{Net sales}} \times 100\%$$

The profit and loss account provides another useful angle on the capital structure. Is there a healthy margin of safety in the profits to meet the fixed interest payments on long-term debt? An over-geared company (see the next section) may show signs of running out of profits to pay this fixed burden.

$$\text{Times interest earned} = \frac{\text{Profit before taxes}}{\text{Interest charges}}$$

To be sure that their dividend is safe, shareholders will want profits compared with the dividend payable.

$$\text{Dividend cover} = \frac{\text{Profit for the financial year}}{\text{Dividend payable}}$$

Activity and efficiency

The ratio showing stock turnover and average collection period helps managers and outsiders to judge how effectively a company manages its assets. The figure of sales is compared with the investments in the various assets.

$$\text{Stock turnover} = \frac{\text{Sales}}{\text{Stock}}$$

$$\text{Average collection period} = \frac{\text{Debtors}}{\text{Sales per day}}$$

Managers should aim to extend the period of credit taken to pay suppliers. However, too long a period will lead to poor trade relations with suppliers, and may even be an indication to the outside world of cash flow problems.

Profitability

This ratio shows management's use of the resources under its control.

$$\text{Profit margin} = \frac{\text{Profit before taxes}}{\text{Sales}} \times 100\%$$

Extraordinary items are excluded from this ratio because they do not represent normal operating profit.

$$\text{Return on total assests} = \frac{\text{Profit before taxes}}{\text{Total Sales}} \times 100\%$$

Profit is closely related to the assets employed by the company. Some analysts calculate the return on specific assets, e.g. the inventory

$$\text{Return on owner's equity} = \frac{\text{Profit before taxes}}{\text{Owner's equity}} \times 100\%$$

If a quoted company fails to earn a reasonable return, the share price will fall and prejudice chances of securing additional capital or long-term debt on beneficial terms.

Capital gearing

The terms *capital gearing* or *capital leverage* are used to describe the relationship between the fixed interest capital and the equity capital of a company. A highly geared company is one which has a high proportion of fixed interest capital compared with equity capital. A low-geared company has little fixed capital compared with ordinary shares, including reserves that form the shareholders' interest. The finance that is provided by the investors themselves is referred to as their *equity* in the company. Tables 4.2, 4.3, 4.4 and 4.5 illustrate the importance of capital gearing. Table 4.2 provides an example of three different firms showing low, medium and high gearing.

Table 4.2   Capital gearing

	Firm		
	A (low)	B (medium)	C (high)
		(£m)	
Ordinary share capital	40	25	10
Reserves	10	10	5
Fixed interest capital	10	25	45
	60	60	60

Table 4.3   Gearing ratios

Firm	Proportions	Ratios	Gearing
A	10:50	0.20	Low
B	25:35	0.71	Medium
C	45:15	3.00	High

Table 4.4   Profitability

Firm	Profit	Interest deductible	Available profits for dividend
A	10 000	1000	9000
B	10 000	2500	7500
C	10 000	4500	5500

Table 4.5   Earnings per share

Firm	Ordinary share capital	Profit	Dividend
		per share	
A	40 000	9000	0.22
B	25 000	7500	0.30
C	10 000	5500	0.55

The most common method of measuring gearing is that of comparing the fixed interest capital with the total shareholders' interest. The gearing ratios can therefore be calculated for the three firms identified in Table 4.2 as shown in Table 4.3.

If the fixed interest capital is raised, and can be used to earn a return in excess of the interest charge then this goes to the ordinary shareholders of the company. This is the rationale for raising the gearing in a company's financial structure.

If each of the companies in Tables 4.2–4.5 make a profit of £10 000 and the fixed interest capital is 10%, then with the same profit, but different gearing, each of the firms will produce differing profits for the ordinary shareholders of the company, as shown in Table 4.4.

However, the higher gearing will produce better earnings per share for the ordinary shareholders, as indicated in Table 4.5.

Gearing tends to have advantages for property companies when rents and values are rising. For example, suppose a company borrows £5 million at 5% to meet half the costs of purchasing a £10 million property showing £600 000 income per year. Its own equity or investment in the property is £5 million. Where the rents rise rapidly and after five years these have doubled, then the value of the building will also probably have doubled. The company still owes £5 million but the property is now worth £20 million. The value of the company's equity in the property has risen from £5 million to £15 million. Due to the principle of gearing the company's interest in the property has risen 200% for a rise in the value of the property of 100%. The gearing ratio has also changed from 1:2 to 1:4.

In the second place, initially the interest charged in the above example was 10% of £5 million, £500 000, leaving the company with a rent return of just £100 000. On the basis of its £5 million investment it was therefore achieving just 2%. This was insufficient on the basis of interest rates of 5%. However, its rent has now doubled to £1.2 million, leaving a residual profit of £700 000 after interest charges have been deducted. This is a more realistic long-term return of 14% on the capital invested.

This scenario would of course be reversed, where property values and their associated rents were falling. In practice it would be usual for the borrower to put up at least a proportion of the full cost. Only in times of rapidly rising property values will finance companies be prepared to put up the 100% full cost. Additionally with property developments, the developer has to fund the venture from inception, including site purchase, through to occupation. The interest charges are often rolled up on to the original loan. Like many other sorts of investment it takes some time before the project begins to provide a real return for its investors.

The MM theory

Another school of thought suggests that capital structure has no effect on the overall cost of capital, except at extreme levels of gearing. This is the argument associated with the names of Modigliani and Miller (1958) who first suggested it. The MM theory, as it has become known, states that the total value of a firm depends on its expected performance and its commercial risk and is independent of the way in which it happens to be financed. Any other position is one of equilibrium which will be eradicated by investors playing the market through the process known as *arbitrage*.

It is not easy to make a realistic comment on what is an appropriate level of gearing for a property company, since this depends upon a number of different factors. These include:

- Whether the property assets are to be held for long-term investment or sold
- The levels of floating rate debt against fixed rate debt
- Whether the company is borrowing secured or unsecured
- The structure of any *off-balance sheet* financing and development. (The Accounting Standards Committee (ASC) of the United Kingdom Accounting Institutes, have prepared information that specifically addresses this issue.)

Interest cover

It has typically been assumed that the operating profits of a company should cover the interest charges of any loans by a ratio of about 3:1. In demonstrating an ability to repay a debt, this ratio is the clearest indication of a company's health. It focuses on operating profit and thereby ignores the distorting effect of provisions in a company's figures included in an annual report. Before the recession of the early 1990s, most companies could demonstrate this requirement. The requirement might be relaxed where there was a virtually guaranteed increase in income within the accounting period, or where exceptionally high interest rates were expected to move downwards in the near future. Of course a sharp rise in interest rates produces just the opposite effect. In the early 1990s, outstanding bank loans to property and construction firms increased rapidly posing problems both for investors and developers as well as bank managers.

During the last recession, many covenants associated with construction and property companies were being reviewed twice a year rather than as previously, where they were typically set at three-yearly cycles. These were not, of course, the only industries to be treated in this way.

Both gearing and interest cover are used extensively by financial lenders, to determine whether a company's borrowings are at a reasonable level and whether a company should be allowed to increase the amount it has borrowed. All investors are concerned with a company's capacity to absorb a downturn in profits without the possible need to sell valuable assets under unfavourable conditions.

Depreciation of assets

When new plant or equipment is purchased, its value from the time of purchase will begin to reduce. The life of such plant may be, for example, five years for some items of equipment, possibly a much shorter period (three years) for equipment that is rapidly changing such as computers, or perhaps a longer period of time for some heavy items of plant installed in a factory.

Buildings, however, have tended to appreciate in value over time, and are one of the few items of capital expenditure to do so. Much of this increase is, of course, attributable to the land on which the building is placed, rather than to the building itself. However, the majority of buildings do not last indefinitely. In *Obsolescence in Buildings: Data for Life Cycle Costing* (Ashworth, 1997), characteristics of building life are considered. Some buildings have a remarkably short life, such as some of the multi-storey blocks of flats that were constructed in the middle of the 1960s and which lasted barely 25 years. There are also examples of buildings that have lasted only for a matter of a few years. The Millennium Dome, at Greenwich, may fall within this category, due to a growing dissatisfaction with this project. However, for the majority of buildings constructed today, they would be expected to have a life approaching upwards of 100 years. During their life they may undergo many changes and regular refurbishment and upgrading. In a proportion of cases one might expect a change from their original designated use.

Depreciation is the term given to a reduction in value over time. It may be necessary to allow for this in the company's balance sheet. It is also important to understand this concept in the context of property investment. There are several different ways of calculating depreciation, in order to distribute the appropriate costs over the expected life of the project. Where depreciation occurs over a long period of time it may be necessary to allow for the time value of money through the use of one of the discounting methods that are outlined in Chapter 5. The following are the usual methods of allowing for depreciation and they are compared in Figure 4.1.

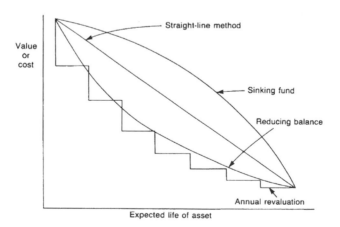

Figure 4.1   Methods of depreciation

The straight-line method

This is also sometimes described as the fixed instalment method. The original value of the asset, less any residual value, is divided by the number of years of its estimated life.

Example

A manufacturer has purchased a new item of capital equipment for £30 000. It has an expected scrap value of £5000 and an expected life of 5 years. What is its annual depreciation?

$$\frac{300\ 000 - 5000}{5} = 5000$$

Whilst the straight-line method is simple to calculate, it has the disadvantage that it does not represent the actual depreciation of an asset. This will be higher during the first few years of ownership and reduces as it reaches the end of the asset's useful life. If these figures were used in a company's accounts, then it would be necessary to amend the final year's figures by a balancing adjustment, to agree with the actual amounts involved. If the equipment was to be replaced at the end of its useful life then it might be necessary to allocate these amounts to a sinking fund for the equipment's replacement, in which case, due to inflation, it would probably be insufficient.

The reducing balance method

Since most plant and equipment involves higher depreciation amounts during the earlier years of their lives, it is desirable that the depreciation calculated models this fact. The reducing balance method reduces the value of an item of plant or equipment by a fixed percentage each year. Whilst the percentage is fixed, the actual amount will vary, reducing each year. The Inland Revenue typically imposes a figure of 25% for taxation purposes. The Inland Revenue in its calculations for depreciation adopts the principles of initial allowances, writing-down allowances and balancing allowances and charges (see later). For items that have a very short life expectancy then the depreciation rate will need to be increased. The example in Table 4.6 is based upon 25% depreciation.

Table 4.6 Depreciation

Year	Depreciation for year £	Book value £
0	0	10 000
1	2500	7500
2	1875	5625
3	1406	4219
4	1054	3155

The depreciation fund method

A fixed proportion of the initial cost is transferred each year from the revenue account to the depreciation reserve. If this is allowed to accumulate with compound interest, it should at the end of the asset's life produce an initial cost less value. An alternative to this method is referred to as the sinking fund method, where the annual sum is then reinvested. The advantage of this method is that it provides the actual cash in order to replace the asset. A further alternative approach is the insurance policy method, whereby a policy is taken out with an insurance company for the amount of the asset, due when the asset is to be replaced.

The valuation method

Depreciation may also be determined by the process of actual revaluation of the asset at fixed periods of time, normally annually at the end of the financial year. Such valuations are also a useful check on the other methods that might be used to calculate depreciation. In contracting, it might also be used to revalue the major items of mechanical plant at the end of a contract, where such costs in use can then be allocated to the overall project costs.

Share analysis

In the *Financial Times*, information is published on a range of comparisons of different companies and their financial performance using a system of financial ratios. These ratios measure only the current performance of a company's shares. They do not seek to measure future risk or return. The assessment of these two factors relies ultimately upon skill and judgement using a wide range of commercial information that is available.

Gearing ratio

This measures the relationship between debt capital and capital employed.

$$\text{Capital gearing} = \frac{\text{Debt capital}}{\text{Capital employed}} \times 100\%$$

The higher the ratio, the greater the risk posed to the company's equity capital by a fall in asset values. It might also lead one to expect that the company's earnings would be more volatile. This might not be the case since a lower interest rate payable on loan stock will to some extent compensate for the greater gearing. It is the income gearing that determines the risk and potential return to equity earnings.

$$\text{Income gearing} = \frac{\text{Interest payments}}{\text{Gross trading profit}} \times 100\%$$

Price–earnings (P/E) ratio

This ratio is one of the most widely quoted share ratios, and is provided along with other information in the publication of share data in the *Financial Times*. It is the ratio of equity earnings, i.e. profit after the deduction of tax, per share.

$$\text{Earnings per share} = \frac{\text{Equity earnings}}{\text{Number of share issued}}$$

It is also necessary to consider the above in the context of the price of the share to obtain a figure that is more of a true comparison.

$$\text{Price–earnings ratio} = \frac{\text{Share price}}{\text{Earnings per share}}$$

Figure 4.2 shows the FTSE all-share price–earnings ratio from the beginning of 1993 to the middle of 2001. The price–earnings ratio is a far from perfect measure. The earnings are the earnings after all costs except dividends, but exceptional costs and credits can distort the picture. By the same token, some element of subjectivity will creep in if adjustments are made to allow for one-off events. But since a company's single most important goal is to generate profit, it is surely sensible to use a per share measure that takes account of all except the most unusual credits. Adjustments to individual corporate circumstances are better made through shifts in the value of the shares, not by redefining the earnings measure. Ordinary earnings and price–earnings ratios have advantages over other measures because data is readily available. Investors may also get a better idea of underlying value by drawing comparisons across sectors, and that is made easier using p/e ratios.

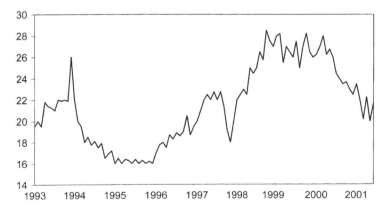

Figure 4.2   FTSE all-share index price–earnings ratio

The price–earnings ratio is a robust measure. But shares that look cheap on a price–earnings basis may not present bargains. House builders, most of which are well financed and are stringing together good results because of the current buoyant housing market, defy logic by remaining at the bottom of the pile of companies. Many trade on forward price–earnings ratios of between 5.5 and 7.0.

Dividend yield

Although earnings are traditionally the return to shareholders, after all expenses have been met, companies often retain a proportion of these earnings to plough back into the business. This is to replace plant or equipment that has become obsolete, to invest in new buildings or to provide for other capital items. While the gross trading profits allow for the depreciation of these capital items, further capital may be required to cover any shortfall from this calculation or to allow for investment in improved technology. This of course may be a disguise to avoid paying the full amount of the equity earnings to the shareholders. The dividend yield per share of the two companies is therefore calculated as follows:

$$\text{Dividend per share} = \frac{\text{Dividend payment}}{\text{Number of share issued}}$$

However, it is the amount of dividend received per £1 spent on buying the shares that is important. The relationship of the share price and dividend is normally quoted as a percentage yield, termed the dividend yield:

$$\text{Dividend yield (gross)} = \frac{\text{Net dividend per share} \times 100\%}{\text{Current share price}}$$

Dividend cover

A ratio that provides an investor with an indication of both the security of future dividend payments and the potential for increase is termed the dividend cover. This calculated as follows:

$$\text{Dividend cover} = \frac{\text{Earnings per share}}{\text{Dividend per share}}$$

This ratio measures the relationship between what is earned and what is actually paid out by way of a dividend. If the ratio calculated is 1, then it indicates that everything earned is actually paid out with nothing being retained for possible future investment or contingencies. Where the ratio exceeds 1, then this suggests that the dividend paid is less than what has been earned, whereas when the ratio is less than 1 then the dividend has used some of the reserves.

Assets per share

Although it is the profit-earning capacity of a company which is the dominant factor influencing the market price of its shares, the value of the assets owned by the company is also of importance. This is especially important in companies where a takeover bid is possible. This ratio represents the total net assets less total liabilities per share. If the company ceased to trade then this is the amount that each shareholder would expect to receive.

$$\text{Net assets per share} = \frac{\text{Total net assets} - \text{total liabilities}}{\text{Number of share issued}}$$

Equity capitalisation

The value of a share is largely dependent upon the company's ability to earn a profit. The share price will, however, not normally be closely related to the value of a company's assets, except in the case of a takeover or liquidation. The usual measure of the value of a company is its equity capitalisation, i.e. the number of shares issued multiplied by their current price.

Other considerations

This concludes the description of different forms of analysis, but in practice it would also be important to examine as much other information as possible regarding a company, such as its annual reports. In normal times, it would be expected the earnings per share in a good company would be as a minimum 15%, perhaps totalling 100% over the preceding five years. There might have been a small setback in one of these years, but earnings in the remaining four should make up this difference. The following are indicators to look for in share performance expectations:

- The chairman's statement in the annual report should be optimistic
- The dividend should be positive (where a company has constantly been either increasing or maintaining a dividend and this suddenly drops (or is even maintained) then this should be taken as a warning signal)

- It is important to look for general indicators, such as the overall performance in a particular sector, taking into account government economic policy and expectations.

The key indicator for all investors is the dividend yield paid on the amount invested, rather than necessarily the number of shares owned since these are likely to be at different prices, unless they are rights issues.

Audit

The shareholders, not the management, appoint an independent firm of accountants to report to them each year on the examination of the accounts. The report will state whether in the auditor's opinion, the financial statements have been properly prepared in accordance with the provisions of the Companies Act and whether in their opinion, they represent a true and fair view of the profit, or loss, of the company and the financial position at the year end.

It is sometimes assumed that the main function of an audit is to detect errors or fraud. This forms only a subsidiary objective of the main auditing function, which is described as follows. The auditors must ensure that the books, records and internal controls have been satisfactorily maintained during the financial year. They will therefore:

- Examine the accounting system to ascertain whether it is appropriate for the business and properly records all transactions
- Compare the accounts with the underlying records to see that they are in compliance with them
- Verify the title, existence and value of the assets, and the amount of liabilities in the balance sheet
- Verify that the results shown in the profit and loss account are fairly stated
- Confirm that the statutory requirements have been complied with and that the relevant accounting standards have been correctly applied.

The final accounts of building contracts are frequently subjected to a technical audit, in addition to the above. This involves an examination of all the transactions and procedures that have been used throughout the contract and the manner in which the accounting work has been recorded. The extent of the examination of the documents will depend upon the auditor's assessment of internal control and the procedures that should have been followed.

Financial auditing is undertaken by someone with an accounting qualification, indeed this is a requirement for all larger companies under the Companies Act. The needs of a technical audit are different. These require someone with a more specialised knowledge of construction practices and procedures, and since it is in part a financial operation, a quantity surveyor is frequently employed for this purpose. For further information on technical auditing refer to *Practice and Procedure for the Quantity Surveyor* (Ashworth and Hogg, 2006).

Audited accounts will carry more authority with the Inland Revenue and are also useful to prove to third parties, including prospective clients, that the financial status of the business has been independently scrutinised. The system is not perfect. There are examples where auditors have been shown to be lax in the execution of these duties, with fines or damages having to be paid to an aggrieved party. The near collapse of Baring's Bank resulted in a claim for damages of over £1 million, against the well known firm of accountants who earned out the audit. The matter was eventually settled out of court. The more recent case of the American company ENRON is an example of auditors failing to verify the true value of a company's assets.

Ultimately questions about financial regularity might be asked in Parliament regarding public authorities, at a shareholders' meeting for a company, or in the seclusion of a study of a private client.

Taxation and property

The influence of taxation and its effects on property are constantly changing due to revisions in taxation principles and the introduction of new measures or rates from time to time by the Chancellor of the Exchequer. The application of taxation through statute law is also governed by case law that is tested in the courts. Taxation may be *direct* or *indirect*. The former is normally collected by the Inland Revenue and the latter by Customs and Excise. Taxes such as the council tax and business rates are collected by the local authority in which the property is situated. Taxes may be further classified as *proportional*, in which case they are a fixed percentage of price, *regressive* where the percentage is smaller with increased income; or *progressive* which takes higher percentages on larger incomes. Government adjusts taxation to provide income to meet its programmes and policies of social, political and economic needs. It normally does this through an annual budget expressed in legislation through the Finance Acts, but also at other times when needs arise.

New developments now attempt to take taxation into account in order to minimise the overall expenditure that is likely to be incurred. This is particularly important in the context of capital allowances, which are discussed on p. 108 and the need to design tax efficient construction and development projects. The main types of taxation that are relevant to property are listed and examined separately below:

- Capital gains tax
- Inheritance tax
- Stamp duty
- Value added tax

- Council tax
- Business rates.

Capital gains tax

This tax was introduced in April 1965. The law has been consolidated in the Taxation of Chargeable Gains Act 1992. Capital gains tax applies when chargeable assets such as property are disposed of.

Inheritance tax

Inheritance tax is charged on certain lifetime gifts by individuals, on wealth at death and on certain transfers in and out of trusts. The tax, originally referred to as capital transfer tax, was introduced in 1975 to replace duty. The law is contained in the Inheritance Tax Act 1984. United Kingdom domiciled individuals are chargeable to inheritance tax in respect of property anywhere in the world. Non-domiciled United Kingdom individuals are charged in respect of property in the United Kingdom only. Note that domicile is a more involved concept than residence and specialist advice should be sought when dealing with it.

Stamp duty

Stamp duty is a fixed or *ad valorem* charge on documents such as conveyances and land transfers, with limited exemptions in certain disadvantaged areas of the UK designated since November 2001. It has been charged since the seventeenth century. Current legislation is based on the Stamp Act 1891, as amended by numerous and subsequent Acts, especially the Finance Acts. It is administered by the Commissioners of Inland Revenue through the Office of the Controller of Stamps. A separate tax, *stamp duty reserve tax*, was introduced by the Finance Act 1986 to charge certain exemptions that escape duty.

Value Added Tax

Valued Added Tax (VAT) is charged on the supply of goods and services in the United Kingdom, and on the import of certain goods and services into the United Kingdom. It applies where the supplies are *taxable supplies* made *in the course of business* by a *taxable person*. Significant changes were made to the United Kingdom system as a result of the introduction by the European Community of the Single European Market on 1 January 1993.

Value Added Tax was introduced to the construction and property industries through the Finance Act 1972. Since this time the extent and rates of tax have been amended several times. The current legislation is covered in the HM Customs and Excise leaflet 708/2/90 dated 1st August 1990. In addition to this leaflet, other provisions cover Protected buildings (VAT leaflet 708/1), Property development (VAT leaflet 742A), Property ownership (VAT leaflet 742B), Aids for handicapped persons (VAT leaflet 701/7) and VAT refunds for DIY builders (Notice 719).

Building work is either standard rated work (currently 17.5%) or zero rated work. Examples of zero rated works are residential buildings which include children's homes, old people's homes, homes for rehabilitation purposes, hospices, student living accommodation, armed forces living accommodation, religious community dwellings and other accommodation which is used for residential purposes. Certain buildings intended for use by registered charities may also be zero rated. Buildings which are specifically excluded from zero rating include hospitals, hotels, inns and similar establishments. The conversion, reconstruction, alteration or enlargement of any existing building are always standard rated. All services which are merely incidental to the construction of a qualifying building are standard rated. These include architects, surveyors and other consultants' fees and much of the temporary work associated with a project. Items which may be typically described as 'furnishings and fittings' (fitted furniture, domestic appliances, carpets, free standing equipment, etc. are always standard rated irrespective of whether the project may be classified as zero rated. The VAT guides provide examples, but throughout the document individuals are advised to check their respective liability with the local VAT office, and seek special-ist advice. The rating of some items is arbitrary and some will need to be tested by the courts.

Council tax

The council tax replaced the community charge, also known as the poll tax from April 1993. Whilst many people suffered large increases in their household bills under the poll tax, compared to their payments under the previous local rates system, they were cushioned by a reduction scheme. A similar scheme was introduced to provide transi-tional relief to households whose council tax bills were significantly more than their combined poll tax.

Council tax is payable on a dwelling, i.e. a house, flat or mobile home. There is a single bill for each dwelling. Properties that are mixed hereditaments will be propor-tionately assessed under each tax. There is a long list of dwellings that are exempt and these include unfurnished property that is empty for up to six months, students' halls of residence, unoccupied managed property, etc. The amount of the bill is calculated on the basis of allocating the property to one of the valuation bands A–H. Properties are normally only revalued when they are sold, even where they may have been substantially improved or extended in the meantime. However, there is the provision for adjusting values downwards at any time if there is a major change in the area, such as a motorway being built nearby or if the property is adapted for someone who is disabled. Like any tax there are procedures that can be used to appeal against a valuation in certain circumstances. There is only one council tax bill for each dwelling. The council tax is based upon a 50% property element and a 50% personal element. The overall bill is reduced by 25% where there is only one householder resident and by 50% if the property is not a main home. Those on low incomes are able to claim benefit of up to 100% of the bill.

Business rates

From April 1990, business rates are paid by businesses at a uniform level fixed by central government on the basis of their rateable values. This tax is also referred to as the national non-domestic rate. This applies throughout the country other than to businesses served by the Corporation of the City of London for which there are separate arrangements. The introduction of this system was preceded by a property revaluation. There was a further revaluation at April 1993 that has affected assessments since April 1995. Future annual increases in the business rate will not exceed the increases in the retail price index (RPI). There were transitional arrangements to limit gains and losses for individual businesses over the first few years. Where property was transferred to a new owner, the transitional relief could be transferred. Property that has remained empty for three months attracts only half the normal rates assessment.

Capital allowances

Whilst capital expenditure is not allowable in calculating income profits, the taxable profits may be reduced in the form of allowances. The law on capital allowances is contained in the Capital Allowances Act 1990. It includes for provisions to allow it to be amended by subsequent Finance Acts. Allowances may be given for plant and machinery, industrial buildings, agricultural buildings, hotels, etc.

The allowances are calculated on the basis of the following:

- *Initial allowance*: this is an initial sum that is allowed against the expenditure of an item in any financial year.
- *Writing down allowance*: the writing-down allowance is a sum allowed by the Inland Revenue that can be offset against income on an annual basis for a specified term of years. These may represent, for example, the depreciation of an asset.
- *Balancing allowances and charges*: a balancing allowance or charge is provided, where upon the item's disposal, the actual amount is calculated and adjusted to take into

account the above allowances that have already been given. If the sale of the asset falls short of the unrelieved expenditure then a balancing allowance is made.

These allowances do not need to be taken in full in order to make the best possible advantage of the reliefs and allowances that are available.

Plant and machinery

The treatment of capital expenditure on plant and machinery is a very complex area. Whilst the definition regarding machinery is generally understood, plant is not and has thus come before the courts on many occasions. The main problem lies in distinguishing the apparatus with which a business is carried out from the setting in which it is carried on. Items forming part of the setting do not attract relief unless they do so as part of the building itself and not as plant. Lifts and central heating systems are treated as plant, but plumbing and electricity systems are not. Specific lighting to create an atmosphere in a hotel and special lighting in fast food restaurants have been held as plant.

Expenditure on computer hardware is a capital expenditure on plant and machinery. Allowances are usually claimed under the short life asset rules. Where software is purchased at the same time the Inland Revenue have suggested that this should also be treated as part of this same capital cost. Licences to operate software are treated as a normal expense against profits.

Industrial buildings

These are treated differently to other types of buildings. They may be broadly defined as buildings used for the processing or manufacture of goods. They also include buildings that are used for the storage of materials before manufacture and for goods after production. They must have a direct link with production and as such wholesale warehouses are excluded. Offices that form a part of the factory are included if they do not exceed 25% of the total cost.

The full costs of construction including professional fees are allowed. Land costs are excluded but the costs of any site preparation may be included. The full costs of the purchase of a building from a builder are allowed. The costs expended on an existing building and plant and equipment costs associated with that building are also included. The allowable costs are 20% of the initial costs and a writing-down allowance of 4% until the costs have been fully written down. In the event of a sale, a balancing adjustment is introduced and a possible claim for relief by the purchaser.

Where an area has been designated as an Enterprise Zone by the Secretary of State, expenditure incurred on buildings qualifies for an initial allowance of 100%. Where fixed plant and machinery are an integral part of the building, this can also be treated as a part of the building for the purpose of claiming Enterprise Zone allowances.

Risk and uncertainty

When preparing a plan for the future, the associated risk and uncertainty increase the further these predictions are projected. We are able to use our own experiences to assess the confidence of our predictions. We can also enhance this confidence by examining data and other information. Both the forecasting of events at some distance in the future and the preoccupation with the issues and events of the present time militate against thorough investment appraisals and subsequent decision making. Because capital investments often span a long period of time they are never made under conditions of certainty. Risk covers attempts to quantify events about which we have some knowledge. Uncertainty is concerned with events that either cannot be measured or perhaps not even contemplated. Both risk and uncertainty may result in outcomes that are better or worse than expected.

There is therefore the argument that to apply any forms of analysis to problems of risk and uncertainty is misplaced. These arguments tend to perpetuate the views that decisions are made without the support of all the facts or a rational evaluation of the information that is available. It must, however, be stressed that the analysis must always be supported by experience. The analysis should in many ways help to confirm what is already supposed, although this might not always be the case, since analysis frequently identifies other issues or events that may have not been considered in the assessment of the project. Techniques for dealing with risk and uncertainty must therefore, always seek to combine both judgement and analysis, each relying on the other for support.

Risk may be defined as uncertainty about a possible future loss. It is however, different to uncertainty since it can be mathematically predicted, whereas uncertainty cannot. For example, on the basis of past performance, a developer could assess the risks involved in building for lease an office block in any major city in the United Kingdom. Whilst risk is involved in making the decision there is sufficient data available by which to assess the risks. Uncertainty occurs where there is no data on previous performance on which to base the judgement. The valuation of property in Eastern Europe provides just such an example. The state, until the demise of communism, owned all property and transactions between previous owners did not exist. Risks may be dealt with in several different ways. Table 4.7 gives examples of the kinds of risks and uncertainties that are associated with the development of construction projects.

Table 4.7 Examples of risk and uncertainty affecting project development

Need and demand for the project
Costs of construction
Gross development value
Economic conditions
Type of procurement arrangements
Timing of the project

Risks can be avoided

The total avoidance of all risks can only be achieved by non activity. However, the risks associated by building on poor load bearing ground can be avoided by choosing to build elsewhere. In practice the site investigation report may recommend such a course of action, since the costs involved in building under these circumstances might not be recoupable from the finished project. Where the risk involved cannot be counter-balanced by the utility gained from the risky action, then the risk should always be avoided. Avoidance is only possible if that choice exists. If it cannot then it must be handled in some other way.

Risks may be accepted

The development of property is such that if one wishes to be involved, then the risks must be accepted in one way or another. Sometimes risks are accepted in ignorance since the different liabilities have not been carefully considered. Contractors, for example, may undertake work for development companies, unaware of the implications of the conditions of contract that are being suggested. Sometimes risks are recognised but the opportunity to either transfer or minimise them is neglected. A contract for a new development project may involve aspects of work of a highly technical or specialised nature. The contractor may recognise this, but instead of subcontracting the work to a specialist firm, may decide to execute the work and therefore accept the risks that are involved. In other circumstances, risks may be accepted unintentionally, believing that the contract documents for a project suggest one thing but they actually mean something rather different.

Risks may be averaged

This involves sharing the risks involved between those involved. The principle should be that risk should be allocated to the party that is best able to control it. Evidence suggests that where a building client places all of the risk with the contractor, a worse deal is struck. This is especially true, since some of the risk will never materialise.

Risks may be ignored

This is not a recommended approach. The belief that the project will work out all right in the end is the precursor to many bankruptcies in the construction and property industries. Another view suggesting that all risks will be controlled and contained is equally unrealistic, because past evidence suggests the folly of this statement. Others will argue that their own intuition is sufficient on which to assess the risks involved and see no point in attempting to quantify risk.

Some techniques for dealing with risk and uncertainty

Probability

Probability is an important concept when dealing with the analysis of risk. It is often helpful when dealing with uncertainties in cost and price forecasting to consider a range of results rather than a single outcome. It might be suggested that the development value of a proposed project is likely to be within £10 million and £12 million. It is also possible to attach to this range an estimate of its probability. Probabilities may be estimated from past results or statistical records of previous events, or they may be obtained by conducting experiments using a sampling procedure.

Decision tree analysis

Many investment decisions are not isolated events but are often a process regarding an overall strategy of development. A decision tree comprises a number of branches that originate from a first question such as 'Should we carry out development?' The analysis is characterised by a series of either/or decisions.

Sensitivity analysis

This is discussed in Chapter 5. Essentially it is a method that is used to test the impact of change in the values of variables in a model. Such variables are capable of having a range of values attributed to them. For example, in a developer's budget, the selling prices of houses are estimated. The selling price is affected by a number of different circumstances. Sensitivity analysis is used to test whether under extreme circumstances, if the development was to proceed, it would do so at a loss. Any of the variables can be changed to provide a worst and best scenario and the likelihood of either of these events actually taking place.

Simulation

This approach assumes that the values of the different variables may be combined with each other on a random basis. It is sometimes referred to as the Monte Carlo method since it uses random numbers to select outcomes. The origins of simulation are threefold. First, there has always been a desire to avoid direct experimentation whenever possible. Direct experimentation may involve developing and testing a particular system and this may be a very costly procedure to manage. Obviously at some stage of the development of the procedures this will, however, become necessary. The second reason stems from the solution of purely mathematical problems. Simulations, unlike these which generally represent steady state behaviour, are observations that are subject to experimental error. This means of course that they must be treated as a statistical experiment and any inference regarding the performance must therefore be subject to the tests of statistical analysis. The third reason lies in the growth area of the subject of operational research and the consequent need to explain variability that occurs in the information that is provided. The art and science of simulation is in designing a model that behaves in the same way as real life.

Portfolio theory

Portfolio theory or analysis is a technique that seeks to identify the efficient set of investment characteristics. Once these have been identified, then the developer will choose those that best satisfy the overall objectives for the project. At the outset the analysis may require a long list of characteristics that may be of some importance. The list is often too extensive to explore fully and some of the characteristics may have a very minor impact upon the final solution. The important characteristics are then evaluated, often numerically but not so in every case. Some of the characteristics are considered as essential, others as required and a third group as desirable. Correlation coefficients can be measured between these characteristics.

Break-even analysis

Break-even analysis is discussed in Chapter 10. The break-even point (or more correctly, break-even circle) can be used as a basis to assess the risk of adopting a particular decision pathway.

Scenario analysis

This technique essentially examines different scenarios as possible options. The aim is to consider the likely outcomes of a solution in a more carefully considered manner by examining the different variables involved in the decision-making process. Scenarios are chosen that represent the most likely, optimistic and pessimistic cases.

Chapter 5
Development Appraisal

Introduction

The development of construction projects arises for several reasons. They may be undertaken in the public sector to meet political, social or community needs. They may be undertaken in the private sector for use, or as projects that can be sold upon completion, be rented or leased to some other organisation. Different techniques are available by which to evaluate the original needs of development. In some cases, these rely upon investment appraisal techniques that assess the expected profitability of undertaking such work. Other techniques may also be used that attempt to form a relationship between the benefits that might be achieved of developing compared with the costs involved with the project. In other cases it is possible to calculate the costs of not undertaking such work and to compare these against the costs of development. Public accountability also requires that funds have been spent wisely on appropriate developments. In every situation the necessity of understanding the full financial implications is very important, since whether the development is private or public sector funded, there are only limited funds available for investment purposes.

If the project is to be effective, then adequate systems of investment appraisal must be adopted at inception, while the concept is still little more than a possible solution to meet either a need or a desire. It has often been suggested that at least 70% of the initial capital costs of construction are already committed to the design, once the project leaves the inception stage and enters the next stage in the development process. Development appraisal is the title given to examining the financial implications of a project at this stage. It is therefore important to have some understanding of valuation methods, valuation tables, the developer's budget and other ancillary matters.

Development value

The development value of a plot of land is the difference between the costs of development, which will include the costs of the land, construction, etc., and the market value

114

of the finished work. The latter is much more subjective and difficult to predict since it is more influenced by the state of the economy at the time of completion, which will be some time away. It will also be in competition with other similar projects that will be available at the time of completion.

There are a wide range of considerations influencing the development of a construction site. These include:

- Type of development envisaged
- Location
- Shape, size, topography, aspect and access
- Ground conditions and site preparation difficulties
- Availability of utility services
- Planning controls
- Legal considerations
- Government assistance that might be available.

These factors are discussed in Chapter 6. In addition, the costs of developing the site and its eventual worth must also be considered.

General determinants of value

The supply of property is relatively inelastic due to a number of factors, particularly the physical nature of land and the length of time required for development purposes. The demand for property is also relatively inelastic. The demand for property arises from four possible motives:

- Occupation
- Investment
- Speculation
- Development.

Some of the major factors that affect demand, and hence value, are as follows:

- *The general state of a country's economy*: in a time of economic well-being there is a desire to invest in property
- *Structural changes in the economy*, such as a movement from manufacturing to a service sector economy, increases the need for office accommodation
- *Significant changes in the costs of ownership*, such as rents and taxes; a significant increase in business rates may cause businesses to cease to trade
- *Location*: the better located offices are able to charge the higher rents
- *Condition*: this factor will have an effect at the margins of value (where the condition is deteriorating, then this will be a more significant factor)

- *Obsolescence*: especially where the building is no longer suitable for modern techno-logy or practices
- *Government*: provision of grant aid and other incentives will affect the worth of the property
- *Infrastructure*: the provision of new or easier means of transport will have a positive effect on value where the provision offers benefit to the owner
- *Population*: demographic trends will influence the size of families and hence hous-ing; age longevity may dictate more sheltered housing schemes
- *Funding*: changing the costs of borrowing or the amounts that lenders are prepared to lend will have significant effects upon value.

Price, value and worth

To avoid confusion, care is needed over the use of the words price, value and worth. Whilst price and value are market-driven terms, worth is subjective and based on a client's particular circumstances, i.e. value in use. The calculation of worth means the provision of a written estimate of the net monetary worth at a stated date of the benefits and costs of ownership of a specified interest in property to the instructing party reflect-ing the purposes specified by that party.

The property market has laboured for many years without a clear understanding of the difference between price, worth and value. This confusion derives in part from the use of the word value as a substitute for price when in many people's minds value also means worth. This problem is exacerbated by the fact that, when asked to provide an opinion on the spot price of a property, a valuer produces a figure which is described as the open market value. This situation is not helped when in most dictionaries, the words value and worth are merely synonyms: value being defined as both worth (desir-ability or utility) and price (value in exchange) and vice versa.

In the language of economics, worth can be considered as value in use, whereas price or open market value can be considered as the most usual manifestation of value in exchange. Price or value in exchange is the outcome of the interplay of the respective values in use of market makers. In an open and free market, no transaction will be likely if the value in use or worth to the putative vendor is greater than the value in use/worth to a putative purchaser. Similarly, potential purchasers will not be willing to deal at a price which is greater than their respective values in use/worth. Hence, where practitioners are providing purchase or sale advice, they should provide calculations of worth and value in use in order to advise as to whether a sale or purchase should proceed at any given level of price. In the property market, what is often called a *valuation* is the best estimate of the trading or spot price of a building or land and calculation of worth is the range of individual assessments of worth/value in use to a range of potential purchasers or vendors. A prospective owner may also undertake calculations of worth of two or more different assets on a like basis to enable investment choices to be made in a rational way.

Bearing in mind the confusion between price, worth and value, it is important to be fully appraised of the relevant definitions adopted by the RICS. These fall under the headings of valuation, appraisal and calculation of worth and are contained in the RICS *Appraisal and Valuation Manual*. They identify the task being undertaken by the valuer as distinct from the definitions of the bases of value, such as open market value, existing use value and the like. The following are useful definitions that are adopted by the RICS:

- *Valuation* means the provision of a written opinion as to capital price or value, or rental price or value, on any given basis in respect of an interest in property, with or without associated information, assumptions or qualifications. It does not include a forecast of value.
- *Appraisal* means the written provision of a valuation, combined with professional opinion, advice and/or analysis relating to the suitability or profitability, or otherwise, of the subject property for defined purposes, or to the effects of specified circumstances thereon, as judged by the valuer following relevant investigations. It may incorporate a calculation of worth as requirements dictate.
- *Worth* means the provision of a written estimate of the net monetary worth, at a stated date, of the benefits and costs of ownership of a specified interest in property to the instructing party reflecting the purpose(s) specified by that party.

Having due regard to the preceding commentary on the difference between price and worth, it is felt appropriate that the following convention should be used by those involved in the provision of valuations and calculations of worth:

- *Price* is the actual observable exchange price in the open market.
- *Value* is an estimate of the price that would be achieved if the property were to be sold in the market.
- *Worth* is a specific investor's perception of the capital sum which would be prepared to be paid (or accepted) for the stream of benefits which are expected to be produced by the investment.

Worth is sometimes believed to be in the eye of the beholder. It is not an abstract concept and is capable of calculation. However, because the value in use of a property will vary from one user to another, calculations of worth will require significant input from the client/user in respect of their individual requirements and preferences.

RICS Red Book

The RICS *Appraisal and Valuation Standards* (The Red Book) contains mandatory rules, best practice guidance and related commentary for all RICS members undertaking asset valuations. It was first published in 1980, and has been updated many times since then.

The current fifth edition was first published in 2003. Changes to the standards are approved by the RICS Valuation Faculty Board, and the Red Book is updated accordingly on a regular basis. The standards are divided into two main parts. The first contains rules and guidance applying to RICS members anywhere in the world and is consistent with the principal rules of International Valuation Standards. The second contains material which relates specifically to particular countries.

The Red Book also has a related suite of Valuation Information Papers which discuss valuation methodology as it relates to specific property types. The Red Book is also available in printed format or via a fully functional premium, web-based version.

The RICS *Code of Measuring Practice*, an RICS Guidance Note, provides succinct, precise definitions to permit the accurate measurement, calculation of sizes and description of land and buildings. The fifth edition has been updated throughout to reflect current practice, and now takes into account the application of the Code to leisure property. Further information on supporting Guidance Notes is available from the RICS.

The Red Book and the European Blue Book both recognise two main bases of valuation for secured lending purposes. *Market Value* which is an estimate of exchange price at the date of valuation, assuming the marketing period to have already taken place, and *Mortgage Lending Value*, which is a more subjective assessment. The former is readily understood and most frequently used by valuers. For residential property, the Red Book also recognises *Projected Market Value* which is an estimate of exchange price on the assumption that the marketing period begins on the date of valuation. The Red Book prescribes the minimum contents of valuation reports. Many within the profession believe that valuation reports have improved in the past few years. The Competition Commission found that lenders on residential property routinely sought to control the selection of valuers through the use of in-house valuers and panels. Lenders on commercial property were far less prescriptive and were often willing to permit the borrower to be involved in selecting the valuer. The information provided by valuers usually includes aspects of the following information:

- Physical features of land and property
- Tenancy agreements
- General suitability for lending
- Specific location
- Demand for property
- Planning situation
- State of lending market
- Valuation method
- General information on comparables
- Suitability of specific property for a loan
- State of economy
- Uncertainty of valuation figure
- Valuation calculations.

Valuation Office Agency

The Valuation Office Agency (VOA) is an executive agency of HM Revenue and Customs (HMRC) with 85 offices throughout England, Wales and Scotland, employing around 4300 people. Its main functions are to:

- Compile and maintain the business rating and council tax valuation lists for England and Wales. (In Scotland council tax and business rates are dealt with by the Scottish Assessors.)
- Value property in England, Wales and Scotland for the purposes of taxes administered by the HM Revenue and Customs.
- Provide statutory and non-statutory property valuation services in England, Wales and Scotland.
- Give policy advice to Ministers on property valuation matters.

There has been no change to the way that homes are valued for council tax since the system was introduced in 1993. There are no extra charges for things such as living in a quiet area, having a nice view, a conservatory, or off-street parking. These characteristics are considered with others such as age, location and type, to assess the overall value, as they would be if you were selling your home, and a council tax band is allocated, based on its value as at 1 April 1991. Characteristics that increase the value of a home are taken into account as much as those that decrease its value. This is to ensure your local council bills you for the correct amount of council tax. Many people commonly believe that they will have to pay more council tax because house prices have risen in recent years, however, this is not the case.

Valuation Office valuers rarely need to visit people's homes. Less than 1% of homeowners receive a visit. Most of the information is derived from questionnaires, by telephone or from the local council. On those rare occasions when it is necessary to visit a dwelling, all the information required can be obtained from the outside, usually from the road. It is extremely rare that photographs need to be taken inside a home and this would never be done without permission.

It was announced in September 2005 that a revaluation of properties was being postponed and there is no indication when the next revaluation will take place. The purpose of revaluation is to help keep the assessments on homes in line with fluctuations in the property market.

Investment appraisal

In its simplest form, an investment decision can be defined as one that involves a firm in making a financial outlay with the aim of receiving, in return, future cash inflows. Different variations on this definition are possible, such as financial outlay today resulting in a saving of expenditure at some time in the future.

Investment appraisal is an aid to decision-making. Its objective is to achieve the maximum return that can be obtained from investment expenditure. It comprises a range of techniques for sorting, organising and presenting information and alternatives to assist decision-makers to achieve the best value for money from the use of resources. The techniques of investment appraisal are relevant for a whole range of capital investment decisions such as buildings and equipment. It may be used to decide:

● Whether to invest in new facilities
● Between alternative methods of achieving a given objective
● Whether to continue to use or dispose of existing assets
● Upon the quality and standards of a design
● Maintenance and service schedules.

Investment appraisal has therefore an application throughout the whole life of the project and the associated costs involved. This ranges from the setting of strategic priorities up to the final details of design and operational practice and the eventual disposal of the site.

It is worth remembering that the different methods used in investment appraisal may produce different solutions. These depend on the objectives that have been specified. It should also be noted that these techniques are only a guide for decision makers: they will not make the decision automatically. However, they will help to make a more informed decision, where judgement then relies upon some form of analysis in addition to other forms of skills and expertise. They provide a more objective solution to the problem. However, such techniques will never replace managerial judgement. The preparation of estimates or forecasts of some future activity will also always include some element of uncertainty and this can only be assessed on the basis of previous performance and expected trends.

The basic steps of investment appraisal normally follow the following sequence:

● *Define the objectives*: it is necessary to define the expected outcome clearly at the outset, and to evaluate how the results are to be measured. The objectives need to be a balance between the general and the specific. If they are too general, then they may lack credibility, whereas if they are too narrow, other viable options may be overlooked.
● *Identify options*: several different solutions may be available, all of which achieve the prescribed objectives.
● *Measure costs and benefits*: the correct choice can only properly be made in the light of the full facts, and these should be measured as accurately as possible.
● *Discount costs and benefits*: in order to evaluate present and future costs properly, these must both be transferred to a common time base by discounting.
● *Consider uncertainties*: some aspects of the proposal will be unknown and it is essential to assess how these may affect the final outcome of the appraisal.

- *Assess other factors*: other factors which are outside the scope of the analysis may have a bearing upon the final decision: for example, political uncertainty and how this might affect the decision will need to be considered.

Relevant techniques are available to assist and help identify the most profitable of a number of options. These can be used alongside professional judgement, to provide some objectivity in the analysis. The different techniques can be subdivided into the two classifications of *conventional* and *discounting* methods which are discussed on p. 142.

Methods of valuation

Valuations of land and property are undertaken for a variety of different purposes, usually by the valuation surveyor. The purpose of the valuation will affect the assessment of its value, and this may differ because of the assumptions made and also because they are only estimates of value anyway. Valuations are required, for example, for statutory purposes in order to assess inheritance tax or when a public body seeks to acquire land or property by means of compulsory purchase. A valuation may also be required when a purchaser such as an insurance company or pension fund wants to invest its capital. It may also be required during the sale and purchase of property, in connection with a mortgage loan or for determining an auction reserve. A number of alternative methods (Table 5.1) can be used to estimate either the capital value or the rental value of an interest in land or property. It should also be noted that values can vary considerably depending upon the location nationally or even within a small area.

Table 5.1   Methods of property valuation

Comparative method
Contractor's method
Residual method
Profit or accounts method
Investment method
Reinstatement method
Hedonic price modelling

Each of the methods in Table 5.1 other than the profits methods (and hedonic price modelling which is considered separately) is useful for estimating capital values, whereas the residual method and investment method are not really suitable for the determination of rental value. The demand for a particular type of landed property will be influenced by changes in the size of population, methods of communication, standards of living and society in general.

The comparative method

This is the most popular method used for valuation purposes. Its main uses are in connection with residential property where direct comparisons can be made against other types of property on the open market. However, the method is only reliable where there are sufficient records of many recent transactions and the properties are in the same geographical area. Other factors that will influence the valuation are the similarity of properties in respect of design, size and condition, and the legal interest. A stable market and economic factors such as lending rates will also affect the reliability of the valuation. A key criterion is the postcode of the property being valued.

The contractor's method

The basis of this method is to suggest that the value of a property is equivalent to the cost of erecting the buildings, together with the cost of the site. It is, however, an unsound assumption since value is determined not necessarily by the component costs involved, but by the amount which prospective purchasers are prepared to pay. Its main use is in connection with valuations for insurance purposes and for buildings such as schools, churches, hospitals, etc. for which there may be little in the way of comparative valuations.

It is necessary when using the contractor's method that allowances are made for depreciation, since a building that is 60 years old is unlikely to have the same value as a modern building of a similar type and quality. Some of these buildings may be ornate and have been costly to construct, but this will not necessarily be reflected in the value.

The reason why this method is unsound can also be attributed to the concept of added value. The cost of the land, plus the construction costs and any profit should equal the sale price. However, the various components of cost when added together frequently do not equate with value.

The residual method

This method is used in those circumstances where the value of a property can be increased after carrying out development work. For example, an old house may have the potential and be suitable for conversion into flats, when its best potential can be realised. The building is valued on the basis of its future worth after conversion, and the costs of this work together with developer's costs are then deducted. The resulting sum is the value of the property in its original state and is known as its residual value.

The profits or accounts method

Almost all types of property are capable of producing an income under certain conditions, and a relationship will exist between this and the capital value of the property. It is more appropriate to commercial premises such as hotels, shops and leisure projects than domestic premises. The usual approach is to estimate the gross earnings, deduct expenses, and the balance remaining then represents the amount available for payment of rents. This can then be converted into a capital sum.

The investment method

This method can also be used in those circumstances where the property produces an income. The income expected must be comparable with that which could be earned by investing the capital elsewhere. In considering alternative investment possibilities factors such as security values, ease of realisation, costs of purchase and sale, and any tax liability will influence competing proposals. The principal investors are pension funds, insurance companies, property companies, historic owners, local authorities and government agencies.

The reinstatement method

This method requires the estimation of the cost of rebuilding a particular property and then adding to it the value of the land on which the property stands. It is a useful method for fire insurance purposes, in order to calculate the premium to be paid. It may sometimes appear that the insurance premium should only be based upon rebuilding costs, since the site will remain, even in the event of a fire. It will, however, be necessary to allow for demolition and site clearance costs where the building is to be rebuilt. These costs will also have to take into account the possible site damage and temporary works that might be necessary before demolition can commence.

Hedonic price modelling

This is a computer-based system for valuing property on the basis of the different variables involved. It uses the technique of multiple regression analysis to find a formula or mathematical model that best describes the data characteristics that have been collected. The technique is normally used in those circumstances where the relationship between the variables is not unique. This is in the sense where the value of one variable always corresponds to that of another. In order to calculate the value of a property, it is first necessary to identify the variables that might be important. In the case of residential property, variables such as location, type, size, number of bedrooms, garages, central heating, etc. are important. Large amounts of data from previous transactions are then required in order to discover mathematical relationships. It is unlikely that a perfect relationship will be found, just as it is unlikely that a number of valuers would all predict the same value for a property. The model will be able to predict confidence limits to the results, and where a good model has been constructed then these should allow the value to be stated within tolerable limits. The location of property is the most significant variable that affects value. Knowing a property's postcode will therefore allow a value to be predicted within the above model formation, as long as the data in the model is representative of the value that is being predicted.

Valuation tables

In order to allow comparisons to be made between money spent or received at different times we need to be able to convert these sums into a common time scale. Valuation tables are used as the means of making this conversion.

In most societies the payment of interest for the use of capital or money is an established part of economic life. The money which is lent is called the *principal*. The sum of the principal and the interest for any length of time is called the *amount*. The money paid for the use of the principal is calculated on a percentage rate basis and this is generally calculated using either simple or a compound interest formula. The basis of valuation tables uses compound interest (see Figures 5.1 and 5.2).

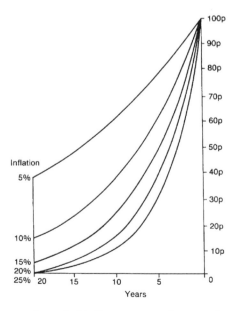

Figure 5.1   The present value of £1 (Source: Barclays Bank)

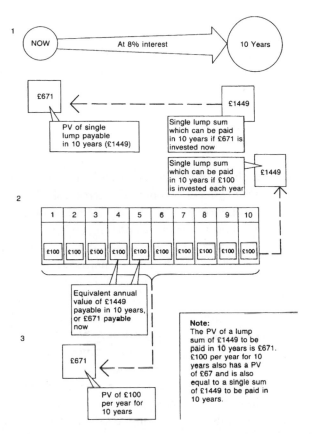

Figure 5.2   The varying value of money at 8% (Source: Ministry of Public Building and Works (1970))

Simple interest

Simple interest arises when only the original capital invested earns interest. For example, if A borrowed £100 from B with the agreement that 5% interest was payable each year, then at the end of the first year if B paid A £5, then the obligation would be met. £5 would be paid each and every year on this basis.

Compound interest

Interest may accrue on interest as well as the capital. For example, C decides to borrow £100 from D at a 5% rate of interest for two years. At the end of the first year C owes D £105. The interest in the second year is calculated as 5% of the £105, which equals £5.25. At the end of the second year C would repay D £110.25. Using simple interest as a basis this would only have been worth £110.

Calculations involving compound interest involve:

- *Compounding*: the way a present sum of money will grow
- *Discounting*: which is the reverse of compounding, and considers how a future sum of money might be worth today, given a particular rate of interest.

Parry's Valuation Tables

In valuation practice, and other studies associated with land and buildings where some aspect of financial analysis is involved, compound interest calculations of a tedious and time consuming nature are often required. The use of valuation tables can be used to reduce the time-consuming aspect of such calculations. There are a number of different books of valuation tables now available. The best known is *Parry's Valuation and Conversion Tables* (Davidson, 2002). These tables were first prepared in 1913 by the late Richard Parry. Although other valuation tables are in common use, Parry's has become synonymous with property valuation, where calculations requiring interest rates are to be taken into account. The comprehensive set of tables provides for the requirements of current practice. The later editions of these tables have been prepared by computer and in order to minimise possible error a computer linked typesetter has been used. Over the years different editions of the tables have seen both the introduction of new material, and the removal of tables now thought to be obsolete. In addition to the actual computational values, an explanation of the purpose or use of the tables is also provided. The first chapter of the book also deals with the construction and use of the tables generally.

The use of electronic calculators and computers has made the tables somewhat dated. Nevertheless, they are still used extensively by surveyors. The eight most basic tables are described below.

Amount of £1 table

This table forms the basis of construction of many of the others. The multiplying factors given in the valuation tables represent the amount to which £1 invested now will accumulate at compound interest over a given period of time. It is represented by the formula:

Amount of £1 $= (1 + i)'^n$
where i = interest rate
n = number of years.

The principal is multiplied by the appropriate figure from the tables for the required interest rate and term of years. The amount of £1 table has multipliers greater than unity. Tables of this type are commonly referred to as accumulating tables.

For example, a builder purchased a plot of land 5 years ago for £15 000. Assuming that land has increased in value by an average of 6% per annum, what would be its value today?

£15 000 × 1.34 = £20 100
  (multiple of 1.34 from amount of £1 table: 5 years at 6%)

Present value of £1 table

In this table, the investor is seeking to find what sum must be paid into the bank today in order that it may amount to £1 at the end of a given period of time, using compound interest. It is a discounting table and is the reciprocal of the Amount of £1 table above, and is thus represented by the formula:

$$\text{PV of } \pounds 1 = \frac{1}{\text{Amount of } \pounds 1}$$

For example, a boiler will need replacing in 20 years time at an estimated cost of £8000. If the average annual rate of interest is 4%, what amount should be invested today in order to be able to make this replacement?

£8000 × 0.456 = £3648
  (multiplier of 0.456 from PV of £1 table, 4% for 20 years)

Amount of £1 per annum table

This is a similar table to the Amount of £1 table above in that it is an accumulating table. The difference is that whereas that table represents a once and for all single sum of money, this table provides for an equivalent amount annually for the required term of years:

$$\text{Amount of } \pounds 1 \text{ per annum} = \frac{(1 + i)^{n-1}}{i}$$

Since the formula for calculating the amount of £1 = $(1 + i)^n$, then substituting this for A, the amount of £1 per annum can be simplified to:

$$\frac{A - 1}{i}$$

For example, what sum will be obtained if an investor puts £200 in his bank account every year for 10 years at a rate of 6%?

£200 × 13.18 = £2636
  (multiplier of 13.18 from Amount of £1 per annum table, 6% for 10 years)

Annual sinking fund table

This table is the reciprocal of the previous table (amount of £1 per annum). It is used to calculate the annual amount to be saved each year at a given rate of interest in order to meet a known expense at an expected date in the future.

$$\text{Annual sinking fund (ASF)} = \frac{1}{\text{Amount of £1 per annum}}$$

For example, extensive modernisations to a client's offices are expected to be carried out in eight years time. What sum need he invest annually at a rate of interest of 5% to cover the future costs of £75 000?

£75 000 × 0.106 = £7875
(multiplier of 0.106 from ASF table 5% for 8 years)

Present value of £1 per annum table (years purchase (YP) single rate)

This table is used to calculate the present value of future payments which are made at regular annual periods. The formula is derived from the basis of the PV of £1 table by adding together the multipliers for each year. It can therefore be represented by the formula:

$$\text{PV of £1 per annum} = \frac{1 - \text{PV of £1}}{i}$$

As the number of years approaches perpetuity, the value of the PV of £1 becomes so small that it really is insignificant. The formula is sometimes referred to as the *years purchase* table and can in these circumstances be abbreviated to:

$$\text{YP (single rate)} = \frac{1}{i}$$

For example, a client wishes to know how much should be invested today at 5% rate of interest to cover the average annual payments of £2000 for energy consumption during the next 25 years?

£2000 × 14.09 = £28 180
(multiplier of 14.09 from PV of £1 per annum table, 5% for 25 years)

Present value of £1 per annum table (dual rate)

The previous table (PV of £1 per annum table (single rate)) provides for the same rate on both the interest on the sum invested, and the annual sinking fund (ASF) to recover the capital value over the term of years.

In practice, the rate of interest on the loan and the ASF may be different. In these circumstances it is therefore necessary to use a dual rate (DR) table, which allows these rates to be different.

$$YP_{DR} = \frac{1}{i + ASF}$$

For example, assume that the cost of capital is 12% and an ASF rate of 5% is required to cover future replacements of a boiler plant during the next 30 years. The PV of these replacements is £33 936. What is the annual charge to cover this amount?

In order to convert a PV to an annual equivalent, we divide this by the YP factor calculated above, and as follows:

$$YP_{DR} = \frac{1}{0.12 + 0.015}$$

(12% inflation; 5% ASF for 30 years)

$$= I = \frac{7.4074}{0.135}$$

$$\frac{£33\ 936}{7.4074} = £4581 \text{ is the annual charge}$$

Annuity £1 will purchase

The term *annuity* is generally used to mean a series of payments that are to be made during a given period of time at fixed intervals. If these payments are only to last for a fixed period of time then it is termed an annuity certain. The rent from property, either for a fixed term of years or in perpetuity, is an annuity certain. When the period of the annuity is perpetual, the annuity is more properly described as in perpetuity. The annuity £1 will purchase is given by the following table:

$$A = i + SF$$

For example, a leaseholder paid £3000 for an interest last month. The lease has a 30 year unexpired term. It is decided to let the property. What is the minimum rent to be accepted if a return of 10% on the outlay is expected with a sinking fund of 2.5%?

Capital cost	3000
Annuity £1 will purchase 30 years @ 10% and 2.5%	0.1228
Equivalent annuity	£368

Mortgage repayment tables

A mortgage is the annual equivalent of a capital sum lent by a mortgagee, often a building society, normally for house purchase. The mortgagor or purchaser agrees to repay the capital borrowed together with interest charged at the society's rate. The parties agree beforehand upon the number of years for which the mortgage will run. The amount of repayment therefore depends upon the size of the loan, the term of years and the interest rate applicable. The annual equivalent is the sum of 12 monthly payments as shown in *Parry's Valuation Tables* giving multipliers needed to repay £100 on a monthly basis. This table represents those of the annuity tables (single rate) multiplied by 100 and divided by 12.

$$\text{Mortgage instalment} = \frac{(i + \text{SF}) \times 100}{20}$$

For example, what is the monthly repayment for a mortgage of £25 000 over 25 years at 12%?

$$\frac{£25\ 000}{100} \qquad = 250 \times 1.0625 \qquad = £265.63$$

(tables based upon   (mortgage instalment)   (monthly payment)
units of £100 table)

Annual payment = £3187.56

Alternatively this could have been calculated from the annuity table (single rate):

£25 000 × 0.1275 = £3187.50

The annuity table is sometimes referred to as the *annual equivalent* table.

Developers' budgets

Developers seeking sites for development purposes will need to consider many different factors. They may, for example, be looking for a site that is suitable for one of several different development proposals. Alternatively, they may already have a particular scheme in mind and are seeking a site which is most suitable for this need. Developers will usually be seeking an overall scheme which is likely to be the most profitable and one that is attractive to potential investors.

When a suitable site has been identified, it will be necessary to ensure that planning permission, for the type of development envisaged, will be granted prior to its acquisition. The developer will therefore need to make enquiries at the offices of the local planning

Table 5.2 Summary of valuation formulae

(1) Amount of £1

$$A = (1 + i)^n$$

(2) Present value of £1

$$PV = \frac{1}{A}$$

(3) Amount of £1 per annum

$$Am = \frac{A - 1}{i}$$

(4) Annual sinking fund

$$ASF = \frac{1}{Am}$$

(5) Present value of £1 per annum (single rate)

$$PVA = \frac{1 - PV}{i}$$

or

$$YP = \frac{1}{i}$$

(6) Present value of £1 per annum (dual rate)

$$PVA = \frac{1}{i + ASF}$$

(7) Annuity £1 will purchase

$$A = i + SF$$

(8) Mortgage instalment

$$MI = \frac{(i + SF) \times 100}{12}$$

authority in order to determine this. Permission at this stage will only be given in principle, and it might include conditions to be met if approval is to be obtained. Where permission is not forthcoming then a notice of appeal can be made, if there are reasonable and likely grounds for its success. The type of development that will be allowed is generally quite clear, but there is always room for some debate and discussion on arbitrary cases.

In order to determine whether a scheme is feasible it will be necessary to prepare a developer's budget. This will then provide answers to the following questions:

- How much should be paid for the land?
- What will be the maximum building cost?
- What should be the selling price or rental value for the property?

The developer's budget considers the following items.

Gross development value (GDV)

The total rental value is estimated by comparing the proposed scheme with the rents obtained from similar properties. The net rental value is used after deductions for out-goings such as maintenance, repairs, insurances, management, etc. have been made. This then provides the net income from the proposed development. The amount before the deduction of any income tax is used in order to compare this with other non-property investments. In the case of a block of flats or shopping centre development, where there may be many different tenants then management costs to cover rent collection, survey-ing, etc. would also need to be deducted. These are currently worth about 2.5% of total rents. The valuer will be able to advise upon these appropriate amounts. These values are, however, more prone to error than are building costs. This is due to the many uncertainties in the property market, not least among which is the difficulty of forecast-ing likely prices and demand some time in the future when the property is constructed and available for occupation. The valuations of two independent valuers could also in-dicate some wide discrepancies, since in addition to the calculations involved, valua-tion is also a matter of considerable skill and judgement. Opinions, which may be based upon wide experience, are known to conflict, as illustrated by case law on the subject.

Investment yield

Whilst a large proportion of residential properties are owned by the occupants (dis-counting any mortgage interest), commercial property is more likely to be rented. The theory is that the profits from a commercial enterprise are probably better employed in running and expanding that business at which the management are expert rather than being tied up in property. This also allows better opportunities to move premises when the business changes shape through expansion or contraction.

The net income from the budget is then capitalised by multiplying by an investment yield. This figure should compare with investment yields from other types of invest-ment and may fluctuate considerably in an unstable economy. An appropriate yield, sometimes referred to as the year's purchase (YP) can be obtained by dividing 100 by the interest rate:

$$\text{YP in perpetuity} = \frac{100}{\text{rate of interest}}$$

$$\text{YP of 8\%} = \frac{100}{8} = 12.5$$

This is also sometimes known as the PV of £1 per annum, e.g. net income £2000 per annum multiplied by YP at 8% = £25 000. This capitalised figure is known as the *devel-opment value.*

It should also be noted that office block rents are based not upon gross floor areas but upon net usable floor areas. Some allowance must therefore be made for non-usable areas such as circulation space, before calculating the development value.

For example, the rental value of an office block is estimated to be £30 per m^2. The total floor area is 10 000 m^2 and the non-lettable area represents 20%. What is the development value if the YP is 6%?

10 000 m$^2 \times 80\% \times$ £30 = 240 000

$$YP \text{ at } 6\% = \frac{100}{6} = 16.67$$

Gross development value = £4 000 800

As a principle, the greater the expected rental growth the lower the initial return an investor would be prepared to expect. Conversely, an investment where income growth is expected to be small, such as in a building society account, will require a high yield to compensate.

Costs of construction

There are several easy to apply methods for calculating the approximate cost of a building. However, whilst the methods rely upon a simple method of quantification, such as the floor area of the proposed building, the skill in selecting a correct current rate by which to calculate cost is much more difficult. This relies upon a knowledge of current prices and being able to interpret these against the designer's brief and outline drawings. For further information on the different methods and techniques that can be used for early price estimating, reference should be made to *Cost Studies of Buildings* (Ashworth, 2004).

During the investment appraisal it is common to use the construction costs at the date of tender, i.e. excluding any increases in cost during construction. This is used because present-day rents are also used, on the basis that any increases in either will to some extent compensate the other.

Fees

Charges for the professional services provided will need to be added to the costs of construction. The various professional institutions publish fee scales which can be used as a guide in assessing these costs. The fee scales are based upon a combination of a lump sum and a percentage of the construction costs. The fees will vary depending upon the type and size of project, and the description of the service provided. Professional fees are now calculated on the basis of competition, using the fee scales as a guide. The larger the project and the more repetitive its components, the smaller will be the overall fee that is charged. For unusual, complex or difficult projects which might include specialist professions such as archaeologists, etc. then the fees will increase accordingly. An addition of 10% will typically cover design, costing and supervision fees. Value added tax will also be chargeable, but may be recoverable by the client depending upon the type of project and type of client.

Legal fees will also be required for the purchase of the site, and the preparation and agreement of leases or the conveyancing documents. Property agents may also be required in connection with letting and management of the property, or for its disposal to potential owners. Their fees may be typically 2–3% of the selling price, depending upon the service provided and the number of units involved.

Developer's profit

Property development involves the taking of considerable risks. Where these risks fail to materialise then increased profits are made. Where the risks are greater than expected then profits are reduced. The developer is paid a return on the development to cover the skills, time and risk that will have been incurred in the development as well as for expected profits. This is typically estimated to be about 10–20%, but is dependent upon a wide range of factors such as the type and size of development, length of development period and possible competition when completed. The risks to be assessed by the developer and the professional advisers involved include rising costs, speculative nature of the development and inability to either sell or lease on completion.

For example, on the apparently ill-fated Canary Wharf development, even the offer of free leases for up to 10 years were insufficient to attract some blue chip companies to take up occupation. The objective of this idea was that, if successful, then other firms would also have been prepared to relocate to be close to such important and prestigious companies. The greater risk than anticipated of letting the property caused profits to be wiped away, leaving a trail of debts and bankruptcy for the initial developer.

Finance (site and buildings development)

The developer will need to have already purchased a site prior to commencing with the construction work. This might have been acquired through retained earnings, in which case there will be bank interest accruing, although a loss of income from such funds will be incurred. Alternatively, the funds may need to be borrowed in which case there will be interest charges to be added. Land is often purchased at least 12 months prior to starting work on site, to allow for planning permission and the security of the site.

Payments to the constructor will be made monthly, the amount to be paid being determined by the quantity surveyor. Payments for professional fees are often paid in two parts. The first is to cover pre-contract design work and the second is for supervision and administration of the project through construction. The time between completion and letting or selling will vary depending upon the local market and the demand for the type of property that has been constructed. Housing developers aim to complete dwellings in line with sales and will deliberately accelerate or delay construction activity to meet the demand for houses.

The interest added to the developer's budget is often on the basis of the full amount for half the time. Whilst this is only an approximate amount, it is adequate for the purpose of the calculation. In the case of long contract periods then compound interest rates should be applied. The interest rates selected will be based upon the opportunity costs of capital, this being a few points higher than the base rates from where the finance has to probably be obtained.

Example

A speculative developer is considering purchasing a site for the construction of 40 detached houses. The selling price of the houses is £65 000. The cost of the land, inclusive of legal charges, is £150 000. The developer requires a profit of 16% of the gross development value. What is the allowable amount for building costs?

Developer's budget
Gross income
40 houses × £65 000   = £2 600 000
                                                 (= GDV: Gross Development Value)

Developer's costs
Land
Cost inclusive of legal fees   = £150 000
Short-term finance, required for say 2.5 years at a compound interest rate of
12% (valuation table could be used)   =   49 450

Fees
Legal, agent's and advertising (3% of GDV) =   78 000
Profit 16% of GDV   =   416 000
Total   = £693 450

By deducting this amount from the GDV, the building costs, professional fees and finance for construction can be calculated.

GDV   =   2 600 000
Development costs (above) =   693 450
Total   = £1 906 550

Building costs
Let B = building costs.
Assume that finance will be required for 1.5 years at 12%.
Professional fees assumed to be 10%.

Building costs	= B
Finance	$= B \times 0.12 \times 1.50$
Fees	$= B \times 0.10$
1 906 550	$= B = (B \times 0.12 \times 1.50) + (B \times 0.10)$
	$= B + 0.18B + 0.10B$
	$= 1.28B$

$$\text{Building costs} = \frac{1906\ 550}{1.28B}$$
$$= \pounds1\ 489\ 492$$

Check

Land costs and finance	=	199 450
Legal/agent's fees	=	78 000
Developer's profit	=	416 000
Building costs	=	1 489 492
Finance	=	268 108
Fees	=	148 949
Total	=	2 600 000

The allowable amount to cover the costs of building is therefore £1 489 492. This represents £37 237 per house. If the size of each house was known, say 110 m², this can be converted to a rate per square metre of gross internal floor area of £338.50. Present-day building costs would be used to determine whether this would be adequate for the type of quality of building that is envisaged.

The developer during the early stages of any development will need to provide answers to the following questions:

- Is there a market for the proposed development?
- What will be the likely selling price?
- How much can be afforded to be spent on the scheme?
- Can the necessary finance be raised?
- What will the finance cost?
- Will the scheme be granted planning permission?

Sometimes the developers are able to predetermine their costs and need to know the likely selling price of the development and whether or not this is achievable. The questions can then be approached in the reverse order:

Land cost and finance   =   199 450
Fees (3% of GDV)   = ?
Profit (16% of GDV)   = ?
Building cost, finance and fees = 1 906 550
GDV   = ?

Let GDV = x = 199 450 + 0.03x + 0.16x + 1 906 550

$$0.81x = £2\ 106\ 000$$
$$x = £2\ 600\ 000 = \text{GDV}$$
$$0.03x = £78\ 000 = \text{Legal/agent's fees}$$
$$0.16x = £416\ 000 = \text{Developer's profit}$$

The cost of each house is then obtained by dividing the GDV by the number of properties.

Contingencies

Some developers may consider that it is prudent to allow for contingencies, to cover unforeseen costs, i.e. amounts that cannot be properly estimated or as a margin to the costs identified above. It is common, for example, for contingencies to be included in bills of quantities for unforeseen items of expenditure. Sums of money will be required in circumstances such as the costs involved in capping a mine shaft that is only discovered when the work starts on site. Building works contingencies are typically about 2–5%, depending upon the degree of certainty of the design and the knowledge of the site. On refurbishment projects it is often much higher since there are likely to be more unknown factors in both the design of the project and the condition of the existing building.

Letting and agents

It is normal practice to assume that the completion of the project will not entirely coincide with the letting or selling arrangements and it is therefore sensible to make allowances for such delays in the budget. Where these delays are considerable because of a downturn in the market, then the developer may need to consider offering induce-ments in order to dispose of the property. If this is considered to be a possibility at the time of inception, then it may be desirable to postpone the project. This happened with a large number of designed office blocks during the early 1990s.

Alternatively developers may consider that improvements will occur in the market for the property by the time they come to dispose of the finished projects. In any case construction costs are likely to be much lower during a recession and it is possibly a good time to build, if one assumes that the business cycle has not stopped. In prime locations, property is always likely to find a buyer. If the developer considers that extra costs will possibly be incurred then these should be included in the budget.

Letting and sale fees usually occur towards the end of the development. It is common practice on larger developments to appoint two or more agents, with a consequent increase in fees. Letting fees are typically based on 10% of a single year's rent where one agent is employed, and 15% in the case of multiple agents.

Example

A speculative developer has provided the following details of a proposed speculative office development and has asked you to calculate the allowable building costs:

Gross floor area 10 000 m^2
Non-lettable area 20%
Estimated rent £60/m^2
Capitalisation of rents 7%

All outgoings are to be recovered by a service charge. The building contract details are as follows.

Period 18 months
Professional fees 15%
Short-term finance 15%
Developer's profit 12% of GDV
Land cost (including fees) £100 000

This solution depends upon an evaluation of future rents in order to establish the amount that can be available now for building purposes. It should be remembered that rents are determined on the basis of the lettable floor area. Therefore the rental received will be as follows:

Rental received

Lettable floor area = 10 000 m^2 × (100 − 22) = 7800 m^2
The net income to the developer is therefore:
7800 m^2 × £60/m^2 = £468 000

Gross development value

This amount must then be capitalised, i.e. converted to a current capital value. It is assumed for the purpose of this question that the rent will be received in perpetuity. It is therefore multiplied by the year's purchase in perpetuity at the given percentage.

$$\text{YP in perpetuity at } 7\% = \frac{100}{7} = 14.286 \times £468\ 000 = £6\ 685\ 848$$

No adjustments are to be made for any outgoings, such as repairs, insurances, etc. as these will be recovered by means of a separate service charge. If these, or any other management charge was incurred then the effect would be to reduce the annual rents, and hence the capitalisation amount.

Developer's profit

The gross development value is the same as the capital value. The developer's profit is therefore calculated as:

$$12\% \times £6\ 685\ 848 = £802\ 302$$

Land costs

The cost of the site which includes professional fees associated with its acquisition is £100 000.

Short-term finance will be required until the development is complete, and then presumably let or sold. This is required for at least the contract period, assuming that the site is purchased at the start of the contract. This may be a conservative assumption, since the land is likely to be purchased much earlier and therefore incur additional interest charges.

Assume 24 months at 15%	=	30 000
Add site cost above		100 000
		£130 000

Summary

Gross development value	= £6 685 848
Developer's profit = 802 302	
Land cost = 130 000	
	£932 302
Amount of allowable building costs =	£5 753 546

Building costs

Let B = building costs, including any allowances for inflation. Finance will be required at 15% for the 18 months contract duration. Note that the finance will be required as the work progresses. This is equivalent to the full percentage for half the time. Professional fees would not be paid until the project was completed. Therefore:

$$B + (B \times 0.15 \times 1.5 \times 0.5) + (B \times 0.15)$$
$$= 1.2625B$$
$$= £5\ 753\ 546$$

$$B = \frac{5\ 753\ 546}{1.2625}$$
$$= B = £4\ 557\ 264$$

This is the amount available for building costs. Dividing this by the gross floor area will provide a rate per square metre. This will then suggest a type and quality of construction that may be possible.

Residual cash flow valuation

A client is considering purchasing a row of large Georgian terraced houses for conversion into luxury flats. At the present time, similar flats in similar locations are selling for about £200 000. Confidence in this market expects that prices will increase by 0.5% per month over the next year. Building costs associated with the conversion amount to £75 000 per flat, but these costs are also expected to increase by 0.75% per month for the foreseeable future. Building costs are assumed to be evenly spread over the 12 month contract period. The expected purchase of the flats will coincide with the start of the work on site. Flats will be sold when complete, i.e. three in each of the months 9–12. Professional fees are 15% of building costs, payable as follows: 25% in months 1 and 6 and 50% in month 12. Agent's and solicitor's fees will be chargeable at 3% when each flat is completed. Bridging finance is 1% per month and the developer requires 20% of the agreed sale price. The acquisition costs of the existing building are 4%. Tables 5.3 to 5.5 give a diagrammatic representation of the various stages of the project.

How much should be paid for the existing row of terraced houses?

Site value

£839 143 is the capital sum that is available in 12 months time. This must allow for site finance, acquisition and the time value of money:

£839 143 at 12% the start of the project = £749 235

Table 5.3   Residual cash flow valuation

Month	Sales	Building costs	Agent's fees	Design fees	Total flow	Cash charges	Interest cash flow	Total
1		−75 000		−35 000	−110 000	−110 000	−1100	−111 100
2		−75 563			−75 563	−185 563	−1856	−187 418
3		−76 129			−76 129	−261 692	−2617	−264 309
4		−76 700			−76 700	−338 392	−3384	−341 776
5		−77 275			−77 275	−415 667	−4157	−419 824
6		−77 855		−35 000	−112 855	−528 522	−5285	−533 808
7		−78 439			−78 439	−606 961	−6070	−613 031
8		−79 027			−79 027	−685 988	−6860	−692 848
9	624 423	−79 620	−18 733		526 070	−159 918	−1599	−161 517
10	627 546	−80 217	−18 826		528 503	368 584	1843	370 427
11	630 684	−80 819	−18 921		530 945	899 529	4498	904 027
12	633 837	−81 425	−19 015	−70 710	462 687	1 362 216	6811	1 369 027
Totals	2 516 490	−938 069	−75 495	−140 710	1 362 216		−19 775	

Note: Interest charges are 1% to borrow and 0.5% to invest

Table 5.4   Sales analysis

Month	Selling price	Developer's profit	Agent's fees
1	200 000		
2	201 000		
3	202 005		
4	203 015		
5	204 030		
6	205 050		
7	206 076		
8	207 106		
9	208 141	41 628	6244
10	209 182	41 836	6275
11	210 228	42 046	6307
12	211 279	42 256	6338
Totals		167 766	25 165
12 flats =		**503 298**	**75 495**

This figure includes site acquisition of 4% and interest charges of 12% per annum − existing building = x; acquisition 0.04x; interest charges 1.12x (total 1.16x) − as follows.

Existing building = 647 387
Acquisition (4%) =   25 895
Interest (12%)   =   77 686
                    750 968

Table 5.5   Summary of residual valuation for the row of terraced houses

	Income	Expenditure
Building costs		938 069
Agent's fees		75 495
Design fees		140 710
Interest charges		19 775
Developer's profit		503 298
Sales	2 516 490	
Totals	**2 516 490**	**1 677 347**
Residual amount (Income – expenditure) = **839 143**		

This is the residual amount and is the capital sum that will be available in twelve months time. This must allow for site finance, acquisition and the time value of money.

839 143 at the end of the project @ 12% = 749 235 at the start of the project

This figure includes site acquisition (4%), interest charges (12% per annum).

Let existing building = x; acquisition 0.04x; interest charges 0.12x = 1.16x

Summary (check)

Existing building	645 892
Acquisition (4%)	25 836
Interest (12%)	77 507
Total	749 235

The developer should be prepared to offer around £650 000 for the existing row of terraced houses.

Conventional and discounting methods of valuation

There are two approaches to valuation of a project, the conventional ones and discounting ones, as shown in Table 5.8. The latter take into account the time value of money. They will be examined in turn.

Pay-back method

This is the crudest form of investment criterion, but nevertheless one of the most widely used. It is defined as the period it takes for an investment to generate sufficient incremental cash to recover its initial capital outlay in full. A cut-off point can be chosen, beyond which the project will be rejected if the investment has not been paid off. The pay-back method appears attractive because it is extremely simple to apply.

Since the pay-back method takes cash receipts into account, it helps to assess a company's future cash flow. This is particularly advantageous in times of liquidity crisis. However, it fails to measure long-term profitability since it takes no account of cash flows beyond the pay-back period. It is therefore difficult to make comparisons between projects with different life expectancies using this criterion. The technique also falls short in its application within the pay-back period since no account is taken of the timing of the cash flows during that period. The use of the method is sometimes justified by claiming that it is a dynamic criterion, since projects are adopted only if they are paid off quickly, but this argument does not allow for the fact that highly profitable investments do not necessarily pay off in the initial years although large gains may be reaped later. Table 5.6 shows how the pay-back period is calculated, comparing three projects.

Table 5.6   Pay-back period calculation

	Year	Projects		
		A	B	C
Expenditure	0	60 000	100 000	140 000
Income	1	10 000	50 000	50 000
Income	2	20 000	25 000	50 000
Income	3	40 000	25 000	25 000
Income	4	20 000	50 000	45 000
Income	5	20 000	50 000	35 000

The pay-back periods for each of these three projects are:

A   2 30/40 years = 2 years 9 months
B   3 years = 3 years
C   3 15/25 years = 3 years 7.2 months

It can be seen from Table 5.6 that the pay-back period is quick and simple to calculate. However, clear objectives need to be formulated in assessing the competing alternatives. In the above example and using the pay-back criterion, project A would be selected since it has the shortest pay-back period. However, there are other criteria that need to be measured that might have an influence upon the decision to be made. In the example provided, over the 5 year period, project C provides the highest cash profit (£65 000), whereas project B offers the largest percentage profit (200%).

Average rate of return method

The average rate of return is the ratio of *profit* (net of depreciation) to *capital*. The first decision that must be made is how to define profit and capital. Profit can be taken as either gross of tax or net of tax, but since businesses are mostly interested in their post-tax position, net profit is a more useful yardstick. However, net profit can be either that which is made in the first year, or the average of what is made over the entire lifetime of the project. Similarly, capital can be taken as either the initial sum invested or as a form of average over the time of all the capital outlays over the life of the project. This method takes no account of the incidence of cash flows so that projects with the same capital costs, expected length of life and total profitability would be ranked as equally acceptable. The method can, however, be extended by calculating the net average yield. This is done by subtracting the stream of cash outlays from the stream of cash benefits and expressing the answer as a percentage of the initial outlay.

Table 5.7 shows four projects, all with similar capital costs but different income streams. In Table 5.7, the average rate of return on project D is 115% of the initial capital expended. The life of this project is one year. In project E, the sum of the positive cash flows is £200 000, but over four years. This is worth on average £50 000 per year giving a rate of return of 50%. In project F, the total income is £120 000 or £30 000 per annum giving an average rate of return of 30%. As can be seen, this method does not take into account the timing of the cash flows. Therefore projects with the same capital costs, expected life and total profitability would be ranked as equally acceptable. In Table 5.7, project G has the same average rate of return as project E, even though no account is taken of its income flows that appear at different years.

Table 5.7   Average rate of return comparison

	Year	Project			
		D	E	F	G
Capital cost		100 000	100 000	100 000	100 000
Income	1	115 000	50 000	60 000	
	2		50 000	60 000	
	3		25 000	−25 000	
	4		75 000	25 000	200 000
Total income		115 000	200 000	120 000	200 000

Necessity/postponability

This criterion is essentially a negative one. The rationale is that the more postponable an investment is, the less attractive it appears and so the basis of investment decision-making is the urgency of requirements. Thus, if a particular project was one which could only be carried out now and could not be initiated at a later date, then it would be chosen in favour of a project which could be undertaken in the future.

Table 5.8   Methods of investment appraisal compared

Conventional methods	Discounting methods
Payback	Net present value
Average rate of return	Internal rate of return
Necessity/postponability	

Discounting methods

A vital factor ignored by the conventional methods of investment appraisal is that money has a time value. A pound today is worth more than the same pound tomorrow.

A sum of money is worth more today than an equal sum of money at some time in the future even ignoring inflation. This is known as the *time value* of money. This is because it allows for the possibility of investment or consumption to take place in the intervening period. The present value of a future sum is dependent upon two factors: the rate of interest and the term of years. The further into the future the sum is or the higher the rate of discount used, then the less will be the present value of that sum. There are two major discounting techniques and these are described as follows.

The net present value (NPV)

In order to determine the NPV of a proposed investment, the forecast net of tax cash flows is simply discounted to the tune of the initial capital outlay (at a rate chosen to reflect the company's cost of capital) and the value of the initial capital outlay subtracted. The company's cost of capital is generally set at a level which would give the shareholders a rate of return at least equal to what they could obtain outside the company. The discounting technique can be readily adapted to take account of real life complications such as cash flows arising in the middle of a year, investment grants, capital allowances, inflation and delays in corporation tax payments. With the help of appropriate tables, the volume of calculation and analysis resulting from these complications is not nearly as weighty as might be supposed.

Internal rate of return (IRR)

This is the commonest discounting method of investment appraisal. It can be defined as that rate of interest which when used to discount the net of tax cash flows of a proposed investment, reduces the NPV of the project to zero. This discount rate which will reduce the NPV of the project to zero can be found by trial and error: if a negative NPV results, the rate chosen is too high, if a positive NPV is obtained, then the rate is too low. Although it appears to involve a large number of calculations, in practice it should never be necessary to carry out more than two trial discounts, the true IRR then being determined by interpolation. The IRR depicts the annual rate of return on the capital outstanding on the investment. Thus, in common with the NPV method, the IRR will generally be higher if the bulk of the cash flows are received earlier rather than later in the life of the project, reflecting the fact that more capital will have been recovered in the first years of the project so that the flows remaining represent a higher rate of return.

Optimal investment criteria

Although some of the conventional methods of investment appraisal provide a useful measure of the vulnerability of investment proposals to risk and liquidity constraints, as gauges of the profitability of projects they must be regarded as extremely inferior to the discounting methods because of their failure to recognise that money has a true value. There are occasions when IRR is meaningless. If, for example, a particular project involves heavy net capital outlays towards the end of the project's life, the IRR could be non-sensical. When appraising independent projects, where the only decision to be made is whether to accept the project or not, then both the NPV method and the IRR method will give the same answer. However, when trying to decide which is the most profitable of two mutually exclusive projects, then the two methods can give very different answers. The risks associated with a project are largely dependent on the quantity of capital involved and the length of the project. By showing a rate per unit of capital per unit of time of the project, the IRR can show the margin over the cost of capital that is being obtained in return for any risk taken.

Conventional methods of dealing with risk, such as sensitivity analysis, probability analysis and game theory can of course be used in conjunction with discounting techniques.

Sensitivity analysis

During a residual valuation calculation, a large number of assumptions are made regarding, for example, the costs of construction and the income that might be generated from the development. It is necessary to test whether the assumptions made are likely to have any effect upon the overall viability of the proposed scheme. There is the need to provide the decision maker with all of the relevant information that may influence the outcome of such a decision. A way of testing the analysis is to repeat the calculations, by changing the values that have been allocated to some of the variables, such as discount rates to be used, expected construction costs, profit expectancy, etc. This might be a tedious process, but the use of a simple computer program or a spreadsheet will easily allow such calculations to be repeated and tested with ease.

The first calculation is assumed to be the most likely, but by changing the values in the equations in this way will be able to demonstrate just how likely it is that the solution might be affected if things do not turn out in the way that is expected. It is possible therefore to produce worst and best scenarios, in addition to what is believed to be the most likely. The use of sensitivity analysis will help to determine the possible risks associated with the development.

Cost–benefit analysis

Cost–benefit analysis (CBA) is a technique used to evaluate the economics of costs incurred with the benefits achieved. It is mainly used in the public sector in connection with investment decisions where some account needs to be taken of those considerations which are not of a purely financial nature. As such it is an investment appraisal technique. It has its origins in a paper presented in 1848 by a French economist, Duput, on the utility of public works. Since then the technique has been further refined and developed in several other countries for a variety of purposes.

Obvious areas of relevance are health and medical provisions, education and defence. One of the early important areas of application was for water resource development in the United States, with the introduction of the Flood Control Act in 1936. This Act stated that the control of flood waters was in the interests of the general welfare. The construction of a number of dams in a river had multiple objectives relating to power supply, provision of water supplies and improved navigation as well as flood control. In such cases it was important to take all these wide repercussions of the dam development scheme into account in deciding the viability of the projects.

In 1971, the UK Department of the Environment undertook a study of office block schemes. Part of the study encapsulated the initial capital expenditure and also the costs-in-use. A part of the study was also devoted to other benefits that could not easily be quantified, such as, aspects of the buildings' design, their flexibility to meet changing requirements and benefits accruing to both employer and employee. This latter group of items could only realistically be evaluated by cost–benefit analysis.

Cost–benefit analysis has also been used in the road building programme, where some of the benefits listed include the saving of lives through fewer accidents and a reduction in travel time for commerce and industry. It has been used to assess the need and value of an oil pipeline across Alaska. This cost–benefit analysis study took two years to prepare and resulted in a 4600-page analytical report. CBA has also been used in connection with hospital building, urban renewal, the provision of leisure facilities and many other types of project.

Cost–benefit analysis was used extensively in the construction of the Victoria Line underground railway during the 1960s, where one of the benefits listed included the removal of traffic from street level to below ground and therefore benefiting the movement of traffic such as cars, buses and taxis. Those people actually diverted to the Victoria Line would also benefit, otherwise they would not use it. Their gain included time, convenience, possibly comfort and maybe lower fares. In addition, travellers switching to the Victoria Line from other underground lines would help ease the travel conditions of those passengers continuing to use the other lines. These are commonly referred to as direct and indirect benefits.

These direct and indirect benefits can also be illustrated in the example in Table 5.9, concerned with the building of a dam in a Welsh valley to provide water for a midlands town.

Table 5.9   Costs and benefits

Costs	Benefits
The scheme	Employment during construction to
The loss of homes and livelihoods to those whose land is flooded	local people
The loss of their productivity to the national economy	The midlands town's water supply
Compensation costs	Watersports facilities and angling
Possible ecological damage	

In the 1960s a Government White Paper gave formal recognition to the existence of cost–benefit analysis and assigned it a limited role in the nationalised industries. However, the efficacy of cost–benefit analysis was seriously in doubt after the advice given by a statutory commission on the proposed location of London's third international airport. The decision in favour of Stansted was severely criticised by many people, largely on the grounds that insufficient attention had been given to the analysis of alternative sites and that only partial attention had been given to many of the wider repercussions of the project. The government set up the Roskill Commission in 1971 to investigate the proposal with the specific instruction that cost–benefit analysis techniques were to be used to evaluate the alternative sites. The analysis suggested an inland site, whereas politically a coastal site, on the Thames Estuary had been preferred. The disapproval focused on the marked differences between the measured costs of noise nuisance and the costs of time lost by air passengers both in the air and on the ground. The access costs had been calculated by a simulation model and noise nuisance using the principles of compensation.

As a result of this apparent confusion, cost–benefit analysis fell into disrepute during the 1970s, although it was still used extensively in the USA and elsewhere. Much of the criticism was misplaced, being based on a misunderstanding of the role that 'money' played in the application of the technique.

Cost–benefit analysis has been refined since the middle of the twentieth century and is designed to provide answers to the following questions:

- Who are to be affected by the decisions which are to be made?
- How much are these people likely to be affected by the decisions?
- When are the effects likely to occur?

Costs are defined in their widest possible sense to include all resources such as labour, materials, land and foregone opportunities. For example, when agricultural land is used for building purposes, the agricultural output from such land is lost. The price of the land will reflect a number of factors, other than the agricultural output, such as the needs of the local planning authority, the demand for buildings, etc. The real value of the land is therefore equal to the value of the opportunities that are foregone.

The approach to the problem arises from the economic proposition that the community at large has relatively unlimited wants, needs and desires when compared with the resources available to satisfy them. These requirements are not all of equal importance and there are therefore choices to be made in the use of resources and some method of establishing their priority. Cost–benefit analysis does not seek solely to justify an alternative on the basis of immediate costs and benefits, but seeks to evaluate these for the life span of the project. The argument therefore is for the best use of limited available resources by the public sector. A public body does not need to balance its income with expenditure in the same context as a private company. It is classified as a nonprofit-making organisation and relies upon subsidies from one source and another in order to make ends meet. Many of its services are for social need and are often charged for below accounting cost. Politicians of all shades of opinion, however, may argue that this is not the real or total cost and other factors therefore need to be considered.

Benefits are usually assessed in terms of the value of goods or services that a person would be prepared to give up in order to be able to enjoy the facilities which the decision-makers are contemplating should be provided. Benefits may be divided into two parts. The amount a consumer does give up when enjoying a service, i.e. the price paid, and the extra amount over and above that which would if necessary be prepared to be given up.

For example, the construction of a new bypass is estimated to save its users three minutes on a journey. The number of journeys undertaken on this stretch of road is estimated to be 1 m per annum. This then provides us with a total saving of 50 000 hours. If the average occupancy of the vehicles making this journey is two then a total of 100 000 man-hours can be saved. These are worth something and can be priced, and the benefits of the scheme can be compared to costs and then assessed in relation to the national economy. The biggest single problem in applying the technique is, of course, to price the benefits. How much is a man-hour worth? Is a highly paid executive worth the same as a lorry driver, someone who is unemployed or even someone who is retired? In practice, there is no real answer to such a question, it rather depends upon one's own assessment and personal judgment. The worth of an average person is there-fore used, whatever that might be.

In another example, a road to be constructed between two towns may have a choice of several different routes. The different routes will incur different attributes, such as the length of road to be constructed, the loss of certain amenities, the need to demolish property and the different costs and legal difficulties involved in acquiring the land. In addition, it is necessary to consider the users of the new road. A longer route will increase the travelling time of the users and this will also need to be taken into account in the way described above. The various costs and benefits can then be assembled, com-pared and evaluated.

After identifying the problem and the alternative solutions, a cost–benefit analysis evaluation can be made using the following seven basic steps:

(1)   Determining the objectives
(2)   Establishing the extent of the effects of the projects
(3)   Valuing these effects
(4)   Fixing a time scale
(5)   Discounting the value of the effects
(6)   Evaluating the alternatives
(7)   The final decision.

Cost–benefit analysis can be viewed as another tool in the decision-making process. The following points should be noted:

- It does have a real use if used professionally, but it is also open to manipulation for political ends.
- Cost–benefit analysis does not stipulate that a project should only go ahead if the gainers actually compensate the losers.
- Wide divergences of opinion have been expressed about the role and usefulness of this management technique.
- The benefits to be achieved are sometimes little better than a guess, and the less quantifiable they are the more they become questionable.
- The money values placed on the intangible elements, such as the value of the 'environment' and the value placed by future generations are highly speculative.
- Many observers have complained that the methods used are often artificial or arbitrary and provide for no means of checking.
- There is, however, at the moment no better alternative technique to be used under these circumstances.
- If the constraints and limitations are understood then this provides for a more objective approach to be used in the selection of projects, rather than relying totally upon opinion alone.

Chapter 6
Urban Land Economics

Introduction

The United Nations recently released a report on the continuing urbanisation of world population. According to the *State of the World's Cities Report 2001*, urbanisation is continuing at a quick pace. Every year the percentage of the world's population living in cities grows by 0.8%. In highly industrialised countries such as the United States, up to 80% of the population live in cities, and the developing world is beginning to catch up to those figures. Moreover, the number of megacities keeps expanding. By 2010, there will be 30 cities worldwide with populations of 10 million or more people. Currently the population of the five largest cities in the world are: Tokyo (34 m), Mexico City (23 m), Seoul (22 m), New York (22 m) and São Paulo (20 m). By comparison, Greater London is the twenty-first largest with a population of 12 m.

This expansion is projected to continue into the near future. In some towns and cities around the world, notably third-world countries, an uncontrollable urban expansion is already emerging and placing greater demands than ever on the urban infrastructure. There are a great many cities around the world that are already having difficulty coping with their population expansion and many of these cities are expected to grow even further well into the twenty-first century. As the report notes, many of those living in cities are living in poverty, with up to one billion people living in slums and squatter settlements in urban areas around the world.

To many, the city appears to be synonymous with civilisation and the heights of man's achievements in the arts and sciences. To others, it represents urban decay, poor quality housing, crime, misery and disease. Deprivation exists in some form in some parts of almost every city. In Britain, the way of life of its people changed rapidly during the second half of the twentieth century. The underlying causes of these changes included lower birth rates, longevity of life, widening educational opportunities, technical progress and higher standards of living. The following are the major reasons that have determined the pace of the growth of towns and cities throughout the world towards the end of the twentieth century.

151

Population expansion

The world population continues to expand and currently numbers just over six billion souls (Figure 6.1). More than 50% of the world's population live in Asia; here and Africa are the largest population growth areas. In much of Western Europe the population is only continuing to expand through immigration.

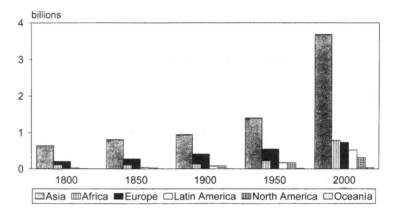

Figure 6.1   World population growth

The population of Great Britain in 1990 was 55 million. This compares with 9 million at the end of the seventeenth century and almost 38 million at the turn of the twentieth century. By the year 2000, the population was estimated to have risen only marginally to about 57 million. The population is the fifteenth largest in the world and is comparable with that of France (58 million), Italy (60 million), the Philippines (61 million) and Vietnam (65 million). Figure 6.2 charts the comparison between birth and death rates throughout the last century. This indicates an almost steady state in

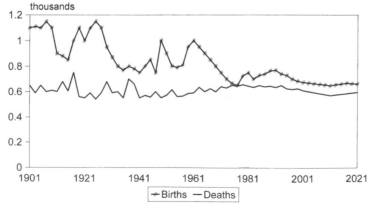

Figure 6.2   Births and deaths in the United Kingdom

respect of these statistics during the latter part of the twentieth century. This feature is unique since records first began.

Figure 6.3 shows a more detailed trend in birth rates in Great Britain since the 1950s. The trends in birth and death rates have an important impact on the need for buildings and development. For example, the traditional age at which many people purchase or rent housing is their early twenties. The trough in twenty-five-year-olds occurs in 2002. While living patterns are changing, with more single households and greater number of people at post-retirement age, the demand for housing is likely to increase, according to forecasts by housing developers. This may be an overly optimistic forecast on their part, although housing improvements and the replacement of derelict stock will always continue to be a feature of the industry.

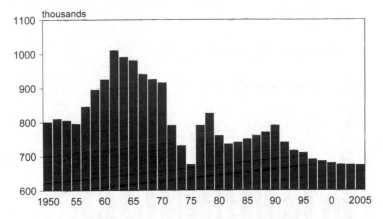

Figure 6.3   Birth trends in Britain

In Britain, 19% of the population is under 15, 65% between the ages of 15 and 64 and 16% is over 65 years of age. The population grew steadily up to 1971, but has since then grown more slowly. The population decline in the mid 1970s occurred for the first time since records began, other than during periods of wartime. However, the age structure has changed with a lower proportion of children aged under 15 and a higher proportion of adults aged 65 and over. This trend in an ageing population is continuing to take place, placing new challenges for planners. By the early part of the twenty-first century, 20% of the population will be over 65 years of age. Whilst more males than females are born, their greater infant mortality and shorter life span of 72 years against 78 years will result in the latter outnumbering males by approximately 5%. Figures 6.4 and 6.5 indicate male and female age profiles from the beginning of the last century and include a forecast for the first quarter of the twenty-first century.

The average population density is in the order of 236 inhabitants per square kilometre (65 in Scotland and 365 in England). This is about the same as Western Germany and India but well above the European Community average of 170 inhabitants in 1988.

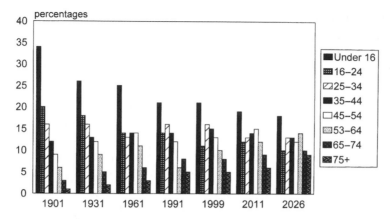

Figure 6.4   Male age profiles

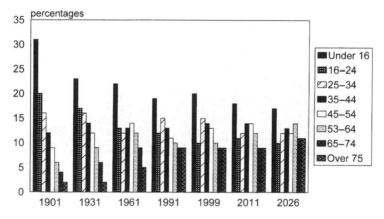

Figure 6.5   Female age profiles

The population density of Japan and Belgium is about one third higher, whilst that of France is less than half and the USA only one-tenth. Since the nineteenth century there has been a trend away from the congested urban centres into the suburbs. There has also been a geographical redistribution of the population from Scotland and northern England to East Anglia and the south.

Economic growth and development

The drift towards towns and cities is largely due to the improved employment opportunities that are available and the wider range of amenities that are to be found in rural areas. Economic growth and development have been synonymous with towns and cities, where the goods created and the business deals made provided a better return for employers, and thus to some extent for their employees. The urban cycle offers increased opportunities for specialisation and thus the increased production of goods and services. This represents an increase in real national income and hence a higher standard of living. This increase in living standards in turn creates a demand for more goods and services, providing firms with larger markets, encouraging more specialisation and increasing living standards yet still further. The rates of industrial growth and urbanisation are closely correlated. The economic opportunities then attracted labour from non-urban areas who helped to produce a surplus of manufactured goods to trade with the agricultural areas and other communities further away.

Employment trends

Since 1955, the number of people employed in agriculture in the United Kingdom has declined by over 50%. In 1955, 5% of the population were employed in agricultural occupations compared with just over 2% in 2000. This picture is mirrored elsewhere in developed societies as mechanisation and new methods of farming have helped to increase productivity. In countries such as Italy, the decline in agricultural employment has been more dramatic, decreasing by over 70% from 41% to less than 10% of the population. In the 1950s, almost 40% of the population from developed countries were employed in agriculture. By 1990, this had declined to 8% and by 2025 is expected to be only 2%. Even in developing countries, these figures show a sharp decrease from over 80% (1950), 60% (1990) and 40% (2025).

Initially the decline in agricultural employment was towards industry and especially manufacturing industry. These were mainly located in the urban areas where labour was available. In some cases, such as the textile mills of Lancashire and Yorkshire, mill owners established communities around the main source of employment. However, by the 1950s, employment in industry was also in decline amongst the developed nations. This was due primarily to two factors:

- The increasing mechanisation and automation in industry
- The lower labour costs in developing countries

Only the presence and considerable investment in high technology has helped to keep such countries competitive. In the 1950s, about half of the working population of the United Kingdom was employed in industry of one kind and another, of whom 40% were in manufacturing. By 1990, these figures had declined to 30% and 20% respectively.

Towards the beginning of the twenty-first century all countries of the developed world were showing increasing employment patterns towards the service sector. In the United Kingdom in 1990, the number employed in the service sector had increased by over 50% to almost 65% of the working population. The United States showed a similar trend, although Japan still retained a higher percentage of its workforce in agriculture, and in Germany, a larger percentage remain employed in manufacturing. This is usual for a developed country. Figure 6.6 shows a comparison of manufacturing outputs in the United Kingdom and United States. Each show a continuing decline, although it is more pronounced in the former than in the latter.

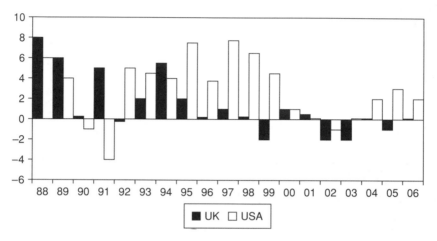

Figure 6.6   Percentage change in manufacturing output

Site identification

The development of land and buildings is promoted by a variety of concerns.

- Property developers who have identified the need for entrepreneurial projects
- Commercial and industrial corporations developing for their own use
- Public and statutory authorities developing for their own functions
- House builders and housing associations
- Charitable organisations
- Government sponsored development agencies.

The most important factors that influence development are the developer's perception of market conditions and the return expected from development. This knowledge is important in identifying the type of development, the region in the country in which the development is likely to take place and in finding a suitable site that satisfies its determined criteria. The developer's ability and knowledge are important in identifying areas of potential growth for development, considering that it will take some time before such a project can become operational. Although there are always likely to be risks involved and imperfect market knowledge, the developer uses a range of market research techniques in order to identify future levels of supply and demand, trends in expectations and measures to stay ahead of possible market competition. The different characteristics of sites for possible development are given later in this chapter.

Land use patterns

Land is unique and its availability is by and large fixed, even after taking into account the relatively insignificant land reclamation schemes. Each piece of land has its own peculiar characteristics in terms of shape, size, altitude, aspect, location, etc. The supply of land for different uses can be either increased or decreased. Change occurs when, for example, land transfers from one use to another; from agricultural to urban use, from residential to office or retailing or from private to public use.

There are about 242 000 square kilometres of land in Great Britain, of which approximately 8% is devoted to urban use. Almost a half of this is used for housing as shown in Table 6.1. Recent estimates of the likely future demand for land raises this figure by a fraction of a percentage point.

Table 6.1　Typical composition of urban land (%)

Housing	45–50
Industry	7–15
Offices	12–25
Retailing	18–20
Education	6–8
Open spaces	17–25

Although much has been written about the need to conserve land for agricultural and other non-urban purposes, the losses of land at the present rate do not seem to present any real economic problem. The main anxieties occur where a loss of agricultural land occurs in a single location.

The main issue is not really concerned with the loss of agricultural land, but the use of former virgin sites for building development and the actual location of these developments. Whilst progress in agricultural practices has improved productivity, it seems sensible to use the best quality land for this purpose rather than for building purposes. Agricultural productivity on the best and poorest land can vary by as much as a factor of ten. However, good farmland is often also the best building land. The area of agricultural land transferred to urban areas per year since 1900 is about 1500 hectares per annum. Recent government initiatives are aimed at focusing development on brown field sites, where buildings may already have been demolished.

There appears to be a good spread of land of all qualities in most regions of Great Britain, and the amounts required for all foreseeable urban development are small compared with the total amount that is available. Overall therefore, the limitation of development through non-availability of land is unlikely to present any serious long-term problem for development, although planning constraints in particular locations in most regions are always likely to be problematic.

There is also a considerable amount of land throughout Britain described as urban wasteland. Whilst governments have sought to remove this, it has been a relatively slow process in terms of regeneration. The examination of any industrial town or city will reveal sites of this kind. Some derelict land has been damaged to such an extent by previous industrial developments that it is incapable of beneficial use without extensive treatment. Contaminated land consultancy is now extensive.

Government has wide powers of control over the development of construction works. It seeks to resolve the conflicting demands of industry, commerce, housing, transport, agriculture and recreation by means of a comprehensive statutory system of land use planning and development control. The government's aim is for the maximum use of urban land for new developments, with the intention of protecting the countryside and assisting urban regeneration.

There are of course wide variations in the use of urban land. Some of this is affected by legislation and legal decisions but more commonly location is the over-riding factor for its use.

Town and city centres

The central focus of any urban area is the town or city centre, which is composed of shops and offices. There is competition for this limited amount of space and often an excess of demand over the availability of suitable buildings and sites. In times of a recession, however, it is common to find many empty shop and office premises. The competition raises land values and development becomes intense, often requiring the construction of multi-storey buildings. Transportation from outlying areas is usually easier than between outlying districts themselves, although the growth of new conurbations is frequently accompanied by increases in road and rail networks and public transport. The demographic increase in population, together with the associated economic growth that was the norm until the recent times, increases the densities within the central business districts. The focal points of town and city centres also show movement over time due to changes in fashion, working and shopping patterns and the influence of the automobile. This together with the development of out of town shopping centres has caused a shift in the site and rental values of land within urban areas. The intensity of development occurs in those areas that are the most sought after. These are also the most expensive in terms of sites and rents, being in prime locations.

Transition zones

There is often no clear dividing line between one urban land area and another, but a merging that is often reflected by an absence of lateral growth of shops and offices. The transition zone lies between the town and city centres and the suburban areas. It contains mixed development of housing for the low and middle income groups, although in times past much of the existing housing stock that is now decaying would have been housing for the middle income groups. These houses were of a sufficient size that many have been converted into flats for multiple occupancy. There are very few areas in towns and cities where the transitional zones have retained their previous high status image. These zones also include areas for industrial and manufacturing purposes and for warehousing.

Evidence exists from some cities in the north of England that where residential developments sit cheek by jowl with industry, pollution can have a severe effect upon health, life styles and life expectancies of the inhabitants. The transitional zones have gained a reputation for urban decay and social and economic deprivation. They are neither prime centre locations nor the more affluent and desirable suburbs. In some areas planning blight has also aggravated the problems, and where proposals for major redevelopments have been suggested or new roads planned, there has been an enforced delay of several years before government funding could be provided. The owners and occupiers also often invest little in their buildings, other than essential maintenance for safety and security matters, due to either poverty or blight. This further reduces the attractiveness of the area. This zone is colloquially referred to as the twilight zone.

Suburban areas

The predominant use of these areas is for residential land and buildings, often segregated by socio-economic class and ethnicity. There is a clustering towards major roads which are often transport routes with an emphasis upon easy access to areas of employment, central areas for shopping, and other general amenity areas. These areas include a range of other types of buildings found in residential areas such as schools, churches, public houses, shops etc. Whilst they may include some leisure facilities, the cinemas, theatres, bowling alleys, etc. are more likely to be located in the town and city centres, where public transport is likely to be easier. With the advent of wider car ownership, new leisure facilities are now being located in out of town areas, where car parking and access are easier.

Where residential development has been extensive, subsidiary commercial and shopping facilities have also been provided, but on a much smaller scale to those in the central areas. Land costs are considerably below those in the central areas. Because of this and a proximity to an available supply of local labour, some manufacturing or light industry has grown up in these attractive locations. However, the higher the social scale of the neighbourhood, the less likely it will be to contain either commerce or industry. The areas also provide an easy access to routes away from the town or city centres. Within these areas there are likely to be parks and other public open spaces.

Rural urban fringe

These areas are largely reserved for low density residential development interspersed with open spaces, farmland and green belt restricted development areas. These are the commuter belt areas where housing is generally occupied by high income earners.

Green belt

A green belt is an area of land, near to and sometimes surrounding a town, which is kept open by permanent and severe restriction on building. There are four reasons for the existence of green belts according to the HMSO publication *The Green Belts* (1995). These are to:

- Check the growth of a large built-up area
- Prevent neighbouring towns from merging into one another
- Preserve the special character of a town
- Assist urban regeneration.

Not all open land surrounding a town or city is green belt, only that which has been given the designated status in the local authority plans. Whilst the government approves their broad outlines, the exact boundaries are defined by local authorities. There are 15 green belts in England covering almost 4.5 million acres or 15% of the country.

Whilst it is not impossible to develop in a green belt, it is nevertheless difficult to obtain planning permission. Anyone who does so must be prepared to show either that the building is required for purposes appropriate to a green belt, e.g. for agriculture or that there is some other reason why it should be allowed. The intention of a green belt is to maintain the rural character of the land, and restrictions on building are therefore similar to those applying to the ordinary countryside which lies beyond.

The severe restrictions on building in the green belt still need to be matched with adequate provision for new development outside these areas. The present trends are to preserve the countryside and encourage developments and the regeneration of urban areas. This policy is implemented by encouraging the fullest use of land in urban areas, including the reclamation of derelict land, the release of unused land owned by local authorities and other public bodies, and by grants and other incentives to developers who are willing to tackle derelict or difficult inner city areas. This has included a range of different measures and these are discussed later, in the section on Government intervention and assistance (p. 180).

Environmental and social factors

The development of property is an emotive issue, with reactions to property development bringing out the not in my backyard (NIMBY) philosophy. During the boom years of the 1960s, many developers were discredited because of the inferior types of development that took place, often destroying familiar and loved environments, creating schemes of inhuman proportions, and having little regard for the existing neighbours living close by the area of the proposed development. A part of the housing property boom was sometimes more concerned with political dogma, associated with the quantity of dwellings completed rather than their longer term quality attributes. The property booms of the 1970s and 1980s were also responded to in a similar fashion but often with protesters becoming more vehement in their responses. Much, however, has been learned of the social consequences of development, particularly in respect of housing and the redevelopment of inner urban areas. Residents will now be consulted about proposals before final decisions are made. They are allowed to comment on the proposals and on the design of the redevelopment. However, developers like other producers respond to market demand, and it is the demand for housing, offices, shops, and industrial buildings that largely determine the shape, location and occurrence of development. Progress and the preservation of the past seem to be incompatible. Laws and procedures exist for the orderly control of development, but often they are limited by political and economic will, with social concerns being considered of less importance. There will always be conflicting demands and it is necessary to attempt to achieve a compromise. Different solutions have been attempted to satisfy those involved but they are often limited due to a lack of adequate research and understanding.

In Britain, from the end of the Second World War up to the present day there has been a substantial increase in the value of landed property due to the effects of an increased demand on a relatively slowly changing pattern of supply. The increased demand is the amalgam of four different factors:

Inflation

Property became attractive to developers, investors and owners, not only because it increased at a level at least in line with inflation, but because it was seen to represent a *hedge* against inflation and therefore a good commodity to possess.

Credit availability

The interest rates on mortgages have barely kept pace with the increased prices of property. This, coupled with taxation relief (although this has recently been abolished), on these interest payments meant that in real terms the costs of purchasing property was very small.

Population growth

The increase in the population, after taking birth and death rates into account in Great Britain since the early 1960s, has been about 12%. This growth created a demand for property of all of its respective kinds.

Affluence

Affluence has been maintained or increased at most levels in society. This is addition to the other factors described above also caused a demand for property.

The changing scene

During the last decade of the twentieth century, structural changes in society occurred which resulted in the state ownership of housing being less desirable. The characteristics are now different. After the Second World War there was an urgent need for property development due to the ravages of war, a lack of development throughout this period, the need to upgrade property to modern standards and to demolish slums and to provide additional property to cope with the rising increase of a growing and affluent population. The property scenario in Great Britain now is different. Population is now at best static or declining and employment is high and likely to remain so for the foreseeable future. Whilst inflation is under control, property price increases are expected at best to be only in line with inflation, over the longer term.

Planning controls

For the purposes of planning control, the Secretary of State at the DTLR publishes the use classes order which allocates land with similar uses to the various land classifications. Normally a change of use within a classification will not constitute development. A change in the use of land from one classification to another will constitute development for which planning approval will be necessary. Buildings are often capable of a variety of different types of use. It will be necessary to seek planning consent where the use of the building is proposed to be changed within the above classification. The value of such a building may be substantially affected where the change of use is possible under the planning regulations.

The Use Classes Order (1972) was replaced with the Town and Country Planning (Use Classes) Order 1987 and its subsequent amendments. The reason for the change was the recognition by the government that the previous order was outdated and too inflexible to meet the changes and needs of modern commerce and industry and their use of buildings. The Use Classes Order is, however, not so precise that it avoids disputes occurring between developers and local authorities.

The system of land use planning in Britain involves a centralised structure under the Secretary of State. The strategic planning is primarily the responsibility of the county councils, while the district councils are responsible for local plans and development control, the main housing function and environmental health. The development plan system involves the structure and local plans. The structure plans are prepared by the county planning authorities and require ministerial approval. They set out the broad policies for land use and ways of improving the physical environment. Local plans provide detailed guidance, usually covering about a ten-year period. These are prepared by district planning authorities but must conform with the overall structure plan. In 1989, the government announced its intention to simplify the two-tier planning system in the non-metropolitan areas by replacing it with a single tier of district development plans.

Before a building can be constructed, application for planning permission must first have been obtained. This is made to the local authority in the form required. In the first instance outline approval is sought to avoid the expense of a detailed design which might fail to secure approval. Full planning permission must still be obtained. If a scheme fails to obtain approval, then a planning appeal can be made to the Secretary of State. The Building Act was introduced in 1984. The Building Regulations (2000) are framed within this Act and allow two administrative systems to be applied: one through the local authority building control department and the other via certification.

Such planning controls are necessary to:

● Secure improved standards of design and construction
● Ensure the safety and health of the occupants
● Provide for the proper location of buildings and industry
● Make the best use of the land available
● Provide for the safety, health and welfare of those engaged in the construction process and those affected by it.

Contravention or disregard of the laws relating to building and construction, whether intentional or not, will render the offender liable to prosecution and in some cases to imprisonment. Some of the more common Acts of Parliament are:

● Town and Country Planning Act 1990
● Housing Act 2004
● Local Government and Housing Act 1989
● Highways Act 1980.

Going green

The concept of a green building is an elusive one. The definition is broad and being green in a professional sense may merely come down to a change in attitude. Most buildings are designed to cope with the deficiencies of a light loose structure, designed to meet the Building Regulations thermal transmission standards and no more. The fact that our Building Regulations are a long way behind those of other European and Scandinavian countries is not seen as a cause for concern. The fact that about 56% of the energy consumed both nationally and internationally is used in buildings, should provide designers with opportunities and responsibilities to reduce global energy demand. There is a need to make substantial savings in the way that energy is used in buildings, but there is also a need to pay attention to the energy used in manufacture and fixing in place of a building's components and materials. For a new building this can be as high as five times the amount of energy that the occupants will use in the first year.

In the 1950s and 1960s, building maintenance and running costs were largely ignored at the design stage of new projects. Today the capital energy costs which are expended to produce the building materials and to transport them and fix them in place is ignored in our so-called energy efficient designs. In any given year the energy requirement to produce one year's supply of building materials is a small but significant proportion (5–6%) of total energy consumption, which is typically about 10% of all industry energy requirements. The building materials industry is relatively energy intensive, second only to iron and steel. It has been estimated that the energy used in the processing and man-ufacture of building materials accounts for about 70% of all the energy requirements for the construction of the building. Of the remaining 30%, about half is energy used on site and the other half is attributable to transportation and overheads. Although the energy assessment of building materials has still to be calculated and then weighted in proportion to their use in buildings, research undertaken in the United States has shown that 80 separate industries contribute most of the energy requirements of construction and five key materials account for over 50% of the total embodied energy of new build-ings. This is very significant since considerable savings in the energy content of new buildings can be achieved by concentrating on reducing the energy content in a small number of key material producers.

Site characteristics

Every construction site has its own individual and peculiar characteristics as listed in Table 6.2. These will influence the value of the site and whether it is suitable for eco-nomic and viable development.

Table 6.2   Site characteristics

- Location
- Location of the project on the site
- Shape and size
- Topography
- Aspect
- Access
- Ground conditions
- Public service utilities
- Historic remains

Location

The location of the site will be of major importance to the developer and the type of development that is envisaged. It may be a congested inner city site with problems of development associated with space, access and the proximity of adjacent buildings and structures. Alternatively it may be in the heart of the countryside presenting its own distinct characteristics and difficulties, perhaps in terms of the delivery of construction materials or the availability of labour.

The value of the site will be influenced by its location in different parts of the country and the type of development that is permissible. Construction costs in areas also vary. In Britain, they are highest in London and the South-East and the lowest in Northern Ireland.

The geographical location of the building usually depends more upon its purpose than on other factors associated with the planning and design of the building. Some buildings will serve only a local community. Perhaps it is only buildings of a national significance or international purposes that have a wide choice in terms of their location. The development of more efficient forms of transport and telecommunications tends to reduce the importance of site location in these respects.

Location of the project on the site

The location of the development on the proposed site will also affect the project's overall costs. Buildings that are set well into a site may, for example, necessitate long haul roads with their consequent initial and removal costs and the requirement for maintenance throughout the contract period. Such projects may also require permanent access roads and the other substantial costs, such as long drainage connections and the necessity for landscaping. Sites that are developed to their extremities also have their own construction problems, especially in inner city and town areas.

Shape and size

Development sites are costed at £ per hectare. The size of the site and its shape will have an effect upon the site value. Because of the high costs of building land, it is essential to make the best possible use of the site by constructing the type and quantity of buildings that the planning regulations permit. Where land costs are expensive the only real solution may be to allow the erection of multi-storey structures in order to allow economic development to take place.

Topography

The terrain of the site will have an influence upon the design, layout and construction economics. A reasonably level site is desirable from the point of view of construction economy, particularly where projects with large floor areas are being constructed. Steeply sloping sites may be desirable from a point of view of aesthetics, but may also result in large quantities of cut and fill and more expensive stepped foundations that add no value to the project. The choice between providing earth retaining walls or sloping banks will depend upon a number of factors, although the former are always the more expensive. The costs will need to be balanced against special requirements and the land availability and its associated costs.

Aspect

Ideally the site should be on a gentle slope facing south to secure the maximum sunlight and protection from the cold northerly winds. The siting of buildings can also have recurrent implications for the user's future energy consumption and hence costs-in-use. The use of the site also needs to be compatible and in sympathy with its adjoining neighbours, in order to reduce the possibility of any conflict.

Access

The site must provide for adequate access for construction purposes and for its eventual users. Restrictive site access and limited storage facilities may create problems with site planning and programming and involve phasing constraints on the contractor. Consideration must be provided in the majority of cases to allow for access by automobiles and for adequate car parking facilities. Consideration should also be given to the proximity of highways for access to the site.

Ground conditions

Bearing capacity

The bearing capacity of the soil conditions is a factor that can have a considerable effect upon the costs of construction. The different conditions of running sand and rock create their own cost conditions. The increased costs of an expensive type of foundation necessitated by poor ground bearing capacity may be coupled with overall poor working conditions for men and machines as they become bogged down and thus result in additional and sometimes unforeseen costs. Such difficulties are further aggravated by inclement weather conditions. Foundations may typically account for between 4% and 7% of the total cost of a multi-storey building. Where expensive foundations, such as piling, are required, this proportion may increase to around 10%.

Water table level

Construction sites that are prone to possible flooding are undesirable. Sites that are on water bearing ground will need to be either drained or elaborate and expensive methods of ground water control will be required during construction. Permanently wet sites slow down construction operations and are thus expensive. Sites that remain permanently damp can cause a rapid deterioration of buildings and may create unhealthy conditions for the building's occupants. Special land drainage may be required and this may restrict the design, construction and use of the building.

Site preparation

The preparation of the site prior to construction operations needs careful consideration. Redevelopment sites invariably involve some form of demolition work. The salvage of reclaimed materials may to some extent offset the costs associated with demolition, but this is limited. While it will be desirable to deploy the most economic demolition methods, this may be restricted because of safety aspects and the rights of adjoining owners. Artificial strengthening of the ground, the redirection of watercourses and demolition of existing structures will all increase the costs of development, reduce profitability and may bring the viability of the project into doubt. The draining of ponds and the subsequent filling may also be a further factor to consider. The redirection or realignment of watercourses that cross the site through culverting is expensive. The felling of trees and the removal or relocation of other obstructions may also need to be considered. Where these preliminary site operations are not undertaken with care then future problems for the owners and users may arise. In some cases it may be necessary to provide an advance works contract to deal with the demolition of existing structures.

Public service utilities

The availability of essential services utilities such as electricity, gas, water, telecommunications and sewage will need to be considered. The last of these may be less important since private sewage facilities may be permissible, although these may prove to be expensive in terms of their provision and maintenance. The location of underground cables or pipelines on the site may create problems with design and construction, prohibit development in the way intended and prove costly to divert. The lack of such facilities creates additional problems and consequent costs for the developer.

Historic remains

These are protected by a number of different Acts of Parliament. Their discovery will have at least the effect of causing delay to the development and in some cases the project envisaged may have to be temporarily postponed. In some cities, such as York, an archaeological survey is required prior to any construction work taking place.

Sites for different uses

The type of development envisaged will be determined to some extent by the nature of the site that can be acquired and by the type of development that is permitted. However, it will be more related to the work undertaken by the developer and the demand for a particular type of property that will have been identified through research and marketing. The following are some of the factors concerned with the different sectors of development.

Industrial location

Traditionally, much of the development for manufacturing industries was undertaken concurrently with the need for housing, since industry was often labour-intensive. Development was thus often undertaken in inner urban areas. However, due to the obsolescence of premises and the cramped nature of some of these sites, further development often meant relocation. The current approach to the development of industrial and manufacturing premises is to segregate them away from residential areas and to locate them in industrial zones away from the central areas to locations on the periphery of cities. In some towns and cities this has resulted in industrial parks, sometimes with speculative built factory units. Some firms may however, have a preference for remaining or developing on a central site. There may also be a reluctance to relocate with all the disruptions that will be required, to new out of town premises, even where split site working becomes necessary. There may also be difficulties of disposing of a firm's old factory premises, especially where the site is located m an area that is run down but is not scheduled for general redevelopment. Industries are often reliant on markets that are both national and international and so tend to gravitate towards those areas that are likely to assist in maximising profits. Where there is a choice of site they will locate on the one with the anticipated lowest overall costs, taking into account production and investment options. The location of industrial premises relies upon the following characteristics:

- Access to good road, rail and other transportation networks
- Low cost and spacious sites
- Accessibility to materials, components and labour (skilled and unskilled)
- Availability of suitable power supplies
- Suitably located for markets for products
- Government incentives
- The environment.

In England and Wales there are currently about 230 million square metres of available industrial floor space. The general contraction and long-term demise of manufacturing industry may provide for little in the way of real development in industrial premises, other than new small industrial units and the replacement of the more obsolete and now poorly located premises. However, whilst there has been a high rate of decline of manufacturing employment, and at an accelerating rate, this is not necessarily mirrored by the same decline in premises that have been partly offset by automation and a reduction in factory employment densities. Some would also argue that manufacturing employment peaked as early as 1955. The long-term decline in this sector, which is expected to reduce still further, is partly explained by the emphasis upon the more profitable and growing service sectors of employment. Any growth in provision is likely to be seen in the high technology industries.

A high proportion of factory premises, both vacant and occupied, are now outdated, in terms of the facilities that they provide. About one third of them were constructed prior to the beginning of the Second World War. Many are multi-storey buildings that are unsuitable for many of today's industrial processes and have limited external space for storage, car parking or expansion. They are often the wrong size, in the wrong location and the wrong design. Whilst there is typically about 3–5% of the total stock for sale at any one time, this may be inadequate to meet demand despite the decline in industrial activity. The construction of industrial premises has typically accounted for about 9–12% of the total output of the construction industry, acknowledging that this figure has fluctuated considerably. Recent investment reports are projecting a similar proportion of new developments in this sector at the beginning of the twenty-first century.

Generally the depressed regions of the country have the greatest supply of land available for industrial development. London and other areas, where land costs are typically much higher, are disadvantaged in respect of the quantity, quality and cost of industrial sites. Any industrial upturn will therefore have only a limited effect in these areas other than attracting small firms with modest site requirements. Any increase in industrial activity is also likely to result in the industrial land available not being in the right place, at the right price, the right size or available at the right time.

In respect of both industrial rents and land prices, there remains a north–south divide. The recession of the early 1990s did go some way towards remedying the inequality of them, but this was reversed in the early part of the twenty-first century. The attractiveness of a community of high-tech industries, where specialist skills have been developed in the workforce, has created an attractive image for further industrial developments of a similar kind, in areas such as the M4 corridor and 'Silicon Glen' in Scotland.

There are more than 65 000 small firms, defined as those occupying less than 500 square metres of floor space, and these account for over half of the industrial premises in Britain. Many of these small premises have been provided by speculative developers or local authorities in partnership with the private sector.

Different governments in the latter part of the twentieth century formulated policies in an attempt to create wealth, reduce unemployment and eliminate regional imbalances.

They made loans and grants available, increased tax allowances for buildings, machinery and plant and offered employment inducements. These policies have resulted in a shift of industrial activity away from some of the more traditional areas to those offering assisted area status. Both the quantity and quality of industrial development in the former areas has been less than if market forces alone had been permitted to operate freely.

The proportion of land devoted to industrial purposes shows a wide variation between different urban areas. The industrial areas of the north–west, north–east and the midlands do devote a relatively large amount of land to industrial use. Other cities that may be represented by agriculture or farming have a much smaller proportion of industrial land. However, in all urban areas today, there will be some demand for land for industrial purposes, even if to satisfy only local area need.

Office location

In Britain, office jobs as a percentage of the workforce increased from 24% to 32% between 1960 and 1970, with the growth in employment being five times that of employment generally. Offices may be classified in several different ways: by size, local or national, public or private, independent or attached to some other function. In the earlier part of the twentieth century the public service offices of local or central government were located in central locations to signify the importance of the role of government. Where redevelopment has occurred such offices have tended to be relocated in the less costly areas of urban areas. Where aspects of central government departments have been relocated to the provinces, often to locations with assisted area status, inexpensive sites have often been selected.

In the private sector, office accommodation supposedly benefits from a prestige address, where firms have required their offices in the larger cities and in the expensive inner urban areas of those cities. London is a prime example. Office rents in central London areas in the 1950s were about £15 per square metre rising to £165 by the early 1970s and by the late 1980s to as much as £500 per square metre. By comparison, in other cities around Britain, these respective costs were much lower. In some northern cities, for example, prime city centre rates were less than 10% of the London rates. However, even taking these wide disparities into account, typically 70% of all new office investment has been located within the London area. When occupancy costs (rent, rates, service charges) are taken into account, the London rates reached as high as £650 per square metre. Only Tokyo of the major world cities had higher costs. New York and Paris were about two-thirds and Frankfurt about one-third the London rates. Of course rates vary within a city, the west end of London currently being the most expensive.

Traditionally offices have tended to locate in the central business districts because of the following:

- The interrelationship of businesses require ready access to other offices
- Dispersion is uneconomic
- Convenience of access to their staff and the public in respect of public transport systems
- Access to local and central government offices.

Such high costs often mean that much of the central areas of British cities are given over to offices with perhaps the ground floors being used for retailing, banking, etc. where the public need to gain easy access. The earning capacity of some offices is sometimes dependent upon being sited in a specific area, where face-to-face contact can be made and there is easy access to the required information and data. Since these sites are the most valuable it becomes practicable to construct multi-storey structures to gain the maximum benefit.

Due to the need for more space, expiring leases, premises due for demolition or refurbishment and to the rising costs of rents and rates, higher salaries and travel costs and inconvenience, some firms moved away from central London in a process of decentralisation. It was realised that the general administration, routine accounts and technical departments work could just as efficiently be undertaken elsewhere in the country. The towns and cities in the south-east were the preferred location of these firms, not least because of the pressures of the staff concerned. However, in 1983, the government produced a White Paper, *Regional Industrial Development*, and encouraged some firms through financial incentives to move to areas with assisted status in the north and west. These incentives included subsidised rents, interest free grants and loans, redundancy pay commitments by government, removal costs and other forms of regional aid. The Location of Offices Bureaux had already been established in 1963 for this purpose. By 1977, the economic plight of the inner cities, including inner London, was so bad that the Department of the Environment altered the terms of reference of the bureaux in order to attract offices to areas other than the decentralised locations.

Retail location

The total floor area of retail space in the United Kingdom is approximately 90 million square metres. This can be broadly categorised as shown in Table 6.3.

Factors influencing the location of shopping development include:

- Catchment area
- Purchasing power
- Amount of floor space available
- Competition from other centres
- Transportation and car parking
- Shopping habits and preferences.

Table 6.3   Shopping classifications

Community shops: These include the newsagents, confectioners, grocers, bakers, greengrocers, etc. These shops have been increasingly unable to compete with the large retail outlets and their share of business has fallen by a half. Many of these shops remain open for long hours and are often located as focal points in residential areas.

Town and city centres: These include departmental stores and the speciality and variety shops such as well known shopping chains that specialise in a variety of household goods.

Supermarkets and hypermarkets: The supermarket represents the biggest change in shopping patterns since the middle of the twentieth century.

Market stalls and kiosks: These are found adjacent to road, rail and airport centers.

Trades and professions: These include banks, post offices, estate agents, repair shops.

Many shops are located in a town's central shopping areas in order to attract as many customers as possible. This is largely historic in nature, predating the automobile, when shoppers had to rely upon some form of public transport. Within the central shopping area there is a high correlation between retail turnovers and rents per square metre. Apart from the large department stores, many shops change in use or trader over a relatively short period of time. Over a 20-year period, as many as 80% of shops may either change in use or trader. Some of this is dictated by fashion and the response to change by shop owners. Corner sites are especially sought by traders who need to display their wares as widely as possible.

New town centres which included the extensive redevelopment of shopping areas were a response to city centres that had been bombed out during the Second World War. Even those towns and cities that escaped the ravages of war had to redevelop in order to maintain their competitiveness against neighbouring towns and cities. The larger regional shopping centres, located in the large regional cities, had floor areas of about 46 000 square metres. These developments had a significant effect upon the independent retailers who found that after the abolition of resale price maintenance, they could no longer compete with the larger stores and national organisations that had capitalised on some of the new town centre redevelopments. These developments also had an adverse effect both upon areas of close proximity that were not redeveloped and some shops in the inner suburbs.

By the mid–1980s, the out of town shopping centre developments had overtaken the town centre shopping areas in terms of popularity. The decentralised locations are constructed on less expensive sites with consequent lower rents and rates and provide adjacent free car parking. In addition, some of the savings made by the huge retailers are passed on to the customers. The institutional investors in town centres are experiencing a decline in the value of their investments and are thus having to respond to these pressures by redesigning or remodelling these areas. The important objective is to retain the household names and anchor tenants such as Marks & Spencer, John Lewis or Woolworths. Some of these are also locating a part of their business in the large decentralised shopping malls, evident in places like Gateshead and Sheffield in order to keep a presence in both locations. Decentralised shopping areas include superstores, hyper-markets, retail warehouses and edge-of-town shopping centres (see also Chapter 3).

In London, for example, the centre of shopping moved westward from St. Paul's churchyard in the seventeenth century, to the Strand in the eighteenth century, to Piccadilly and Regent Street in the early nineteenth century to Oxford Street by 1900. In each period, the centre was the point of maximum accessibility, and as means of transport changed there was a relocation to suit the current form of transport. As in the past, shopping is likely to be influenced by the pattern of transportation. The only dif-ference is that the pace of change is likely to be more rapid and far reaching, being influenced by government policy, economic growth, local planning and decentralisa-tion based upon an even wider market.

The efficiency of a retail outlet is highly dependent upon location, and within loca-tion highly dependent upon the site. There may be, within a short distance of one site, a superior shopping area. The correct selection of site may be the difference between business success and failure. For example, a dilapidated building sandwiched in the high street between Marks & Spencer and Mothercare is preferable to a modern shop at the wrong end of town. The number and type of shops that can locate profitably in a single geographical area are dependent upon a number of determinants of that revenue. These include the size of the shopping catchment area in terms of population and visitors, more importantly their purchasing power and the degree of competition amongst the different retailers for that portion of the revenue. Retailers may wish to analyse their market in different segments and to respond to and target the variations in age, social class, lifestyle, etc. of the customers in different ways. Retailers may also want to under-take niche marketing in an attempt to satisfy the specialised needs of different consumer groups, such as the high income, credit worthy, etc. Low income shoppers are being catered for less and less in new shopping developments.

Some are beginning to suggest that the traditional town centre is in near terminal decline as a shopping area due to high rental values, car parking difficulties, inner city crime and competition from out of town shopping areas. In some towns and cities one in eight of the shops now lies empty. Research by Verdict, the retail analyst, has revealed that 25% of retail stores are now in out of town locations compared with just 5% in 1980. The change has been led by the nation's top store groups. In the mid-1980s, 80% of Sainsburys were in the high streets; today this has been reduced by half. Marks & Spencer, once the barometer of the country's retail mood, now takes one-eighth compared with one-eightieth in 1986, although its fortunes were severely dented at the end of the twentieth century. The famous names to have joined in the exodus away from town centres include British Home Stores, Burton, Currys, Littlewoods, Marks & Spencer, Next, Tesco and Sainsbury. Their replacements include amusement arcades, charity shops, discount shops and indoor markets selling discounted and some-times shoddy goods. Currys, for example, in the middle of 1994 indicated that it would be closing down a third of its 330 high street shops. It has, however, increased its total floor space by renting 40 out of town superstores at a much reduced cost.

Residential location

Housing comprises the largest urban land use. In some towns and cities it represents over 50% of the total area. There are a wide variety of house types including flats, maisonettes, town and terraced houses, semi–detached and detached houses, bungalows and mansions standing in spacious grounds. Figure 6.7 illustrates the current composition of residential properties.

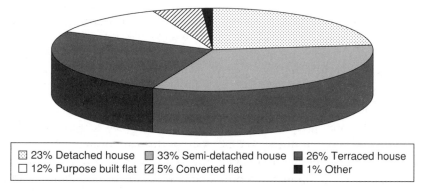

⊞ 23% Detached house	■ 33% Semi-detached house	■ 26% Terraced house
□ 12% Purpose built flat	▨ 5% Converted flat	■ 1% Other

Figure 6.7   Type of housing accommodation (Source: DCLG)

A Classification of Residential Neighbourhoods (ACORN) has attempted to classify the different sorts of housing areas. It is a marketing segmentation system which enables consumers to be classified according to the type of residential area in which they live. The ACORN classification takes into account the different variables encompassing demographic, housing, employment, etc. characteristics. It is based upon census data (Table 6.4). There are 38 neighbourhood types and these aggregate into 10 neighbourhood groups. ACORN is linked to postal geography, thereby permitting survey

Table 6.4   ACORN neighbourhood group classifications (Source: ACORN)

Group	Type
A	Agricultural areas
B	Modern family housing, higher incomes
C	Older housing of intermediate status
D	Poor quality older terraced housing
E	Better-off council estates
F	Less well-off council estates
G	Poorest council estates
H	Multiracial areas; high status non-family areas
J	Affluent suburban housing
K	Better-off retirement areas

respondents to be assigned to their appropriate type by the analysis of the postcode. Although there is a relationship between personal income, type of employment and place of residence, this relationship is subject to conflicting and different interpretations.

The costs and types of housing reflect to a large extent the earning capacity of their inhabitants. Low-income earners, for example, tend to live within the inner cities of urban areas close to where they are employed. They may live in rented accommodation where their rents are regulated, housing densities are high and the costs of travelling to work are minimised. As incomes increase, there is a tendency to move up market to suburbia, further away from the place of work and the inner urban areas, in locations of lower densities, close to open spaces and in housing that is more expensive. In many British towns, lower income residential areas are found in locations that may differ from those suggested because they are local authority housing or properties managed by housing associations that provide subsidised accommodation. However, this type of housing is rarely found on the side of the urban area catering for higher income groups because land values are higher and access to work more difficult and expensive.

The actual choice in housing may be very limited, being dictated by mortgage availability and distance of travel to the place of work. An individual seeking to maximise utility must weigh the desire for access to the place of work against the various possible combinations of commuting costs and accommodation prices with the other desires for urban contacts and amenities. There is often a time-lag delay before householders respond to their changing financial circumstances, and some degree of immobility caused by non-economic reasons outweighing economic considerations in determining choice of location. The main reasons for a change of residence are change of job, marriage and a change in family size. Higher income groups do have a wider choice of residential location since they are able to select from amongst the existing stock of housing as well as having the ability to pay for the newer, more desirable residential facilities.

There is nothing intrinsic in land that creates an upper class residential area. The attractive physical qualities found in such areas include well drained wooded slopes, high ground and separation from other types of land use. However, such features also exist in other types of urban areas, but these areas may be not as conveniently located away from less attractive amenities such as factories. Once established, a high class residential area is able to prevent intrusion by other unacceptable uses.

The total stock of dwelling units in England is approximately 22 million of which currently 70% are owner-occupied (see Figure 6.8). In 1950, less than 50% were owner occupied. The growth in owner occupation has largely been at the expense of the privately rented sector, and through, for example, the selling of local authority-owned houses in the late 1980s. It is estimated that approximately 850 000 dwellings are empty at any one time.

Table 6.5 shows the distribution of dwellings in England allocated to the respective Council Tax bands for rating purposes. Of the 22 million dwelling units, 4% have been placed in the highest band H. This percentage is almost double for London and the

Table 6.5 Number of dwellings on valuation list 2006

	Band A	Band B	Band C	Band D	Band E	Band F	Band G	Band H	Total
North east	660 412	164 934	164 804	88 168	45 004	19 519	11 300	1225	1 155 366
North west	1 313 668	589 478	533 391	305 719	178 943	87 205	59 283	6020	3 073 707
Yorkshire and the Humber	1 006 148	439 680	368 149	205 028	129 001	61 763	36 005	3062	2 248 836
East Midlands	720 522	421 575	338 331	200 489	118 681	57 574	33 083	2908	1 893 163
West Midlands	732 069	576 249	441 834	253 212	159 408	86 394	52 966	5172	2 307 304
East	348 055	513 912	638 485	423 770	257 470	140 248	95 381	11 444	2 428 765
London	109 872	438 795	866 593	822 965	493 693	247 308	200 449	55 546	3 235 221
South-east	309 809	583 822	915 754	711 647	476 258	287 484	230 981	31 742	3 547 497
South-west	393 676	557 874	528 921	361 568	243 288	123 647	71 573	6914	2 287 461
									22 177 320

(Source: Valuation Office of HM Customs and Revenue)

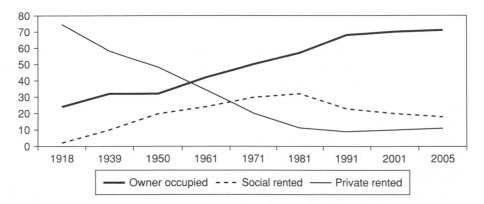

Figure 6.8   Owner-occupied, social rented and private rented housing 1918–2005

South-east of 7.9% and 7.4% respectively. Conversely about 25% of dwellings have been allocated to the lowest Band A. In the three Northern Regions these dwellings account for some 46% of the total number of dwellings in these regions Just over 30% of all the dwellings are located in London and the South-east. It has recently been recommended that a new higher Band I should be introduced to reflect the very high price of some of the country's dwelling houses.

Slum clearance in Britain declined significantly during the 1980s but has recently increased by 6%. The census indicated that there was a surplus of over 500 000 dwellings for the whole of Great Britain. However, this does not take into account that the distribution of available housing throughout Britain does not correspond with the distribution of families requiring household space. It is estimated that 60 000 to 90 000 homes must be built each year to meet needs at the low cost end of the market, but by the middle of the 1990s a shortfall of a third of this estimate was expected. The suitability of private and social housing in 1996 and 2004 is shown in Figure 6.9.

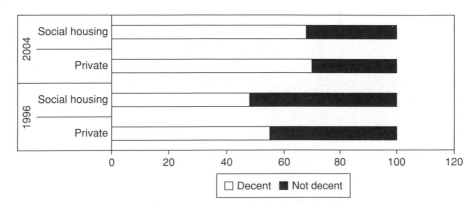

Figure 6.9   Suitability of housing (Source: English House Condition Survey)

Between the 1950 and 1970, a new type of housing was being developed throughout Britain: the tower block. This was one of the remedies to solve the problems of slum clearance. The majority of these high-rise structures were engineered using systems of prefabricated construction. These were off the peg solutions, typically using systems that had been developed in Scandinavia. Their designs failed to take into account the difficulties that they would create in respect of social and environmental issues. They were constructed largely as a quick-fix solution to meet the political dogmas concerned with housing starts and completions. The emphasis was therefore directed towards quantity rather than quality. Their main attractions included their speed of erection and the fact that better grants were available from central government for high-rise dwellings than low-rise housing. Other matters of consideration included the elimination of architects' design fees and the fact that a relatively small land area was required. This would be less expensive to purchase and easier to compulsory purchase. The Ronan Point disaster in 1968 marked the end of such types of industrialised high-rise housing. This was not before almost all towns and cities had recreated their slums in the sky, into which many of the schemes soon developed. Vandalism, crime, squalor and the high costs of repair and maintenance helped us all to realise what a bad decision had been made then.

Legal aspects

English land law has its origins in English history. Originally the land was owned by the King, and the King would allow lords of the manor to hold the land in return for certain services. The lord of the manor, in turn, allowed the inhabitants of the manor to have plots in return for services rendered to him.

Freehold

These are essentially estates of indeterminate duration. The most common sort of freehold estate is the fee simple, which is an estate that may be inherited by any heirs under a will or intestacy, and may thus continue in perpetuity. On very rare occasions someone may die without making a will or without any heirs at all. In this case the land reverts back to the crown.

Leasehold

A leasehold estate may be for any length of time from one day to 999 years. The essential requirement is that the term of the lease will be specified for a fixed and definite duration. During this time the lessee must be given exclusive possession of the land. The grant of a leasehold estate is formally drawn up in a deed called a lease which specifies the details of the lease and any requirements such as covenants that might be imposed.

Easements

An easement is a right that allows the holder to use, or restrict the use of, the land of another in some way. The common examples are as follows:

- *Private rights of way*: these may be presumed if they have been enjoyed for 20 years or absolute if enjoyed for 40 years. Public rights of way are not easements.
- *Rights of light*: the right of light to a building becomes legally protected after it has been enjoyed for a period of 20 years. This right prevents a neighbouring owner from erecting anything on his land that will interfere with this right.
- *Rights of support*: the land carries the right to support adjoining land but for the support of any buildings subsequently constructed on it.
- *Right of drainage*: this allows one owner the right to dram across the property of another.

An easement can exist only in relation to other land which is nearby, which is said to *benefit* from the easement. A quasi-easement is a term that is used to describe a habitual right exercised by a person over a part of his or her own land. This would become an easement if the two parts were in different occupation, such as the right of support between adjoining buildings. A quasi-easement may become a real easement upon the sale of one or both parts of land.

There are a number of other rights which are similar to, but do not comprise easements. A *profit à prendre* is a right to take something from the land of another person, such as grass or sand. A *licence* is a private right to go onto another's land. Either of these rights may exist without the holder owing land that is benefited. An easement of support relates only to a building and is distinct from the natural right to have unweighted land supported by adjoining land.

Restrictive covenants

Restrictive covenants are a contractual provision, imposed by conveyance, that seek to constrain the way in which the holder of land may use it. The nature of these vary widely and they arise where an owner imposes legal restrictions on the use of land. Restrictive covenants created since 1925 on freehold land must be registered if they are to be enforceable. For example, a type of covenant may exist where an owner selling a piece of land restricts its use to certain types of property or restricts the amount of development that will be allowed to take place. It will also be binding upon successive owners, but only if it eventually constitutes an easement. The covenant may also be specified to exist only for a limited period of time.

Where restrictive covenants are reasonable in their requirements, they can assist in maintaining some order to development that may be allowed in an area. They may thus help to sustain property values. In those circumstances where the character of an area changes they may retard normal development from taking place. Restrictive covenants on leasehold land are more common and numerous, since one of the reasons for granting leases rather than selling is that the freeholder can retain some control over the use and development of land. Among the more common restrictions are:

- Use, such as the restriction or prohibition of certain uses on property in residential areas e.g. for offices or trades
- Alterations not to be made without consent, with plans needing to be approved by the freeholder
- Assignment or subletting, not to be permitted without consent.

Covenants may also be positive in nature, by imposing obligations on the purchaser to do something, such as to erect a building within a specified time scale.

Party walls

There are problems that frequently occur in heavily built-up areas over the respective legal rights of adjoining owners in a party wall or party fence wall. These problems are more acute in London and the respective rights have been codified in the London Building Acts. There is no absolute definition of the term, but it is used where the boundary line between the two properties is built on, or where one adjoining owner has acquired rights in a wall built up to the boundary on neighbouring land. The issue is to attempt to define where the boundary lies, often amidst alterations and extensions over the years. Once this has been agreed, it is invariably found that the walls vary in line and thickness and that there is no simple solution regarding ownership and responsibility. Assuming that ownership can be resolved, there are rights at common law to undertake the work to a party wall, notwithstanding that part of the wall is not owned by the party wishing to carry out the work.

Government intervention and assistance

There are several different ways by which government can either encourage or discourage building development activity. It has, for example, a series of planning guidelines and regulations that prohibit the various sorts of development. Local authorities are able to identify in an orderly fashion the types of development that they themselves would like to see taking place through their structure plans. Particular projects in designated locations may be prohibited as being not in keeping with the structure plans. Decisions to prohibit development can, of course, always be challenged through the courts or directly with the Secretary of State for the Environment.

Particularly since the mid–1970s governments of different political persuasions have attempted to intervene in the market by offering incentives to encourage building development. Financial assistance, for example, has been required to encourage industry to relocate in areas of higher than normal unemployment or to address issues associated with urban regeneration and the improvement of the quality of housing. They often seek to encourage development by injecting public finance in order to attract further private investment. The kinds of development grants that are available are constantly under review. Some represent long-term initiatives. Others are introduced, but perhaps do not fulfil the ambitions of ministers and are then replaced by other measures to solve particular problems. The following sections describe the major sources of development grant funding. Other funding is available, for example, from English Heritage, the Countryside Commission and from different government departments. All grants attach certain conditions that a developer in the normal course of business might not usually undertake. Table 6.6 summarises funding initiatives in the 1980s–2000s.

Table 6.6   Urban development activities

1980s	*Urban initiatives*
1981	Urban Development Corporations
1981	Enterprise Zones
1982	Urban Development Grants
1982	Derelict Land Grants
1984	Garden Festivals
1985	City Action Teams
1986	Task Forces
1987	Urban Regional Grants
1988	City Grants
1990s	*Comprehensive approach*
1991	City Challenge
1992	English Partnerships
1993	Urban Partnerships
1993	City Pride
1994	Single Regeneration Budget
1994	Rural Challenge
1996	Capital Challenge
1996	Sector Challenge
1998	Local Challenge
1998	New Deal for Communities
2000s	*Sustainability*
2000	Our Towns and Cities: The Future
2002	Regional Economic Performance
2003	Sustainable Communities
2006	The State of English Cities

Enterprise zones

The first Enterprise Zone was established in 1981 in the Lower Swansea Valley. These are zones to allow the testing of innovative, social, economic and political arrangements. The zones may be one or a number of sites each subject to their own planning regulations. Eleven zones were first selected but more have been added to encourage industrial and commercial development by removing certain fiscal and administrative burdens. The Second Interim Evaluation of Enterprise Zones was published in 1995. The following benefits are available, for the period of the designation of the Zone, to both new and existing industrial and commercial enterprises in an Enterprise Zone.

Tax breaks

Enterprise Zones attract 100% allowances for corporation and income tax purpose for capital expenditure on industrial and commercial buildings. Elsewhere in the UK, industrial buildings may not qualify for capital allowances, greatly increasing the tax burden. There is also exemption from the national non-domestic rate (uniform business rate) on industrial and commercial property. This local property tax can be substantial based upon usable space.

Regulatory breaks

There is a greatly simplified planning regime in Enterprise Zones. Developments that conform to the published scheme for each zone do not require individual planning permission. The statutory controls that remain in force, e.g. planning for non-conforming developments, are administered more speedily.

Employers are exempt from any industrial training levies and from the requirements to supply information to Industrial Training Boards, although few of these now remain. Government requests for certain statistical information are reduced.

Applications from overseas firms operating in Enterprise Zones for certain custom facilities are processed as a matter of priority and certain criteria are relaxed.

Each separate zone sets out its own scheme for the kinds of development which result in the automatic granting of planning permission. Only a limited number of issues then require clarification and approval by the Enterprise Zone Authority. Other schemes that do not conform with the conditions of this deemed permission require planning permission in the normal way.

The main advantages, therefore, of development within an Enterprise Zone are speed in dealing with planning applications, exemption from rates for occupiers and the tax incentives that are available for commercial and industrial buildings. Taxation relief is available to investors at the highest rate. The investment may be disposed of after 25 years with no clawback on the tax allowances. These tax allowances refer only to the costs of construction and the associated professional fees and not towards the purchase of the land.

Urban Development Corporations

London's Docklands and Merseyside were the first Urban Development Corporations (UDCs) to be established in the United Kingdom in 1981. Their objectives are similar to those of the former New Town Development Corporations. Their objective is urban regeneration within their designated areas. They reclaim derelict land, refurbish existing buildings, provide the required infrastructure and encourage the private property development for housing, industry and commerce. UDCs act as the local planning authority for the area for which they have been designated. They may sometimes provide developers with financial assistance towards the purchase of sites or assistance in their acquisition. The extent of their work can be illustrated by the following.

The London Docklands Development Corporation (LDDC) was the second UDC to be established under the Local Government and Planning Act 1980. Its objective was to secure the region of the London Docklands urban development area comprising 8.5 square miles of East London in the Boroughs of Tower Hamlets, Newham and Southwark. It was established in response to the huge decline in the economy of the area. The UDC had power to acquire land by agreement or through compulsory purchase. It took over from the respective London boroughs their planning but not their plan making process. It had powers and the resources to provide new or refurbished infrastructures.

The task was daunting. The area had been subject to catastrophic job losses as the docks closed. The skills of the blue-collar workers, especially, were inappropriate for the growth of the London economy. The extent of dereliction was so severe that the costs of development would be both very high and uncertain, lowering the incentives to investors. Many sites were poorly serviced by the local infrastructure and there were poor strategic links between Docklands and the rest of London. The market alone was unlikely to provide any environmental improvements. There were also gaps in information that were hindering the operation of the market. There was, for example, an almost complete absence of private house building. It would be difficult to reverse the steep cycle of decline.

The LDDC was in existence for 17 years until 1998. During that time it secured £1.86 billion of public sector investment and £7.7 billion from the private sector. It developed 25 million square feet of commercial and industrial floor space, built 144 km of new roads and developed the Docklands Light Railway. Over 24 000 new homes were constructed and it helped to establish 2700 businesses. There are now over 85 000 people at work in the area and its developments have acquired 94 architectural, conservation and landscaping awards.

City Challenge

City Challenge is a national initiative implemented in different local circumstances. Its aim is to bring sustainable and integrated regeneration to areas of widespread and multiple deprivation. It is intended to be different in terms of its values, organisation and priorities and in the scope of the regeneration sought and the way in which it is delivered.

City Challenge emphasises not just physical regeneration but improving economic and social infrastructure and local quality of life. All areas have the needs associated with Urban Programme authorities. Bids have to demonstrate the opportunities and strategies to tackle problems on a significant scale. They must define the intended beneficiaries.

City Challenge is believed to be the most promising regeneration scheme so far attempted. There is widespread support from those involved across the range of sectors for most aspects of City Challenge's design. They see it as an improvement on previous urban regeneration programmes particularly because of its partnership basis, community and private sector involvement, strategies and targeted approach and its implementation by dedicated multidisciplinary teams.

City Challenge allocated £37.5 million each over five years to 31 Urban Programme authorities to achieve self-sustaining regeneration of their designated City Challenge areas on the basis of two competitions. In the first round, 17 local authorities covering 15 areas were invited to complete for City Challenge status. They were chosen to represent the wide range of circumstances across the country and their ability to work up imaginative plans quickly. Bids were submitted by cross-sectorial partnerships.

City Challenge was innovative because:

- it was competitive
- it adopted comprehensive and strategic approaches
- it targeted upon specific areas
- it was time-limited
- it was output-driven
- it was based upon partnership.

Creating better communities

The Department of Communities and Local Government's main aim is to help people and local agencies create cohesive, attractive and economically vibrant communities. It is working to ensure communities are capable of fulfilling their own potential and overcoming their own difficulties, including community conflict, extremism, deprivation and disadvantage. It seeks to empower communities to respond to challenging economic, social and cultural trends. To deliver this, the Department of Communities and Local Government is integrating and building on the approach set out in People, Places, Prosperity, its five year plan that was published in 2005. This identifies a number of areas of activity which include the following:

- *Sustainable communities.* This is about the Department of Communities and Local Government's plans to create thriving, vibrant, sustainable communities which will improve everyone's quality of life.

- *Cleaner safer greener communities.* Creating cleaner, safer and greener communities by improving the quality of planning, design, management and maintenance of public spaces and the built environment.
- *Community cohesion.* Building strong, cohesive communities.
- *Neighbourhood renewal.* This aims to improve the quality of life for those living in the most disadvantaged areas.
- *Civil renewal.* This is about the work of the Department of Communities and Local Government's civil renewal unit, including the *Together We Can* plan.
- *Together We Can.* This sets out the government's commitment to empower citizens to work with public bodies to influence public decisions.
- *Respect.* Respect is fundamental to functioning places and communities.
- *Social exclusion.* The Department of Communities and Local Government leads on policy and delivery in addressing social exclusion and deprivation in deprived areas.
- *Place matters.* This is concerned with the role of the Department of Communities and Local Government in the management of place.

The Carbon Challenge

The Carbon Challenge, which will be run by English Partnerships, calls on developers to raise standards of design, construction, energy and water use and waste disposal so that these techniques can be used in the future as a benchmark for mainstream development. It also seeks to meet rising expectations from the public for more sustainable communities which offer them reduced bills and a higher quality of housing design.

The Carbon Challenge will spearhead the move towards zero-carbon development in a radical package of new measures for greener housebuilding, including the Code for Sustainable Homes and the first ever planning policy on climate change. In future, most new zero carbon homes will be exempt from stamp duty. Building the new homes that are needed is a prime opportunity to harness new technology and drive up environmental standards.

The first two English Partnerships sites have been named as Hanham Hall near Bristol and Glebe Road in Peterborough. Three further public and private sector sites are expected to be added to the Carbon Challenge in the near future. The Carbon Challenge will be open to developers and construction firms from across Europe with a target of delivering several thousand zero or low carbon homes. English Partnerships will work with the construction industry to meet the challenge of climate change. It will work with the house building industry and local authorities to shape the future of development in this country. The successful bidders for the first two sites will be announced in autumn 2007, with work starting on the new communities in summer 2008. The government is also encouraging other public land owners, including local councils, as well as private land owners, to put their own land forward as part of the Carbon Challenge and be zero carbon trailblazers.

The Carbon Challenge is the successor to Design for Manufacture (DfM) and will build on lessons learnt from that competition. Zero carbon means no net carbon emissions from all energy uses in the home. The amount of energy taken from the national grid is less than or equal to the amount put back through renewable technologies. Currently the energy used to heat, light and run our homes account for 27% of all the UK's emissions at around 40 million tonnes. The key features of a zero-carbon development could include technologies such as:

- combined heat and power
- district heating and cooling systems
- aquifer thermal energy
- ground source heat pumps
- passive heating
- solar and wind energy.

The State of the Cities Database

The State of the Cities Database was created to provide access to updated data that underpins the *State of the English Cities* Report, published by the Department of Communities and Local Government in March 2006. The database contains all the key indicators of urban performance used in the report and allows users to identify in greater detail recent changes in English cities and will allow future monitoring of cities and the impact of policies. The indicators are available for 56 cities in England, that is built up or urban areas with a population of 125 000 or more.

The database can be accessed at://www.socd.communities.gov.uk/socd/

European grants

The European Union has a number of structural funds available which work towards the goal of achieving economic and social cohesion. Resources are targeted at actions which help bridge the gaps between the more and the less developed regions and which promote equal employment opportunities between different social groups. Structural funds are organised through negotiated multi-annual programmes. There are four structural funds:

- The *European Regional Development Fund* (ERDF) aims to reduce regional disparities in the Union, while at the same time encouraging the development and conversion of regions
- The *European Social Fund* (ESF) aims to prevent and combat unemployment, as well as developing human resources and promoting integration into the labour market

- The *European Agricultural Guidance and Guarantee Fund* (EAGGF) supports rural development and the improvement of agricultural structures, within the framework of European and social cohesion policy
- The *Financial Instrument for Fisheries Guidance* (FIFG) aims to contribute to achieving a sustainable balance between fishery resources and their exploitation. It also seeks to strengthen the competitiveness of the sector and the development of areas dependent upon it.

The activities which may be funded are described in the Operational Programmes and the Programme Complements.

Chapter 7
Property Development Economics

Development since 1945

Prior to the Second World War, investment in property by institutional investors, such as insurance companies, was confined largely to mortgages and ground rents. The few property companies that were in existence at that time acted little more than landlords, collecting rents and carrying out essential repairs. The only direct investment in property was restricted to the property being used by the company itself. Some of the larger insurance companies, such as the then Legal and General Assurance Society, began to take a longer term interest in property. However, during this time property was generally looked upon primarily for its use or function rather than as an investment medium. In business and industry it was considered largely as just one of the elements of production. The Second World War introduced changes in the economic, social and political structure, not only in Britain but throughout the developed and developing world. Some of these changes impinged upon the use and ownership of property.

The ending of the Second World War in 1945 followed a period of high unemployment during the 1930s. The British economy was in a serious condition, many of its overseas assets had been used to pay for the war and consequently the country was heavily in debt. There was a backlog of building projects and much needed developments which had been postponed for the six years of war. There was also a new government's social policy. The 1944 Education Act had to be implemented which, with the increase in school leaving age required many new schools. In 1948, the National Health Service Act came into operation, which resulted in a programme of much needed new hospitals and health buildings. The Robbins Report of 1963 on higher education created a demand for new universities with the subsequent requirement for capital spending on new buildings. During the immediate post-war years, there were three major factors that helped to create conditions of an economic boom in the construction and property industries:

- The desire to renew much outdated property, much of which had fallen into decay through neglect
- The need to replace the considerable amount of property that had been damaged by the ravages of war
- The need to provide new buildings for homes, employment, education, health etc. for a rapidly growing population and the growing needs of the welfare state.

Coupled with these factors were changes in manufacturing industry, with the newer industries requiring investment in new premises to suit changes in technology, particularly automation and mass production. There was also the switch that was taking place in society with a greater proportion of jobs then being created in offices rather than on the shop floor. In addition, there was the major development of the automobile and the necessity to build better roads and highways. The Preston bypass, one of the first stretches of new motorways and now a part of the M6, was completed in 1956. These all helped to create the need for building development and a relatively prosperous and expanding construction industry.

The 1940s

However, the construction industry, during the war and immediate post-war years of the late 1940s, had been had hit both in respect of the availability of labour and the capacity for the manufacture of materials and products. Skilled craftsmen, for example, were in short supply and were to remain so until the late 1960s. Many materials and product manufacturers had ceased to trade, had outdated technology and manufacturing processes or were unable to obtain raw materials in sufficient quantities for production. Major building materials, such as steel and timber, were rationed. Bricks, for example, needed to be pre-ordered at the design stage to ensure that they would be available when required in the construction process. In addition, the country could not afford to purchase goods and materials from abroad, since it already had a serious balance of payments problem due to the war years. A system of building licensing was therefore imposed that gave priority to the repair of war damaged property, to aid industries involved in manufacturing goods (wealth creation) especially those for export and in providing accommodation for public authorities. Building in the private sector was strictly controlled. It was necessary, for example, to obtain a licence from the Ministry of Works to build a house and this was not permitted to exceed 1200 square feet in area (111 m^2). Offices and commercial buildings could not be built without a permit from the Board of Trade. The economy lurched from crisis to crisis. Public housing programmes were suspended for twelve months, the schools programme delayed for six months and even by as late as 1966 there was a six-month moratorium on university building.

The 1950s

The 1950s heralded a building boom of almost unprecedented levels for the reasons indicated above. There was huge demand for skilled labour and a large expansion amongst the professions. It was possible to move easily from firm to firm, with job opportunities crowding the trade publications and professional journals. The procedures employed for site development were also much less onerous than at present. Planning permission was simpler with a general lack of restrictions on development. Finance was relatively easily available from banks and building societies.

Developers were also able to avoid many of the possible pitfalls that are common today. Most of the building contracts were awarded on a fixed price basis, almost irrespective of the project's duration. Any unexpected increases in the costs of construction were automatically borne by the building contractor. The developers were also able to raise their own project funding on fixed rates of interest, so that any increases in the costs of finance would then have to be borne by the developer's finance company. The developers were further able to benefit from rising rental and capital values that occurred because of increasing economic growth and rising inflation. The developers' business activities were relatively straightforward, with a need to employ few office staff and to rely upon consultants to secure developments in return for their fees. Eventually some developers were forced into partnerships with the landowners where prime city centre sites were to be developed.

The 1960s

The boom years of the early 1960s were followed in the latter part of the decade with difficulties for developers due to squeezes on credit and this had the effect of slowing down economic growth and reducing the need for office accommodation and other kinds of property. This also made the letting of developments more difficult, as the recession affected demand. This lack of demand was felt throughout the public and private sectors. Mortgages became more difficult to obtain on residential property. The uncertainties surrounding public sector finances caused many projects associated with health, education and social service to be either postponed or abandoned.

The 1970s

The 1970s started with a recovery in the construction industry but this was to be short lived. The year of 1973 marked the end of the golden age of sustained economic growth, full employment and reasonably stable prices. In 1974 and 1975, the GDP at constant factor cost fell for the first time since 1946, by 3.5%, and the 1973 level of output was not exceeded until 1977. The initial rise in unemployment from 2.6% in 1974 to 6.2% in 1977 was far worse than anything experienced since the 1930s. An even greater shock was the height of inflation that peaked in 1975 at 26%. The Prime Minister, James Callaghan, was to tell the nation, that the idea of spending one's way out of a recession was no longer an option. Inflation had taken over from unemployment as the most pressing economic problem. This represented severe difficulties for construction and property with many household names in construction and property ceasing to trade. The construction industry was, however, sustained by large amounts of work overseas, particularly in the Middle East, fuelled from the profits of oil.

After 1973, Britain, like most countries, suffered from a world recession. The most obvious change in the international economy was the rise in the price of crude oil. Such increases had been engineered by the Organisation of Petroleum Exporting Countries (OPEC) at the end of 1973 and again in 1979. In 1974, the price of petroleum was four times what it had been in 1972, and by 1980 it was 2.25 times what it had been in 1978. Oil, in 1973, represented about half of the energy consumption in western countries, so that the sudden change in prices had a major effect on the pattern of consumer demand. Capital stock that depended upon a cheap source of energy was now being rendered almost obsolete. The term 'stagflation' was coined to describe an economy facing rising balance of payments deficits and uncertain futures.

The 1980s

The 1980s were characterised by the privatisation of many public services such as gas, water and electricity. Hospital trusts were formed and many schools transferred to grant maintained status. These were later followed by polytechnics being redesignated as universities (1992) and further education and sixth form colleges being funded and controlled by the Further Education Funding Council. These bodies provided both opportunities and threats for construction and property. They were allowed greater freedom to manage their own affairs, often becoming businesses. This allowed some to invest heavily in new capital projects. In other cases, particularly schools, there was a growing resistance to invest in much needed capital development works due to an anxiety about their viability and the possible future levels of underfunding. Consequently there were considerable capital receipts held in small amounts by each school, which individually were insufficient for major development projects. Whilst the schools were under local authority control, such sums of money were pooled to allow major works to be undertaken at designated sites.

The 1990s

The early 1990s were characterised by a recession, claimed by some to have bigger implications than the recession of the 1930s. Construction output fell in real terms, firms went into receivership and the value of property plummeted (see Figure 3.2). The NEDO forecasts made gloomy reading, predicting that output would continue to fall further until the late 1990s. The commercial sector, for example, declined by 20% in 1991, 27% in 1992 and a further 8% in 1993. Even the normally reliable repair and maintenance sector, typically representing 40% of the output went into decline. The glimmer of hope was the employment of British construction firms and consultants overseas, particularly in the Middle East and on the Euro Disney project outside of Paris. There was also much interest in both development and construction in Eastern Europe after the demise of communism and Soviet control.

The 2000s

At the beginning of the twenty-first century, property values were experiencing a boom. This was also relative to other forms of investment, especially equities which had been in the doldrums for about three years and were showing no signs of improvement by the end of 2001. The increase in property values was across all sectors, including the domestic property market, which had declined during the early to mid-1990s and then had remained flat. However, the first signs of a property downturn was being forecast towards the end of 2001, although with little signs in evidence. Some high profile developments had been put on ice and national vacancy rates for commercial buildings were increasing for the first time in a decade. In 1990, these had reached 15%, but had fallen consistently throughout the 1990s to just over 6% (Figure 7.1). In the Thames Valley, effectively the centre for the United Kingdom IT industry, rents were beginning to plummet as demand for its products evaporated. In Canary Wharf, financial institutions were subletting space that they signed up for earlier. In London's West End, demand was virtually non-existent. Headline rents, which touched £90 per square foot, had dropped back to £65, and were predicted to fall yet further. The activity is

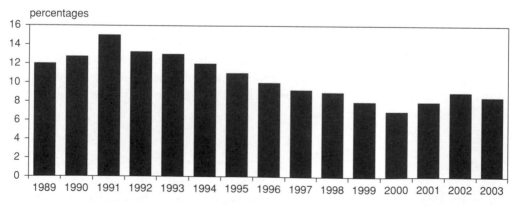

Figure 7.1   UK national office vacancy rates (Source: Knight Frank Research)

reminiscent of the bad old early 1990s. For example, plans to build an extra 500 000 square feet at Stockley Park close to Heathrow Airport were put on indefinite hold, although infrastructure upgrades were under way. In Canary Wharf, more than one million square feet of prime office space came on the market in the middle of 2001. Bank of England data on bank lending to property companies showed an increase in debt to £64 m.

However, if the slowdown did turn into a crash, it was being forecast not to be as severe as that during the early 1990s, a decade earlier. The debt then was secured on floating interest rates, which rocketed to 15%. Most developers now have fixed their borrowing and rates were at historic low levels. However, developers feared that rents in the Thames Valley could halve and leave many empty buildings. This would have repercussions for Docklands and London house prices, which are now so disproportionate to houses prices elsewhere in the country.

It appears that there is no simple correlation between vacancy rates and economic success (Figure 7.1). Other factors are in play to produce a set of statistics that are not simple to interpret. The average for England as a whole has risen from 7% to 9% during a time of sustained economic growth. The level of new completions, turnover of vacant sites and the problem of properties that no longer meet market requirements combine to present conflicting pressures and a confusing picture. As the economy develops, wholesale shifts occur away from manufacturing units and office space that is not equipped to cater for modern IT networks. New property comes on to the market whilst some older property becomes vacant and difficult to let. Other units may be demolished, refurbished for new industrial or commercial use, or may be deleted from the figures entirely by being converted into city centre flats or for other uses.

Whilst property continues to remains one of the best performing investment sectors, towards the beginning of 2007, it was being suggested that dwellings, in particular, were over-valued and that a correction factor was likely to be applied before the end of that year. This was not envisaged to be as serious as the one that created the term *negative equity* in the early 1990s (see Figure 3.4).

There is no doubt that the overall buoyancy of the construction was being sustained at the commencement of the twenty-first century through PFI (Private Finance Initiative) projects. There can be little doubt that without this idea there would have been fewer new hospitals, schools, university buildings and other publicly funded infrastructure projects. There were others, of course, who believed that PFI was mortgaging the future and critics who felt that PFI resulted in high tendering costs which appeared to add little value to the project.

The development of Terminal 5 (T5) at London's Heathrow Airport was approved by the Secretary of State on 20 November 2001, after the longest public inquiry in British history (46 months). The planning process itself cost nearly £63m over a period of 14 years. This cost was borne mostly by the British Airports Authority (BAA) and British Airways, the two main proponents of the project. Construction of the new terminal started in September 2002; phase one of the project is scheduled to be completed and opened by April 2008 with the second phase opening in 2011. The project is requiring an estimated investment (mostly from BAA) of over £4.2bn. Terminal 5 is a large infrastructure project involving over 60 contractors, 16 major projects and 147 sub-projects on a 260 ha site. The T5 terminal construction was 80% completed by November 2006.

New football stadiums were also constructed at the beginning of this century, most notably at Arsenal with a budgeted cost of £0.5bn and at Wembley where the constructor, Multiplex, handed over the keys to say that it was officially completed in March 2007, two years later than expected.

Added to these are the buildings and infrastructure for the London Olympics in 2012 which have a current budget of £8.6bn, which sadly is more than likely to be exceeded. However, this will provide much needed regeneration which will benefit local communities for years to come.

Land ownership

Before the Second World War, a large amount of urban as well as rural land and property was owned by a relatively small number of aristocratic families, who in most cases had owned land for centuries and this had then been inherited by subsequent generations. Even in the present day, after the splitting up of old estates in order to pay death duties, four large estates: Grosvenor, Howard de Walden, Cadogan and Portman own much of the valuable property in Central London. There are also the Crown, the large Cambridge and Oxford colleges and the Church of England who also own huge amounts of land between them.

During the immediate post-war years there was a sharp increase in the public ownership of land. This was due in part to nationalisation, the Town and Country Planning Act of 1947 and the powers required for the compulsory purchase of land, which in some cases included derelict or war damaged property. This was required for new schools, housing, hospitals and other socially needed projects. It also included the acquisition of land that was necessary for the associated infrastructure of new roads and rail networks. This all helped to transfer land ownership into the control of the public sector.

During the ensuing years, different governments exercised policies of nationalisation and privatisation, together with their associated land ownership. During the 1980s, there was a large-scale privatisation of the former nationalised industries. Even industries such as water and power were released into the private sector and local authority housing, that was originally conceived as social housing, was sold to sitting tenants at prices, at the time, well below their levels of valuation. Many of these activities were to do with political ideology and a distrust of local government, which in many towns and cities, was politically opposed to central government policies. However, even after all these transactions were completed, the amount of land that changed ownership was relatively small compared to the total amount of land that was available.

Office development permits

This was one of a number of different measures employed to control or direct building development. The office development boom in London was eventually brought to an end by the White Paper in 1964, which introduced the Control of Office and Industrial Development Act 1965. Office development permits were required for all offices in excess of 3000 square feet (279 m^2) in the south-east and the midlands and in central London there was a complete ban on all such development. Although these limits were eventually raised to 10 000 square feet (929 m^2), the effect was to raise office rents far in excess of inflation due to the inadequate supply of new office accommodation.

Even so, about 8 million square feet of new office space were still built in London over the following three years, partly because developers had stockpiled planning permissions for new office buildings. There was also the time-lag delay in completing the larger projects. Office development permits were also rarely refused, and by the early 1970s, were granted to about 75% of the applicants. A major consideration that resulted in the legislation was to encourage office dispersal from central London and meet the need for regional development. This idea was coupled with rising land costs in the south-east, high office rents (which actually increased as a consequence of the legislation), escalating wages and shortage of office staff. In contrast, there was a reliance on the need for more senior office staff to have direct access to clients, which meant that more of the executive jobs remained in central London.

Development land tax

This tax was introduced in 1976 on persons who realised development values on the disposal of land and property. The tax applied to individuals, companies and trustees, although there were a number of exempted projects such as private housing, approved housing associations, local authorities and organisations that had charitable status. The tax was charged on disposal or part disposal of an interest in land or property. The granting of a lease was treated as part disposal. The tax was also charged on the commencement of a project of material development. Immediately before such a project began on the land in question, every major interest in that land was treated as having been disposed of and immediately reacquired at its market valuation. The first £50 000 of development value realised in any financial year was exempt. The tax was abandoned in the early 1980s.

Urban regeneration

Since the post-war period, it has been recognised that the inner city areas of most conurbations have been in decline due to poverty, unemployment and crime. Although such areas have always existed, they came to the national attention during the 1960s.

These areas have been the recipients of many different initiatives, launched initially through the 1968 Urban Aid Programme. This provided funds from central government to allow local authorities to provide housing advice centres and nursery schools in deprived areas, and funds for research to establish the causes of such deprivation and action programmes to encourage self help mechanisms. Frustrated by the slowness of local authority action the government in 1979 introduced the Urban Development Corporations for London's and Merseyside's docklands areas. In 1981, Enterprise Zones were announced with the stated aims to help to regenerate the national economy but also to assist those cities that had experienced acute employment difficulties. The initial locations of Enterprise Zones were in areas of classic urban decline, but in 1982, new areas were added to account for the rationalisation in the steel industry. In addition to these there were Educational Priority Areas, Community Development Projects, the Home Office's Brunswick Neighbourhood Project, Inner Area Studies and Inner City Partnership. Whilst each of these schemes have received limited success, they have not had any substantial impact in reducing the problems associated with inner city living. In some cases they have had the opposite effect by fostering scepticism amongst the communities they were seeking to assist.

A successful example of recent urban regeneration was undertaken by the Trafford Park Development Corporation. Trafford Park in Manchester, the world's first industrial estate, was transformed during the eleven-year period 1987–1998. The development corporation attracted 1000 companies, almost 30 000 new jobs and £1.8 billion of private sector investment.

Commercial development

The majority of new office developments until the middle of the 1960s took place in London and the Home Counties. About half of these developments were undertaken by property developers, the remainder by owner occupiers. Limited development only had occurred during the previous half a century, owing to:

- the legacy of previous Victorian developments that had been retained
- the lack of development caused by two world wars
- the economic depression of the 1930s
- the emphasis that was still placed on manufacturing industry rather than office-based activities.

It has been estimated that during the Second World War about 10 million square feet of office space was either destroyed or made unusable through bomb damage. About one-third of the City of London was destroyed by aerial bombardment. In addition to this, the need to replace much of this accommodation was coupled with demand for commercial and financial services that emerged as a result of a growing affluence, particularly in the South-East of England and around the capital. The trend away from industrial activity was most marked in London with its already expanding financial services and markets. The Stock Exchange continues to retain its importance alongside Wall Street in America, even though there are real threats from Frankfurt. There was also a pattern of merger amongst a number of smaller firms, as a basis for increasing their profitability and world competitiveness. These needed headquarters based in London. There was also an influx of foreign firms, notably from America, moving into the expanding markets of Britain and Europe. The comparative cultures and language similarities, for example, favoured Britain as a base rather than mainland Europe. Britain at that time was also not a part of the then developing Common Market, which might pose a threat to American trade. The Labour Party had also recently been elected with the aim of public ownership through the nationalisation of many industries and the expansion of public services associated with welfare and social needs. These organisations also required office space and their huge growth required large amounts of space.

This inevitable demand for suitable offices in the right location caused rents and capital values to soar, since need was unable to be met overnight. The unsatisfied demand for office workers, with consequent high salaries in Central London, also caused some firms and public bodies to relocate to the provinces. Here land was less expensive and employees did not demand the same high salaries of their counterparts in Central London, since housing was cheaper and employment opportunities were not as great nor as well paid. Offices which prior to the war could not be let, now quickly found tenants and the bombed sites were reclaimed by developers and speculators eager to satisfy this identified need.

The demand for office space was strong throughout the post-war period. Whilst there has been some movement towards decentralisation, office employment with the consequent need for space has continued to grow in most towns and cities. The expansion and rapid growth of the service sector has also contributed towards this development. The major period of office development occurred in the late 1960s and early 1970s. Demand was high, returns to developers were good and the decline in manufacturing industry encouraged some financial institutions to switch their funds to office development. Many of these schemes were high-rise, changing the skylines of our cities.

Towards the end of the 1980s, when the deregulation of the financial markets occurred, major new developments and refurbishments were carried out around the City of London. These developments occurred because the existing office buildings were unsuitable for accommodating the electronic cabling and trunking that were required for the new information highways. In some cases, developments such as Broadgate and Canary Wharf in Docklands were of a size that made other projects small by comparison. Canary Wharf at its peak employed over 4000 construction workers and when completed was designed to cover almost 30 hectares. However, by the time of the completion of some of these projects, the world had entered into yet another period of prolonged recession, and the demand that had been forecast at the planning stage had long since evaporated. This caused the financial collapse of the developers of Canary

Wharf (see p. 203) and the need for financial restructuring in the case of the developers of Broadgate. It would be some years before these buildings would be brought into full use. For different reasons, this scenario echoed the Centre Point development that was constructed in London in the 1970s, with its office space remaining vacant for several years. This was believed to benefit the developers, whilst at the same time many individuals remained homeless.

The so-called 'Big Bang', the deregulation of the City's financial markets through the Financial Services Act, created an intensified demand for office space from firms associated with these activities. Rents rose rapidly, often exceeding £465 per square metre, with even small offices being able to charge almost £400 per square metre. There was a fear of shortages in office space in central London, where vacancy rates are never much higher than 8%. In 1983, there were estimated to be 4.1 years worth of new office supply in central London, but by 1986 this had plummeted to only two years. The situation was exacerbated with existing offices being taken out of commission for redevelopment or refurbishment, to meet the demands of new technologies. New developments were already underway at Broadgate and Canary Wharf in what seemed to be at the time an endless appetite for demand. However, the collapse of share prices in October 1987 changed this picture considerably. The glut of empty offices remained until almost the turn of the century.

Changes in use

A report published by English and Overseas Properties in 1994, compiled by planners, investors and academics, highlighted changes expected in the social and economic practices of business. Business organisations would concentrate on fewer more efficient buildings, with all but management and key workers operating from home or satellite offices in the suburbs. The removal of office accommodation would reduce town and city centres, leaving further areas of deprivation, dereliction and crime. Parts of some cities would be abandoned. These areas would be occupied by an economic underclass, a pattern already seen in the United States. The report stated that office buildings were grossly inefficient and after allowing for unoccupied nights and weekends, etc. are only used for 21% (76 days) of the year. The report concluded that e-working would increase and some office blocks would only survive because of a token presence at prestigious addresses. In the future, travelling 50 miles only to sit at a desk would seem a ludicrous idea. Many existing office buildings will be converted for other uses, such as homes for single people. This trend was already established with over 250 planning applications in 1994, to convert office properties in London to housing use. The report urges investment in old-fashioned high streets, where the mix of shops, restaurants, homes and entertainment means a lively community.

Later reports suggested that the need for homes in these inner city areas could be easily met by utilising the underused space above shops and restaurants for domestic use.

There were clearly several broader issues that needed to be resolved before this could become a general reality but there are perceived benefits to both shop owners and their future tenants.

Industrial development

There has been a long-term trend and decline in the United Kingdom from heavy industrial engineering towards light manufacturing. This is now partially a matter for history with the collapse of many traditional industries that were unable to compete with other countries, often due to the latter having much lower labour costs. There was also a lack of investment in Britain, without which it must have been clear that such industries would only become uncompetitive at some future date. The decline is perhaps best evidenced in the textile industries of Lancashire and Yorkshire. This has resulted in some former fine industrial buildings becoming obsolete and falling into decay and dereliction. Only a few now remain, now part of our industrial heritage. In some northern cities huge industrial complexes have been levelled and in some cases replaced with other types of projects. This has been the case in Sheffield where a former derelict steelworks was demolished and replaced by the Meadowhall Shopping Centre. There has also been a further trend away from manufacturing towards importing, assembly and wholesaling.

In order to encourage new industrial development, successive governments have offered a range of financial incentives to encourage new investment from home and overseas (see below). These have helped to provide industrial estates and quasi-business parks in many towns and cities. Some have been successful in attracting overseas investment, notably from Japan, in automobiles and electronics. Others, such as the development of the M4 corridor west of London and Silicon Glen in Scotland have attracted high technology industries that were compatible and supportive of each other. The location of such industrial zones was much less associated with the close proximity to housing, as had existed in the old traditional industries, but more related to easy means of transport for the raw materials and the finished goods. The existing mills and factories were often of a scale, construction and design that made them unsuitable for these new high technology ventures. Some were also obsolete and in a poor state of repair and needed to be demolished. In other cases such buildings have been adapted successfully to meet a range of modern uses.

Housing

There had been a massive inter-war house building programme, particularly in southern Britain, when over four million new dwellings were constructed, generally on the outskirts of towns. This new physical form of urban dwellings was associated with new lifestyles that required relatively low density and comfortable surroundings. These developments were encouraged by government subsidies to private builders, the expansion of building societies, which channelled the savings of white-collar workers into housing development, the availability of inexpensive land and cheap labour. There was also an expansion of the suburban local authority housing estates.

The population of Britain appears to be relatively well housed. There are approximately 23 million dwellings for a population of 55 million, equating to about 2.4 inhabitants per home. However, these basic statistics disguise the inequality in housing provision and shortcomings in the state of the housing stock. It does not take into account the mismatch between family sizes and dwellings, nor that the distribution of available housing units throughout Britain does not correspond with the distribution of families requiring houses. The Department of the Environment commissions a quinquennial report, *The English House Condition Survey*, which assesses the condition of houses in both the private and public sector (see Chapter 6).

There has been a substantial decline in house building since the peak in 1968, when 222 000 private sector and 192 000 public sector homes were completed. In the past few years there has been only a limited number of public sector dwellings constructed, due to a shift in government policies directed towards home ownership and private rented accommodation. The sale of local authority houses in the 1980s to sitting tenants helped to reinforce this policy. The proportion of housing occupied by owners has now risen to 70%, with 20% rented from local authorities and 10% from private landlords.

New towns

The New Towns Act of 1946 was made largely to facilitate the planned decentralisation from London into such areas as Crawley, Bracknell, Stevenage, Harlow, etc. Between 1946 and 1950 14 new towns were designated. Other areas designated as new towns were added to the list later. The newly formed development corporations appointed to manage these new towns had extensive powers to purchase land, finance housing, provide an infrastructure and attract business and industry. During the immediate post-war years only a small proportion of development was undertaken in the new town areas. Under the Town and Country Planning Act 1947, local authorities were given powers to acquire land at its existing use value and then to resell it at full market values for its development price. The difference was referred to as *betterment*, whereby the increased values in land due to development would then accrue to the state rather than a private developer. The control powers in the Act of 1947 were negative in order to curtail rather than initiate development. The betterment provisions were later repealed in 1953. In 1952, the Town Development Act allowed conurbations to make individual arrangements with smaller towns for overspill schemes.

By the late 1950s, the inadequacies of existing urban policies could not cope with the population growth and expansion that was taking place. Several more large new towns were planned throughout the country to provide housing overspill areas adjacent to large city areas such as Skelmersdale New Town for Liverpool and Manchester. Washington New Town was developed to meet the requirements of Tyneside in a similar way.

Much later, in the late 1960s, other new towns were developed such as Northampton, Warrington and Peterborough. These developments were different, being based upon already existing conurbations, rather than green field sites. This tended to lower the overall development costs, which had become an ever-increasing issue in central government. The last of the new towns, such as those in Central Lancashire, based upon the existing towns of Chorley and Leyland had hardly started before it was decided that such urban corporations were to be dismantled.

City centre redevelopments

In the immediate post-war years, shopping centres were rebuilt in many war-damaged cities. Many buildings had also fallen into serious disrepair during the war years and immediately thereafter. There was also the need to keep up with the Joneses in adjoining towns. Two separate models were evolved, one that incorporated car traffic and the other based upon pedestrian precincts. Urban policies, as outlined in the Buchanan Report in 1963, had already assumed that traffic growth in towns and cities was unavoidable and many of the new town and city redevelopments, such as those in Birmingham and Newcastle, undertook major urban road building schemes in the 1960s. The major period of shopping centre redevelopment was in the 1960s and early 1970s. Of the largest cities, (about 155), 105 acquired major (larger than 50 000 square feet) shopping centres between 1965 and 1977. Property developers promoted such schemes since they benefited from both land speculation and high rental yields in the new centres. Such schemes were also supported by local authorities, such as in Manchester with its Arndale Centre and the Eldon Square development in Newcastle. They were prepared to cooperate with the developers since the modernisation that resulted also encouraged commercial and industrial development in the area, which in their absence would have been much less attractive. The local authority planners also favoured such developments, since they avoided the need for out of town hypermarkets. These were narrowly defined as having over 25 000 square feet of selling space. In 1980, only four such developments existed, although this number has increased considerably since that date, although much less than comparable developments on mainland Europe.

Out-of-town shopping malls

A trend in developments that occurred towards the end of the 1980s was the provision of out-of-town shopping areas. These provided a range of shopping facilities similar in some cases to those that might be found within town and city centres. One of the most well known is Lakeside in Essex, described as one of the first American-style shopping malls. Others include the Metrocentre at Gateshead, Meadowhall at Sheffield and Merry Hill in the West Midlands. However, whilst over 50 planning proposals for developments of this type around the country have been made, only the Bluewater Shopping Centre near Dartford in Kent and the Trafford Centre on the outskirts of Manchester have otherwise been constructed. What might therefore have been seen as a trend in development seems to have petered out. These shopping malls offer an attractive shopping experience under cover from the elements in modern state of the art buildings. They are also located on the outskirts of urban areas and provide for easy access with adjacent can parking.

Some of the reasons for a lack of other developments of this type have been due to a failure to obtain planning permission or being withdrawn as not economically viable. A House of Commons Environment Committee on Shopping Arcades and their Future (1994) also issued proposals to tighten planning regulations which may effectively stop any further developments on the scale of those mentioned above. The refusal to approve more of these developments is in response to concerns about the damage caused to the environment and to nearby town and city centres. For example, the Merry Hill Mall development, which in 1993 attracted 22 million visitors who spent more than £350 million, has been blamed for the devastation of the nearby Dudley city centre. A study indicated that Dudley lost 70% of its market share to Merry Hill, while towns such as Stourbridge and West Bromwich also suffered.

Concern has also grown about the increased use of cars, as millions of shoppers head to the out of town malls. Recommendations have been made that no new malls should be built unless they bring demonstrable environmental benefits, a feature that is likely to become necessary for all future major building developments. The Council for the Protection of Rural England (CPRE) and other environmental groups have greeted the House of Commons report with enthusiasm. The big supermarkets and other retailers were more cautious. The committee concluded that the government should change its planning policy to require developers to demonstrate that their proposals for new stores will not harm the vitality and viability of existing town centre shopping areas. Out-of-town centres should also be rejected if there was a suitable site available in or close to a city centre. Many of the large supermarket chains, that have an interest in out of town developments, have already suggested a slow down in their own proposals.

The committee recommended that superstores are best located either in or on the edges of town centres, unless there were strong reasons to the contrary. They further suggested that no proposals for superstores or other large retail developments should be considered in or around market towns, unless they were accompanied by comprehensive studies of how such developments would affect their whole catchment areas. The report also dealt with the problem of planning gain, such as inducements in the form of open spaces or road improvements offered by developers in order to obtain planning permission.

The Big Bang of 1986 and the crash

There were two notable events that had an impact upon property development and property investment in the late 1980s. The first of these was the deregulation of the Stock Exchange in 1986, known as the Big Bang. This was concerned with the removal of restrictive practices that, for example, determined and fixed commissions on dealing. A number of barriers between banks, merchant banks, investment institutions and stockbrokers were removed in order to allow one-stop financial shopping to be available. It was also in response to changes in technology that now allowed for high speed telecommunications so that international deals could be carried out anywhere. A further reason for these changes was a reaction to trading practices being carried out in the USA, that were much simpler, cheaper and more tax effective. These reforms provided immense benefits for building development by putting London as a more attractive place for international location with firms requiring modern and up to date premises. The enormous number of new office developments around London were due in part to the Big Bang. Many more commercial projects were planned but never commissioned (see Chapter 1).

Almost one year after the Big Bang, there was another momentous event: the crash of the financial equity markets in 1987. Whilst most people in southern Britain were busy repairing damage to their homes, caused by one of the worst hurricanes this century, another storm was brewing across the Atlantic. In the first few hours of trading on Wall Street, the Dow Jones Industrial Average had slid 100 points; its biggest ever one day fall. Over the next few days the falls in equity values were to be as large as those of the great crash of 1929, when bankrupt brokers leapt to their deaths from the skyscrapers of Manhattan. One person lost £308 million in 1987. The effect of this crash was to rekindle the view that 'there is nothing as safe as bricks and mortar'. Whereas yields on property investments over the previous ten years had been inferior to those on equities (see Chapter 2), the apparent safety of investing in property had again become popular for both institutional and private investors.

Analysis of Canary Wharf

One of the projects that responded to need for new office developments in London was Canary Wharf. The first impression of Canary Wharf is its enormity. The main tower dominates the skyline even from points on the distant M25 motorway. Its shape is artistic, its stainless steel cladding pierced by a symmetrical pattern of windows that are aesthetically attractive. For many it is a masterpiece of building design and development. The 1.3 million square feet of office space in the tower is only a fraction of the entire scheme that was planned to exceed 10 million square feet when completed. The average area of each floor is 27 000 square feet including two central service columns that contain lifts, lavatories and staircases.

The London Docklands Development Corporation was founded in 1981 to regenerate the area. By the time of the Canary Wharf financial collapse, it had already built 21 million square feet of commercial buildings and 16 000 homes. The Docklands Light Railway connected the area with the City and it was proposed to extend the Jubilee line underground service. Whilst improved road, rail and river connections were promised by government, treasury officials wanted a £400 million contribution from the developers before the work could commence on the underground railway link.

The developer; Olympia and York (owned by the Reichman brothers) negotiated as hard as possible with the government at the strategic planning stage of the project. It also claimed that this property company had invested more in this single project in the United Kingdom than all the Japanese car manufacturers put together. There were, however, plenty of people to blame for the financial collapse of the Canary Wharf project. It is also always easier to be wise after the event than to forecast possible disaster beforehand. The building of the new financial centre at Docklands underestimated several different factors:

- the logistical problem of getting office workers there, due to poor road and rail communications
- the self-preservation instincts of the City of London
- the conservatism of the British business classes.

London was also being equated with New York, where the recently completed World Financial Center had at the time been a huge success. However, the two projects were very different. The New York project was just a few minutes walk from New York's financial centre at Wall Street, whereas Canary Wharf was at least a half hour taxi ride from Threadneedle Street, the location of the Bank of England. Wall Street also did not react by constructing its own skyscrapers, but the City did exactly that by providing office space at Broadgate, Cannon Bridge and Alban Gate. In 1991, it was reported that almost 13 million square feet office accommodation was vacant in the City alone. Rents, due to the recession and the over-supply of accommodation, plummeted from £65 to £35 per square foot. Why then should a company want to relocate from the hub of financial activities to the area of Docklands? The understandable reply from the British financial establishment was that they shouldn't.

The banks too also received their fair share of criticism. Some, like Barclays in the 1980s, became aggressive lenders to the then hopeful property industry. Their loan portfolios in this sector grew by 35% between 1987 and 1991. The banks compounded their errors in the case of Canary Wharf, by lending hundreds of millions (£1.2 billion) on the basis of little more than the Reichman brothers' good reputation. They did not even require cross-collateral on Olympia and York's valuable North American properties. As a result, the security on their loans issued was practically worthless.

Government also cannot escape some of the blame for the collapse. Some of this is reflected in:

- the depth of the ensuing recession and the lack of government policies to create real and sustained growth
- the ideological fervour that did not believe in public and private partnership
- the failure to provide an adequate infrastructure, especially road and rail communications, at the required time.

The developers' own optimism, now known to be misplaced, in the 1980s also cannot escape this debacle. In 1991, £15 per square foot was the likely rental at Canary Wharf taking into account rentals in the City and the glut of empty commercial property. If its 5 million square feet could be fully let, this would only generate £75 million per annum. However, as noted above the £1.2 billion debt required about £120 million just to service it, leaving an annual deficit of around £45 million. Since that date, Canary Wharf has of course remained much less than full with the eleven main bank lenders' debts continuing to escalate.

In more recent times, after the above events, the project has settled down but perhaps not to the same extent as that originally envisaged by the developers.

Canary Wharf today

Canary Wharf today is a success story and is a large business development that now rivals London's traditional financial centre, The Square Mile. Canary Wharf contains the UK's three tallest buildings: One Canada Square that is sometimes known as the Canary Wharf Tower, at 235 m, the HSBC Tower and the Citigroup Centre. Canary Wharf tenants include major banks, law firms, technology companies and major news media and service firms, such as the *Daily Telegraph*, Reuters and the *Daily Mirror*. It has also gained more tenants from the public sector such as the Financial Services Authority, and the 2012 Olympic Games organisers, ODA. Around 100 000 people are employed on the estate, of whom 25% live in the surrounding five London boroughs. Plans are underway for Canary Wharf to more than double in size. It is not just an office scheme. It has had impact at the local level, at the metropolitan level and, to a lesser extent, at the national level.

The most immediate impact of Canary Wharf has been to substantially increase land values in the surrounding area. This means that the Isle of Dogs, which had previously been seen as suited only for low-density light industrial development, has been uprated. At the metropolitan level, Canary Wharf was, and remains, a direct challenge to the primacy of the City of London as the UK's principal centre for the finance industry. Relations between Canary Wharf and the Corporation of London have frequently been strained, with the City accusing Canary Wharf of poaching tenants, and Canary Wharf accusing the City of not catering to occupier needs.

Canary Wharf's national significance comes from what it replaces. The former docks were, as recently as 1961, the busiest in the world. They served huge industrial areas of east London and beyond. Both the docks and much of that industrial capacity are gone, with employment shifting to the kind of service industry accommodated in office buildings. In this respect, Canary Wharf could be cited as the strongest single symbol of the changed economic geography of the United Kingdom.

Regional development

There is an increased economic dominance in activity in the south-east of England compared with the other regions of the country. This can be explained in many ways not least in the area of property pricing. The dominance of London and the south-east region of the country in terms of property valuations, for example, accounts for over 50% of the value of residential property and as much as 85% of the rateable value of all office accommodation. This region also accounted for over 40% of the construction industry's output in the boom of the late 1980s, compared with over 30% in the early 1970s, indicating a growing dominance. This picture represents a distorted view when compared with the population distribution as a whole. The recession of the early 1990s did redress this imbalance in a small way, where property values in the northern regions did not decline to the same extent as those in the south of the country. In fact, property prices in the North were continuing to rise for a time whilst at the same time they were falling in the South (see Figure 3.2).

The rise is partially accounted for by the disproportionate increase in the demand, for example, for office space over that period, coupled with the need for housing and the ancillary infrastructure. Ironically, land on the M4 corridor that was for sale at the height of the construction boom of the late 1980s, was selling for a fifth of its price by the end of 1991. The four most northernly regions, known commonly as the north of England, and Scotland accounted for less than 30% of the nation's construction work in 1990. This compared with over 35% in the early 1970s. The regions at the extremities from London: Scotland, Wales and the South-West also showed a decline in their share of construction development activity, down by a fifth from 25% to 20% of the national share. Any upturn in development activity is likely to occur first in the southern regions, partially due to history, but also due to the fact that this region is better prepared in terms of the anticipated future activities that are expected to occur.

Repair and renovation

Recent trends in construction development activity have seen a shift away from public sector building to private enterprise and from new build towards renovation. The former trend will only be halted due to the need to invest more widely in the public sector owned infrastructure, although there are ideas that new road programmes for example, should be constructed using a mixture of public and private capital. More recent studies have questioned the whole basis for the roads programme as largely an incentive for more car ownership. At the start of the 1970s, about 50% of the construction workload was described as public sector projects, but by the start of the 1990s, this percentage had halved due to a variety of reasons including government privatisations and limits on public borrowing and spending. New build projects have typically represented about 60% of construction related activities and whilst this percentage fluctuated, it has rarely in the past been below 55% of the total output.

In Britain in 1989, 200 000 new homes were added to the housing stock and 6684 were demolished or closed through programmes of slum clearance. By contrast, over twice as many homes as were built were refurbished. Refurbishment was also carried out on many other buildings such as office blocks, departmental stores and industrial buildings.

The majority of buildings are constructed from materials that are capable of exceeding the practical life of the structure. Refurbishment makes good use of this inherent extra value and provides the building owner with a modern, redesigned building fulfilling most of the client's requirements. The result also represents a measure of conservation although this is not usually the primary objective. Refurbishment represents a recycling concept which prevents potentially valuable building stock from being prematurely wasted. However, some buildings do not lend themselves easily to refurbishment, owing to changes in methods of working or usage. Working within the confines of an existing building often requires more flexibility in approach and design. Outdated building shells are also sometimes unsuitable to house intelligent modern buildings of the future limitations due to their physical limitations.

During the 1980s, repair and maintenance work accounted for about 40% of all construction work, excluding DIY. This contrasts with the early 1960s and early 1970s when repair and maintenance was 20% and 30%, respectively. This is not due entirely to the volume of new work being smaller. In real terms, repair and maintenance exhibit in-phase cyclicity with all work. Nonetheless, it has grown by 50% since the early 1970s. In other countries of Europe, this type of work varies between 14–44% of all work. In the USA, it forms a much smaller proportion of construction and development output than in Britain. This identifies a philosophy variance between Britain and the USA with a greater tendency in the latter for demolition and rebuilding rather than repairing or as some have suggested 'make do and mend'.

Environmental aspects of building development

The impact of the construction industry on the environment is substantial. During the extraction and manufacture of construction materials, their transportation, the process of construction and use of buildings, large quantities of energy are used. Major contributions are made to the overall production of carbon dioxide which exacerbates the greenhouse effect. The environmental impact of the construction of new buildings is a global issue since it requires the use of raw materials from around the world. During the construction activities on site, which often last for a number of years, communities and individuals can be severely affected by the process of construction. During more recent years there has been a much greater effort to reduce these disruptions to normal life to a minimum.

There is a growing trend of concern in society with the effect of human activity on the environment. There has been greater pressure on clients and developers to state the possible effects of their projects on the area in which the project is to be constructed. Since 1968, a European Community Directive has required an environmental impact assessment to be provided with all planning applications for major projects and for smaller schemes where the planning officials consider them to be important. The assessment requires a statement of the impact of the project on the surrounding area and details of how this can be limited, for example soundproofing in the case of a noisy transportation system. A further requirement is that clients and developers should wherever possible undertake wide consultations involving the public and environmental groups. Despite the requirements of the Directive, the quality of assessments varies considerably with the DEFRA attempting to formulate appropriate standards.

Assessments also tend to be parochial and not to examine the wider issues involved beyond the confines of the particular project concerned. Alternative proposals should be considered that compare the environmental factors involved in the choice of different sites or locations and the different constructional methods that might be adopted.

Energy issues

In the 1950s and 1960s, building maintenance and running costs were largely ignored at the design stage of new projects. Today the capital energy costs which are expended to produce the building materials and to transport them and fix them in place are often ignored in our so-called energy efficient designs, where the emphasis is placed upon the energy use of the building. In any given year, the energy requirements to produce one year's supply of building materials is a small (5–6%) but significant proportion of total energy consumption, which is typically about 10% of all industry energy requirements. The building materials industry is relatively energy intensive, second only to iron and steel. It has been estimated that the energy used in the processing and manufacture of building materials accounts for about 70% of all the energy requirements for the construction of the building. Of the remaining 30%, about half is energy used on site and the other half is attributable to transportation and overheads. Although the energy assessment of building materials has still to be calculated and then weighted in proportion to their use in buildings, research undertaken in the USA has shown that 80 separate industries contribute most of the energy requirements of construction and five key materials account for over 50% of the total embodied energy of new buildings. This is very significant since considerable savings in the energy content of new buildings can be achieved by concentrating on reducing the energy content in a small number of key material producers.

Most buildings are designed to cope with the deficiencies of a light loose structure, designed just to meet the Building Regulations thermal transmittance standards. Our Building Regulations are also some way behind those of other European and Scandinavian countries. About 56% of the energy consumed, both nationally and internationally is used in buildings and this should provide designers with opportunities and responsibilities to reduce global energy demand. Whilst there is a need to make substantial savings in the way that energy is used in buildings there is also a need to pay much more attention to the energy used in manufacture of materials and components and their fixing in place in the finished building. It has been estimated that this may be as high as five times the amount of energy that the buildings occupants will use in the first year.

Future demand

Since the mid-twentieth century, Great Britain has experienced a significant change in all sectors of society. There are very few activities that have remained unchanged or unaffected. Standards of housing, education and health have improved for the great majority of people. The construction and property industries must be given some credit in fostering positive improvements, even though on occasions building developments have created more problems than they attempted to solve. The average individual living at the beginning of the twenty-first century is clearly more affluent than those living at the start of the previous century. Despite this, there is a global economic crisis with civil unrest in many parts of the world. In Britain, unemployment continues to remain an important issue. Even highly paid financial services staff and information technology specialists are not immune from redundancy in current times. For others, unemployment and redundancy are part of a way of life with job security now being all but a part of the past, even in the public sector. The increasing use of new technologies is a major factor in the equation. There are a number of issues that influence the future demand for construction and property. These can be interdependent and conflicting and include:

● Population demography
● Political systems and decision-making
● Economic and financial performance in the country and throughout the world
● Social attitudes, expectations and requirements
● Legal and justice systems
● Technological developments – exploitation of new resources, materials and methods.

The examination of historical records indicates that the development of construction and property has been the lead sector, providing a trend for the economy as a whole.

When recession has forced down prices far enough, a recovery in the demand for construction has led to an upward demand in trade generally. The recovery in construction development has also traditionally been led by an increase in house building. The end of a recession is signalled by house completions which often increase spectacularly. House sales are now also affected as much by demography and family patterns as they are by employment related issues.

The electronic revolution will continue apace. Its effect on office workers will increase. Manufacturing based industries moved overseas to preferential labour rates in the 1950s and 1960s. In this process of realignment they also became more automated. Office-based employment is following a similar pattern, with the advent of improved and advanced telecommunication networks.

There will be an increased demand for high technology commercial buildings, with many of the older office blocks, constructed in the boom of the 1960s, being either refurbished, demolished or redesignated for other uses where their structural forms will not suit modern day needs. Prime rents, for example, increased to £40 per square foot in the City and £55 per square foot in the West End of London. These figures compare with as little as £5–10 per square foot in some provincial towns. Industrial premises and warehousing are also becoming more high technology orientated, relying ever more on information technology as a means to store, sort, retrieve and reorder products. At the time of writing (early 2002) traditional manufacturing is set to continue to decline, remaining uncompetitive against new product designs and technologies.

In addition to these much needed improvements will be the replacement and maintenance of the infrastructure, especially roads and sewers, suitable for a new age. Many public buildings, such as blocks of flats built to replace the slum housing of the 1960s, will themselves have become uninhabitable and be demolished to make way for more people focused dwellings. Visionary projects concerned with leisure and tourism will also become high on the agenda.

Indicators

The economic state of the construction and property industries is linked to the overall economic position of the country and world economies, especially those of the USA, Japan and mainland European nations. The general state of these economies is measured broadly through performance indicators such as those shown in Figures 7.2–7.6. These offer a picture of past performance and current trends as well as a snapshot of the economy at any point in time.

Figure 7.2 shows inflation in the UK from the middle of the 1970s up to 2001. This emphasises the comparatively low and continuing rates, which have not been seen since the 1950s. These are in line with the world picture, although many countries notably in South America and in Eastern Europe exhibit much higher rates. Compared to Europe and the USA, the UK's absolute and relative position has improved further, marginally

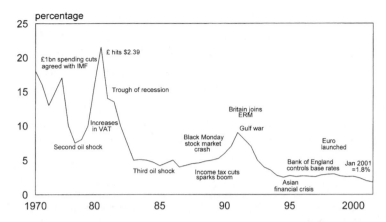

Figure 7.2 Underlying UK inflation (Sources: Office of National Statistics and Times Newspapers)

below the government's own target of 2.5%. Looking forward, all forecasters suggest that UK inflation rates should remain impressively low, compared with most countries around the world.

Figure 7.3 compares UK and USA base rates. There is clearly some trend between the two and their pattern follows to some extent that of the underlying trends in inflation. These indicate a steady fall in percentage points, encouraging industry and commerce to borrow money for future investment purposes. In order to attempt to offset the effects of a recession in world markets, base rates in the UK and the USA have been consecutively reduced, especially following the events of September 11 2001, which was a further trigger towards a probable world recession. Economists are commenting that for the UK, should this occur then it is likely to result in a much softer landing than the recession of the early 1990s.

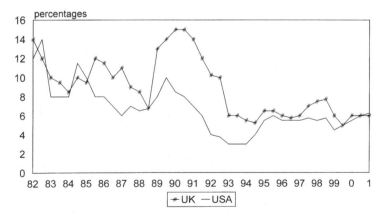

Figure 7.3 UK and USA base rates (Source: HSBC)

The Gross Domestic Product (GDP) measures the value of goods and services produced for final use in consumption, capital expenditure and exports. The construction industry currently accounts for 8% of GDP. Overall the GDP remains well above its trend rate (Figure 7.4). But the growth has been completely unbalanced, characterised by strong private sector demand pulled back by net trade and with the public sector adding little. Given the UK's high exposure to the USA, the global slowdown forecast is likely to detract from the UK GDP growth directly through lower exports. By the end of 2005, GDP stood at 2%, being boosted by the construction industry, the service sector and general household spending.

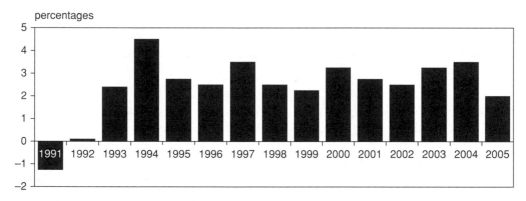

Figure 7.4   Gross Domestic Product growth rates (Source: ONS)

The current account suffered badly in 1999 and did not really show signs of much improvement up to the time of writing (Figure 7.5). Within the current account, the trade in goods deficit was partly offset by a surplus in trade in services (Table 7.1). Indications suggest that manufacturers, in part, are successfully adapting to the new environment and are now winning back export market share. But the news is not all

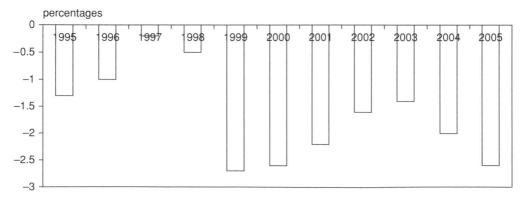

Figure 7.5   UK current account as a percentage of GDP (Source: ONS)

Table 7.1 UK Current Account as a percentage of GDP

	Goods Trade	Services Trade	Net Income Flow	Current Transfers	Current Account
1995	−1.7	1.2	0.3	−1.1	−1.3
1996	−1.8	1.4	0.1	−0.6	−1.0
1997	−1.5	1.6	0.4	−0.7	−0.2
1998	−2.5	1.6	1.4	−1.0	−0.5
1999	−3.2	1.5	−0.2	−0.8	−2.7
2000	−3.5	1.4	0.5	−1.0	−2.6
2001	−4.1	1.4	1.1	−0.7	−2.2
2002	−4.5	1.5	2.3	−0.8	−1.6
2003	−4.3	1.5	2.3	−0.9	−1.4
2004	−5.2	1.8	2.3	−0.9	−2.0
2005	−5.4	1.5	2.3	−1.0	−2.6

(Source: ONS)

good. The improvement in the trade balance can be partially explained by oil prices. As the UK is a key exporter of oil to the EU, this has had a significant impact on the value of UK exports.

Figure 7.6 illustrates unemployment trends for males and females. It is uneven across the age groups, being more pronounced amongst the over-50s. According to the claimant count, the level of unemployment at the first quarter of 2001 stood at 1.04 million, the lowest level since January 1976. Some of the fall in employment was concentrated in self-employment and government supported training schemes. The percentage rate at that time was 5.3%, although its rate of decline had begun to level off for the remaining years of data that was available. The construction industry continued to offer excellent job opportunities with around one person in ten finding their livelihood from this industry. The prospects were equally good at both trade and professional levels.

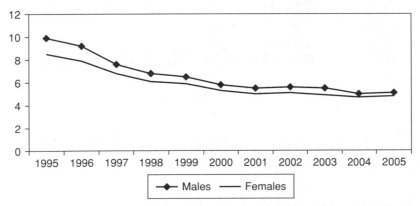

Unemployment rates represent unemployed persons as a percentage of the active labour force.

Figure 7.6 Unemployment by gender (millions) (Source: Eurostat)

The number of tower cranes in our cities was noticeable. Property generally was performing well, typically better than most stocks and shares or government gilts. Whilst house prices were predicted to fall, they have not so far. The overall prospects look good even against the odds that a recession may be on the horizon. Any recession is unlikely to be with us before the end of the Olympic Games in 2012. The objective picture offers encouragement to the construction and property sectors, since these are likely to be at the forefront of future economic recovery, based upon previous patterns of economic activity.

The European Union

The construction industry in the United Kingdom is no longer simply affected by the economics within the UK but also by world markets. These create a demand for property and buildings and an investment in property in the UK. Various markets around the world influence this demand, especially the USA and Japan, which are two of the world's biggest economies. The European Union of member states also represents a large combined economy and the way in which these various economies are managed also influences the need for property and buildings within the UK. Figures 7.7 to 7.9 illustrate the comparative performance of a number of European states against a sample of three different but important economic criteria. Figure 7.7 compares the gross domestic product (GDP) performance. This indicates that Germany, in 2005, was leading the rest of Europe. The UK's performance is at the leading edge when compared with other member countries in respect of GDP.

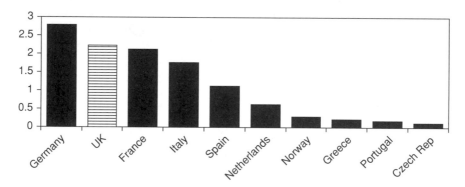

Figure 7.7   GDP percentages of selected European countries, 2005 (Source: Eurostat)

Figure 7.8 compares inflation in the years 1997, 2000 and 2006 across a number of European states. It is notable from these figures how the Czech Republic has moved more into line with other European Union states since joining the Union. Whilst the UK is performing in line with other European states, it is not doing as well as either Sweden or Germany.

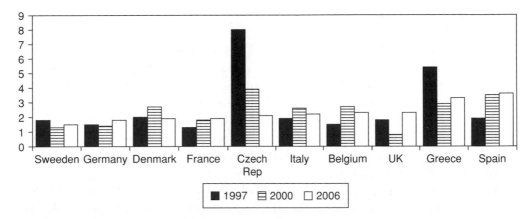

Figure 7.8   Inflation rate percentages of selected European countries (Source: Eurostat)

Figure 7.9 considers male unemployment, based upon published outputs. The published outputs are generally recognised to be lower than the true rate of unemployment. Whilst the UK is not performing as well as some other states, the data progression downwards is in contrast to that of Germany who, since unification has seen rises in unemployment.

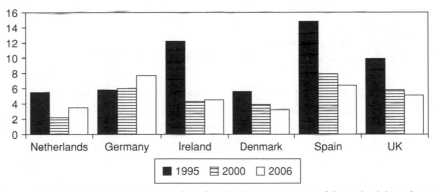

Unemployment rates represent unemployed males as a percentage of the active labour force.

Figure 7.9   Male unemployment rates across selected European countries (Source: Eurostat)

These outputs alone provide an interesting European comparator, which shows that the UK economy is performing at least in line with the other member states and in some cases much better than some of them. The extent of the recession affecting both construction and property in 2002 was relatively short lived and by no means as severe as the one a decade earlier. The European Union appears to be working well in encouraging development amongst the healthy member states. The question of when the UK should join the common currency of the Union will remain a matter for debate for the years to come.

The European Union versus the USA

The UK's economic cycles are closely aligned with those of the USA. Economic history suggests that it is typically the balance or moreover the lack of balance, within an economy which determines the severity of any slowdown. Since the mid-1990s the USA economy has built up huge imbalances: the current account to GDP ratio is huge, companies have over invested and personal savings have virtually disappeared. The reality is that the economy needs a recession to rebalance some of these distortions which have developed. The UK, on the other hand, has not built up such imbalances and this should prove to be our saving grace.

If the EU were a part of the USA, would it belong to the richest or the poorest group of states? At the beginning of the 1990s, there was no need to ask. Europe's economic future was a subject of growing optimism. Productivity growth had for some decades been higher than in other countries of similar standing, and that growth was now going to be hugely accelerated by the elimination of trade barriers and the closer economic integration resulting from the Single Market. The EU as an institution was undoubtedly *seen* as a vehicle for growth and economic liberalisation. In other words, the EU was able to do what politicians in several member countries had wished for but had failed to achieve:

- to increase economic openness
- to strengthen the process of competition
- to harness the political process behind a liberal reform agenda.

Today, the perspectives on the EU, and the outlook on its future, are radically different. Economic growth during the 1990s never became what many had wished for. Some countries performed reasonably well, most notably Ireland, but on the whole the EU was lagging far behind other countries during the whole decade. Productivity growth decreased and by mid-decade the EU was running behind the USA in this respect. The process of convergence in productivity, a much talked-about process since the 1970s, had once again become a process of divergence.

The role and status of the EU in the economic reform process have also changed. Instead of a clear focus on economic reforms and growth, the EU has concentrated its ambitions on other political objectives. Hence, the EU no longer is seen as the great economic liberator of Europe.

The per capita GDP is lower in most of the countries of Europe than in most of the states of the USA. France, Italy and Germany have less per capita GDP than all but five of the states of the USA. America's GDP is far higher than Europe's and has been so for a long time and the American economy has been growing faster than the economies of many European countries in recent decades, not least those of countries like France, Germany and Sweden. The US recession, with GDP growth rates of 1 or 2%, represents almost boom conditions in Germany, for example. Europe may have its Eiffel Tower in Paris, its Colisseum in Rome, fine roads in Germany and social security systems in Sweden, but it will take more than past achievements to cope with the economic challenges which many European countries are facing. Economic challenges which among other things will be brought about by demographic developments and will impose heavy strains on comprehensive, publicly funded welfare systems.

Chapter 8
Sources of Finance

Introduction

When approaching a bank or other institution for a loan it is necessary for a developer to provide a detailed cash flow forecast. Such a budget analysis will enable the funding manager to understand more clearly the needs of the company. This should also be supported by a copy of the latest accounts of the company. The budget will show:

- How much money is required
- How this is to be used
- When it will be required
- Whether it can be repaid
- When it can be repaid.

The finance company or bank will then make a decision about the loan on this basis, coupled with the details of the nature of the firm, its performance and the market in which it is working. The ability to present and analyse the historic data and to couple this with the developer's present and future prospects and its financial performance is very important. Some form of security may or may not be required, but a rate of interest charged will be dependent upon the possible risks involved. From the bank's point of view an overdraft facility is preferable since this is self-liquidating. It is also very flexible up to an agreed limit, is easy to operate and usually the cheapest form of finance available. However, it is not normally appropriate for large amounts of finance that may be required over long periods of time.

Investment

Financial investment of any kind relies upon the analysis of three variables; *risk*, *return* and *time*. A good investment can therefore be described as one which produces the rate of return expected within the time required having regard to the risk involved.

The considerations to be weighed in property investment are as varied as the number of funds seeking to invest. A fund manager's strategy will want to consider freehold or leasehold, geographical spread, type, e.g. shops, offices, industrials, etc., pattern of rent reviews, size and yield, age and quality of buildings, types of covenant and tenants. The perfect balance will rarely be achieved. The main factor to consider with any property is location. For example, a dilapidated building sandwiched in the high street between Marks & Spencer and Mothercare is still a better proposition than a modern shop at the wrong end of town. Most of the investment is also likely to be in the building site, i.e. the land, rather than the building.

The second consideration is obsolescence. Early 1960s office buildings which are now unsuitable for modern open layouts or the installation of electronic communications trunking and networks, or warehouses with eaves height too low for the current generation of forklift truck, are examples of buildings that have become outdated within a very short space of time. These are often of limited use in today's modern and developing world, since they often require adaptation of the shell, which is tantamount to demolition.

The third factor is that of marketability. Investors dislike being locked into a property that will be difficult to value or to re-let. Whilst current fashion and modern technology are important, prime sites or locations will always be preferable.

Debt and equity finance

Whether the capital for the development has to be borrowed by the developer (*debt* finance) or invested from reserves (*equity* finance) is one of the fundamental questions to be answered. Borrowed money has to be repaid at some future date along with the servicing of the debt through finance and interest charges. With equity finance, the return for those who provide the funds, is through the success of the development and its ensuing profits. Where profits are not made then this results in no returns.

Debt finance

The majority of lenders normally require security for the funding that they provide. In the case of any form of default they want at least to be able to recover the funds that they have lent. If a borrower fails to pay the interest at the appropriate times, then a receiver can be appointed to manage the affairs of the company in order to ensure that the agreed payments are recovered as fully as possible. These are described as *secured* loans and are much safer than unsecured loans. In practice, however, established companies are generally able to raise unsecured loans. These types of borrowings are more common. The calling in of a debt is a last resort, since in these circumstances there are frequently insufficient funds available to repay the debt. The various considerations regarding finance are described below and listed in Table 8.1.

Table 8.1   Finance: the considerations

Fixed rate or variable rate interest
Long-term or short-term
Secured or unsecured
Project or corporate finance
Parent or subsidiary company
Domestic or international finance

Fixed rate or variable rate interest

With some loans, the interest rate is agreed at the start of the loan and remains unchanged throughout its life. Fixed rate interest is common for short-term loans. It is also possible to fix the rate of interest for a number of years on long-term loans, although this is much less common. The floating rate is normally determined by the London Interbank Offered Rate (LIBOR). It is also possible to purchase a *cap* or insurance cover to limit the amount of interest being charged on variable or floating rate interest loans. A fixed rate loan is preferred by a borrower when interest rates are rising.

Long-term or short-term

Loans are described as either short, medium or long term. It is not easy to uniformly quantify these periods precisely, but 2–5 years, 3–10 years and in excess of 10 years can be used as a rough guide. The duration of funding is discussed later. Overdrafts, for example, are short-term loans that can be recalled at any time. However, many companies have almost permanent overdrafts, which makes the definition misleading. Also under multi-option facilities, a company might be borrowing for only a few months at a time, but repay the outstanding loan and take out a new one at the end of these periods. This matter is discussed further below in the section on Duration of Funding (p. 232).

Secured or unsecured

Those who lend finance will usually want some form of security for their funds. If a business ceases to trade because its expenditure is higher than its income, then a lender will want to ensure that it is able to recover as much of the loan debt and outstanding interest as possible. In order to achieve this, the lender will insist on the option to sell certain assets of the borrower to repay the debt in the case of a default on the part of the borrower. Well established companies are often able to obtain loans that are unsecured, their name being the guarantee of future repayment. They are also often in addition able to obtain preferential loan terms.

Project or corporate finance

Money may be borrowed on the strength of the company itself or against a specific project. A new company, which has no long-term performance on which to base its funding requirements, will have to borrow finance on the strength of its proposed development. This is described as project funding. Interest rates are normally lower for corporate loans than for project loans. Loans for project finance may also be necessary when the development company is itself quite small or where the project being contemplated is of such a size to make the company's own resources look small.

Parent or subsidiary company

The loan required may be for a subsidiary company. In these cases it may be necessary to get the parent company to guarantee the loan, in which case should a default occur then the parent company becomes liable. In some cases the parent company may guarantee the interest payments only but have no responsibility for the loan itself. This is sometimes termed limited recourse funding, since the lender is unable to recover all of the sums involved where the subsidiary company is unable to service the debt. These loans will not appear on the balance sheet of the parent company and are thus described as *off balance sheet*.

Domestic or international finance

A loan may be obtained in any currency or through international financiers. These became much more common in the UK when it joined what is now the European Union. However, in order to secure such loans companies generally need to be both large and internationally recognised. The key considerations remain the terms of the loan and the interest rates payable. A knowledge of international exchange rates and trends is also important. Borrowing money on the international market adds a further risk, with a subsequent return or loss, to the development.

Equity finance

Equity finance comprises the amounts originally contributed by the owners, together with the profits from previous years' trading which have not been distributed as part of a dividend. Equity is represented by the net assets of the firm that it has helped to create.

The traditional policy of property developers has been to retain as much interest in their completed projects as possible, generally through either the raising of mortgages or the issuing of share certificates. Such a process maximises growth. This was particularly lucrative in the 1960s where accelerating inflation increased capital values and rental growth (see Gearing in Chapter 4).

Duration of funding

Short-term finance

Short-term finance may be required for a few days up to a period of a few years, when full repayment will be expected. The conventional City definition of short-term finance is usually regarded as being up to three years. The most common sources of short-term finance are the commercial or merchant banks or through extending hire purchase agreements or trade credit. The rates charged for money that is borrowed from the bank in this way will vary depending upon the creditworthiness of an individual or company, but they will typically be a few percentage points above the bank's own lending rates. The lender of short-term finance is concerned both about the security of the capital and the interest to be repaid, both of which can be at risk. Such capital may be required to help smooth out the strains in cash flow that may have been caused by seasonal variation in trade, fluctuations in demands for the company's products or the need for bridging finance during the purchase and sale of assets. The cost of such loans relates to the individual bank's lending rates on either a fixed or floating rate. Short-term finance may also be self-generated through the tightening of receipts and payments procedures and close budgetary controls. These will reduce in a loss of some financial flexibility and trading.

Medium-term finance

Medium-term finance is required for periods of time from about five to ten years. These loans may be required to start or expand a business and may be used for the purchase of assets such as buildings or more commonly machinery, plant and equipment. This is a time when finance will be expensive, since at this time the company is trying to establish itself and is at a vulnerable stage in its development. In the case of default by the borrower many of these assets would have depreciated to such an extent as to provide insufficient collateral for the loan.

Long-term finance

Long-term finance is required for periods of ten years or more. Traditionally this was provided by means of a mortgage. Borrowers have the opportunity of selecting from a number of different schemes, notably repayment and endowment types. Under the former, the actual payments made reduce the amount of the capital outstanding. With the endowment type, the mortgage capital is repaid by a lump sum when an endowment policy, often provided by an insurance company, reaches its maturity.

Mortgages may be either floating rate, the actual rate of interest being influenced by current inflation rates and the central bank's lending rate, or fixed rate for a specified number of years. Mortgages are generally restricted to 90% of the valuation of the property, although in periods of high inflation 100% mortgages may be available. The main disadvantage to the lender is that the amounts are lent for considerable periods of time during which the effects of inflation will reduce the real value of the capital when repaid. However, security has been reasonably good, particularly in periods of high inflation where property prices have increased beyond the levels of inflation.

The high number of repossessions of properties in the late 1980s and early 1990s has tended to make lenders more cautious, particularly where negative equity was involved. Liquidations and bankruptcies have had a similar effect on bank lending to commerce and industry. In some cases, borrowers had a negative equity in their property, with the valuation below that of the amount owed on the mortgage. This increased the lender's risk to the borrower. However, the assets of the societies are such that they are well able to weather storms of this type.

The principal source of funds for investment in larger developments has been the insurance companies, pension funds, property unit trusts and other institutional investors. Over 80% of institutional investment in property in Britain is from insurance companies and pension funds.

Sources of finance

Short-term finance

There are several ways in which firms can supplement their long-term finance for short periods. These are listed in Table 8.2. The most common way is by borrowing from a bank on an overdraft facility. The interest rate charged will usually be 1–2% above the bank rate. This is the rate at which the Bank of England will lend money for short periods to high security customers. The lender of short-term finance is interested both in the security of his capital and interest, both of which may be at risk. Capital may be required to smooth out the strains on cash flow resulting perhaps from rapid fluctuations in the market demand for the company's goods. Firms may further supplement their short-term needs by the other methods shown in Table 8.2.

Table 8.2   Sources of short-term finance

Bank overdraft
Trade credit
Subcontracting
Hire purchasing
Leasing
Stock control
Venture capital

Trade credit

The use of credit is a form of short-term finance, which some firms choose to extend by delaying the payment of creditors beyond the normal time. It may be preferable to forgo discounts offered in return for maintaining a positive cash flow. The cost of borrowing from other firms is usually less expensive than borrowing from other sources. However, the firm may develop a bad reputation as a late payer and lose the goodwill not only of these firms but of firms in general. Trade credit is frequently owed by large companies to smaller firms. The construction industry has a poor reputation for making payments on time and hence legislation has been introduced to safeguard smaller companies.

Subcontracting

Where part of a design, manufacturing or construction process can be subcontracted to other firms then this will help to offset the initial cash outlay on goods and labour costs prior to receiving payment for the work completed. However, in most cases this will also result in a reduction in overall profits. Using a specialist firm in this way may provide for a better quality product and a reduction in overall costs due to economies of scale.

Hire purchasing

Hire purchasing allows a company to enjoy the full use of the goods or equipment, but avoids paying for them in an initial full sum.

Leasing

The decision whether to buy or lease assets will depend upon the firm's philosophy, its planning procedures and the projected cash flow of the company concerned. Sale and leaseback is a common form of freeing cash from fixed assets. Under this method the company who owns an asset can sell it to a financial institution and then lease it back at an agreed lease hire rate.

Stock control

The reduction of the level of stock in a company or keeping stocks at a minimum through careful stock control has the effect of freeing up funds for other purposes. The construction industry is perhaps less good at this than is the case in other industries.

Venture capital

Merchant banks and other financial institutions are always looking for companies with growth potential. This is particularly true where the company can use such finance to grow to a size that makes an offer of shares to the public possible. This is the provision of venture capital.

Long-term finance

Long-term capital is often used to start or expand a business and would be used for the purchase of buildings and equipment. The risks to the lender can be high because of the time-scale involved. Long-term capital may be obtained from one or more of the sources listed in Table 8.3.

Table 8.3   Sources of long-term finance

Retained profits
Clearing bank loans
Merchant bank loans
Shares
Debentures
Government grants and loans
Sale and lease-back
Joint ventures
Concession contracting

Types of shares

Shares are the equity capital of a company. The shareholders are entitled to the residual profits in the company, should these be available, and voting rights at annual and extraordinary general meetings. The shareholders therefore can voice an opinion on how the company is being managed. The following sections describe the more usual types of shares.

Ordinary shares

These are often referred to as the equity share of capital of a company. These shares have no special rights or restrictions attached to them. They do not have a fixed or percentage dividend and will only receive a dividend after payment has been made on other types of shares. There is, however, no limit to the rate of dividend which can be paid. A well-managed company making high profits should provide for a good return on an investment. Although dividends are normally declared annually, companies may, if they wish, declare an interim or half-yearly payment. The dividend is declared on the face value or nominal value of the share regardless of the price purchase or quoted on the Stock Exchange.

Preference shares

These shares carry the right to receive a dividend in preference to shares of a lower class, such as ordinary shares. They also carry the right to receive payment of capital in the event of liquidation before other classes of share. Ordinary shares in these circumstances carry the right to any surplus assets once liabilities have met. Preference shares normally carry no such right but their capital is repaid in full in preference to the ordinary shareholders. The dividend paid is normally at a fixed rate, which must be paid in full before any dividend can be paid to other shareholders. These types of shares therefore provide some measure of security to their holders and a regular fixed return on the investment. Their main disadvantage is that they offer no possibility of high earnings in periods of high prosperity.

Cumulative preference shares

These are similar to the above but with the proviso that if the dividend is not paid in full, then the unpaid amount can be carried forward to future years.

Participating preference shares

These shares carry the right to participate in the prosperity of the company when conditions permit. They are particularly valuable since they provide both security and the possibility of high earnings.

Deferred shares

These shares are sometimes known as founders' shares, being issued to the original subscribers of the company. They rank lower than an ordinary share, and only receive a dividend after a declared percentage has been paid on the ordinary shares. They are regarded as a form of payment by results.

Rights issues

In a rights issue, existing shareholders are offered the right to purchase new shares in proportion to their holding of the company's existing issued shares. The price at which the shares are offered is below the current market price of that class of shares in the company.

Debentures

These are loans made to the company. They differ from conventional loans in that they attract a fixed interest rate and must be repaid at a set time. The loan may be secured by a mortgage on the company's property, or simply upon the basis of the firm's reputation. The debenture holder will be paid before any of the other creditors in the event of the firm going into liquidation. Debenture holders take no share of the profits since they are not shareholders.

Convertible loan stock

From a company's point of view, a softer and less restrictive alternative to a debenture is convertible loan stock. This also has a fixed rate of interest and is repayable at a specified future date. Convertibles are not secured on any particular assets of the company, so in a liquidation they rank behind debenture holders but before ordinary shareholders. In exchange for less security and lower interest than a debenture, convertible loan stock holders have the right to convert their stock into ordinary shares. The terms of conversion will vary. Typically this is once per year on a set date.

Sale and leaseback

Sale and leaseback is a technique whereby a property owner raises funds from its property portfolio by selling the property without having to sacrifice the use of the property. The transaction must be fair and reasonable, i.e. market valuations. The situation then appears as follows:

- The property disappears from the balance sheet, since it is no longer owned.
- The sale proceeds are included on the balance sheet as cash received.
- A profit will be shown, or a loss if the sale did not meet its original purchase.
- Rent payable is then charged over the term of the lease.

Provisions are often included to allow the original owner the option of buying back the property at a predetermined price, which might be the current market valuation at the time of the sale. A legal document known as the finance agreement sets out the details that bind the institutions, developers and other interested parties to the methods of funding the development. Accounting for leases is covered in Statement of Standard Accounting Practice (SSAP) 21. These are issued by the Institute of Chartered Accountants in association with the other accounting bodies. A finance lease is defined as a lease that transfers substantially all the risks and rewards of ownership of an asset to the lessee. All other kinds of leases are known as operating leases. Property leases are more likely to be defined as operating leases since rentals will be charged and the title to the property reverts to the landlord at the end of the lease.

There are variations on this theme of financing development that include lease and leaseback and sale and repurchase. The latter method is commonly used by house builders to finance show houses. In these cases the builder sells the completed show house to a finance company. The terms of the arrangement allow the builder to use the house for a fee that equates with the interest charges of the finance company. The house builder upon agreeing the sale of the show house pays off the debt to the finance company and retains the excess amount derived from the sale.

Additional, other methods of development finance include the granting of options to purchase property and joint venture developments.

Joint ventures

On very large projects it may be necessary to utilise the resources of several different companies in order to complete the project. Individually a separate company might be unable to raise the necessary finance or be prepared to accept the risk that might be involved. The arrangements between the parties will depend upon the particular circumstances of the project. A contractor, for example, may provide the funding arrangements for construction with the client accepting the responsibility for land and development finance. The profits from such ventures will then be apportioned in relationship to the different input of the partners concerned. Joint venture schemes can by their nature be complex and require clear guidelines for responsibilities and risks. It is not uncommon for contractual claims to arise between the joint venture parties in order to settle disputes within the project.

Concession contracting

Where a public sector organisation requires a new facility for public use, it has to abide by strict rules regarding borrowing. Such borrowing may additionally be controlled by central government and through Acts of Parliament. However, there are circumstances that arise whereby a private organisation can be engaged to undertake such work through an agreement commonly referred to as concession contracting. In the UK, it is more commonly known as *build, operate and transfer* (BOT). The procedure involves the private sector in selecting and financing projects and designing and operating the facility for public use. The contractor may be paid under a negotiated or competitively tendered arrangement. More commonly the contractor is given the right to operate the facility for a specified period of time. The construction of the Channel Tunnel, completed during the 1990s, was just such a scheme. Eurotunnel plc is the concession contractor. In this case a contract was signed for 55 years, allowing the firm an income generation period of about 45 years, after which the project returns into public ownership. The finance was arranged through Eurotunnel, from a major issue of shares. This kind of arrangement is widely used in mainland Europe, for example, through the motorway networks and the use of toll charges.

A variation of this procedure is being used in the UK to finance public sector housing projects. In these circumstances a local authority may donate the land to the contractor to be used for a commercial project on the consideration that the development of part of the site be used for public sector housing, subsequently being transferred to a housing association for its management and operation.

Needs of the institutions

The different lending institutions are conservative in their attitudes towards investment. They need to ensure that the decisions that they take do not place their funds in jeopardy. They prefer to receive smaller returns from sound investment proposals than to make highly speculative decisions that involve excessive risks. Their lending and investments are therefore influenced by the following factors which are listed in Table 8.4.

Table 8.4   Needs of institutions

- Certainty
- Marketability
- Income preference
- Manageability
- Spread or diversification
- Risk and opportunity

Certainty

This is an intrinsically desirable characteristic of any investment. However, institutions are prepared to accept some level of uncertainty in the anticipation of higher returns. Even venture capitalists, for example, who expect a high degree of uncertainty in the anticipation of a high reward, nevertheless desire to keep the level of uncertainty within specified limits. Whilst this may appear to be somewhat cautious in outlook, it must be remembered that the assets of any financial institution are there to meet liabilities, often within the statutory framework as set out, for example, in the Insurance Companies Act 1982. Few of us would care to think that our own retirement pensions or other investments were at any degree of risk, due to reckless speculation on the part of fund managers. The fraudulent use of pension fund accounts of the Maxwell Corporation has brought this into a sharper focus. Also whilst the managers of the pension funds are able to participate in the benefits of wise investments if successful, our pensions may be in jeopardy should the wrong decisions be taken.

Marketability

Although long-term investors in property have long-term horizons, they are nevertheless interested in the perception that others may have about the sale and resale of their investments. No one, knowingly, will want to invest in either the short or long term in an investment that may be difficult to dispose of at the required time. There is a growing importance attached to the managed fund approach, even though, in many cases, the managing may only be at the margins. However, these margins may be of crucial importance in a competitive scenario. Managed funds are also aimed at changing the structure and balance of portfolios to suit modern day needs, and take into account trends and patterns in fashion in development and occupancy.

Income preference

Income received today is always worth more than the same amount in the future because of the time value of money. Discounting, usually at a gilt rate and an allowance for risk is designed to offset this. The availability of an early release of funds provides the opportunity for further investment elsewhere. The changes in taxation amounts and rules confuses the picture since these may have positive or negative influences on the investment.

Manageability

Institutions prefer long-term savings that require little fund management, since they have themselves a limited management resource for these activities. Property management can of course be placed with managing agents, but these too must also be supervised, depending upon the risk in the funds and the capability and efficiency of such agents. There continues therefore to be a desire for trouble free assets, even where the return is lower than might otherwise be the case.

Spread or diversification

It is generally accepted that a fund manager should attempt to spread the investment across a number of well researched activities. It is also desirable for the fund manager not to have a too large holding of a particular asset, in relationship to either the overall size of the fund or the asset.

Risk and opportunity

Risk is inevitable in all business transactions. It must be controlled and this might be attempted in several different ways through, for example, selecting funds or investments that behave in the opposite manner to each other.

Principal sources of development finance

The following parts of this section describe some of the main categories of development finance for property investment. This is a changing market where investors of different kinds will switch their funds to and fro in order to achieve the best return in terms of the capital invested. During the 1960s, finance for development was in restricted supply. This has since changed and long-term investment in property in Britain has been seen as offering good returns. The mortgage famine that dogged development and pegged prices in the late 1960s has been relieved by a host of foreign investors seeking a place for their pension and insurance funds in this country. There are relatively few individuals or organisations who invest in property who are either able or willing to do so entirely from their own retained earnings.

Tables 8.5 and 8.6 show some of the factors to be considered by both the investor and the borrower. In either case the protection of the capital invested and the return on this capital will be of paramount importance.

Table 8.5   Factors influencing the supply of development finance

- Size, type and duration of loan
- Project characteristics
- Taxation considerations
- Terms on which funding is available
- Reputation and experience of supplier

Table 8.6   Factors influencing supply of finance

- Economic climate
- Political stability
- Spread of investments
- Alternative sources for investment
- Taxation structure
- Risk and return
- Duration of investment
- Liquidity capability

Institutions

Insurance companies and pension funds are often referred to as the institutional investors in property. Whilst they are major investors in property, their importance has diminished since the mid-1970s due to the emerging involvement of the banks and overseas investors. The proportion of funds from these institutions has reduced by a half from about 15% at the start of the 1980s, to around 8% in the early 1990s. This has been due largely to better returns in other sectors of investment, most notably stocks and shares. However, at the beginning of the twenty-first century returns from property were performing quite well by comparison to equities.

Insurance companies

An essential characteristic of an insurance company is to spread the risk involved, either over time, amongst the policy holders or both. Each fund will therefore take into account the nature, mix and term of relevant liabilities. Insurance companies are under an obligation to obtain the returns for their policyholders. Evidence of building development providing a good long-term investment is provided by the large proportions of insurance company funds that are invested in these developments.

During the years after the Second World War, the most important suppliers of long-term finance for property development were the life funds and pension schemes of insurance companies. They invest between approximately 12% and 20% of their total annual investments in property depending upon the performance of these markets with investments elsewhere in other commodities. Currently their investments in land and property are about £60 billion. This amount rose steadily between 1970 and 1990, but during the 1990s remained at a relatively constant figure (see Figure 8.1). In real terms

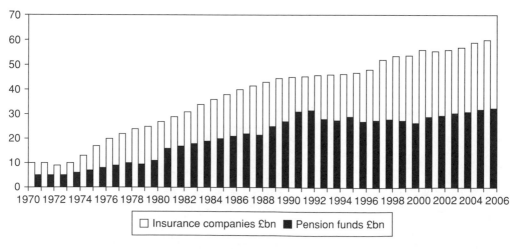

Figure 8.1   Institutions' property holdings (Source: DTZ)

this represented a decline in terms of their overall investments as shown in Figure 8.2. During the 1990s, property represented a poor investment as noted earlier in Chapter 2 (Figure 2.2). However, in the early part of the twenty-first investments in property performed better than, for example, equities (see Figure 3.3).

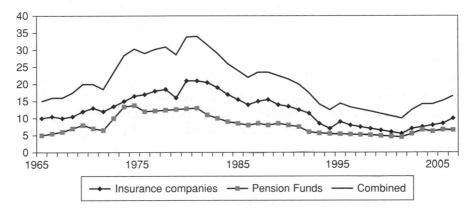

Figure 8.2   Property's share of total institutional assets (Source: DTZ)

The majority of insurance company investments come from the returns on life insurance policy investments, although a proportion of the funds from other insurance dealings such as fire, theft and motor insurance also finds its way into development finance. Like the pension funds, their main concerns in investment are combining the security of investment with equity participation. Property is able to offer just this combination. Typically less than 10% of their general funds are invested in property. The major general insurance companies are listed in Table 8.7.

Table 8.7   General insurance companies

	Total Business 2004	£m 2005
Prudential	8562	8184
Lloyds TSB Group	5635	7797
Aviva plc	7685	7792
Standard Life	7842	7633
HBOS Financial Services	6421	7354
Legal & General	5058	6273
ALICO	3110	4761
AXA	3706	3915
Zurich Financial Services	3243	3576
Aegon	2085	3056
Total market	£72 288	£81 372
Share of top five companies	50.00%	47.63%

(Source: Association of British Insurers)

Life insurance funds are built up from premiums that are then invested by the insurance company in a wide variety of different schemes as managed funds. One of the specialised forms of investment is property bonds. These enable the modest investor to access the returns from prime property investments. There are about 100 property bonds available. The major life funds are listed in Table 8.8. The life funds are dominated by the first three companies (which also may include subsidiary companies within the group). Many have formed larger groups through amalgamations and others are expected to follow this pattern in the future. For example, CGNU was in the mid-1990s at least three separate companies. General Accident and Commercial Union amalgamated to form CGU and within a further twelve months these had merged with Norwich Union to form CGNU. CGNU was renamed Aviva and is the world's fifth-largest insurance group and the biggest in the UK. It is one of the leading providers of life and pensions products in Europe and has substantial businesses elsewhere around the world.

Table 8.8   Life assurance companies

Life Insurance ranking of business	2004	2005
HBOS Financial Services	1	1
ALICO	3	2
Aviva plc	2	3
Legal & General	4	4
Lloyds TSB Group	5	5
Zurich Financial Services	6	6
Standrad Life	7	7
AXA	8	8
Prudential	9	9
Skandia Insurance	10	10
Share of market	45.97%	50.15%

(Source: Associated British Insurers)

Pension funds

These had their origins in Victorian Britain where employers and employees made payments to funds for the provision for pensions beyond the normal retirement age. The principle was that the regular contributions made, if properly managed, expanded in size and would not be required other than in the long term. Occupational pension schemes have grown rapidly since the mid-1960s, with a large proportion of the working population now being members of a pension scheme. Most of the schemes are portable and can be transferred from one employer to another. A wide range of different investments of these types have been designed and marketed, often encouraged by a succession of taxation reliefs to companies and individuals to encourage saving and thus to avoid a reliance on the state in later life. Table 8.9 shows the ten largest pension funds in Britain, grouped broadly between manufacturing, energy and the service sector. The

Table 8.9 The largest UK pension funds, 2006

	£m	Percentage allocations							
		1	2	3	4	5	6	7	8
British Telecommunications	29 229	31.3	29.1	14.2	0.9	8.1	**13.0**	0.1	3.3	100.0
University Superannuation Scheme	21 700	39.3	43.1	5.3	2.1	0.0	**8.3**	0.0	1.9	100.0
Scottish Public Employers Agency	18 800	Data analysis not available								
Coal Pensions Trustees	18 786	46.2	22.6	4.1	1.2	6.8	**12.1**	0.0	6.9	100.0
Electricity Pension Services Ltd	18 329	26.5	22.9	13.6	8.6	17.4	**7.3**	0.2	3.5	100.0
Royal Mail Pensions Trustees Ltd	15 274	50.0	27.0	4.0	3.0	8.0	**8.0**	0.0	0.0	100.0
Railways Pensions Trustees Co	14 729	28.2	37.6	4.7	11.8	8.2	**8.2**	0.0	1.2	100.0
National Grid Plc	13 052	23.9	15.8	31.7	0.0	18.7	**8.0**	0.0	1.9	100.0
Royal Bank of Scotland	11 944	28.0	28.0	8.5	8.5	18.0	**2.0**	0.0	7.0	100.0
Lloyds TSB Group Plc	10 830	32.3	32.2	6.2	0.7	7.0	**8.4**	0.0	13.2	100.0
2006 Average		41.0	23.0	13.9	2.8	8.1	**5.3**	0.1	5.8	
1997 Average		48.5	24.1	5.4	3.1	5.0	**7.3**	0.9	5.7	

1 UK equities
2 Overseas equities
3 UK fixed interest
4 Overseas fixed interest
5 Index linked
6 UK property
7 Overseas property
8 Cash and others

(Source: Pension Funds and their Advisors)

shift in employment patterns over recent years, which is expected to continue, means that some of the funds will reduce as contributions decline and pension liabilities increase to help meet the needs of retired members. In contrast, the funds of the service sectors are expected to continue to grow into the immediate future. It is notable that six of these pension funds were formally nationalised industries. All of these funds and a majority of the smaller ones have some investments in property.

Table 8.9 provides an indication of where these pension funds allocate their investments. The bulk of their funds are invested in equities where evidence shows that this will provide the greatest rewards (see Table 2.1). This is normally a perfectly correct assumption but is not always the case as illustrated in Figure 3.3. Typically pension funds allocate about 5–7% of their capital into property, in one way and another. The reasons for this are largely to do with financial performance but also the ease in the acquisition and disposal of financial assets. Pension funds investment in property has fallen from a high of about 12% in 1986 to just over 5% in 2006. The top ten largest pension fund managers allocate slightly more of about 8%. The investment in overseas property has never figured very strongly amongst these managers' portfolios. At their peak this was only ever just over 1% and has been since the early 1990s almost negligible in the calculations at less than 0.1% on average. This reflects the performance of such investments in the past. The distribution of pension fund investments is shown graphically in Figure 8.3, comparing the relative positions in 1997 and 2006.

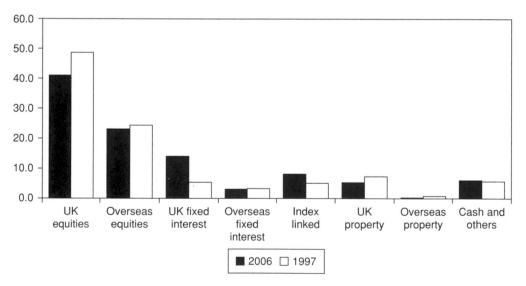

Figure 8.3   Distribution of UK pension fund investments, 1997 and 2006 (Source: Pension Funds and their Advisors)

Pension fund managers still largely allocate their funds in equities, where the long-term returns and other qualities looked for in investment can be obtained (see Tables 8.4–8.6). Equities can also be traded easily and quickly, compared with property which has time and cost disadvantages. Typically, the pension funds allocate about 50% to UK equities and a further 20% to overseas equities. Three-quarters of their funds are thus allocated to stocks and shares of one kind and another. Even the relatively poor performance of the stock markets at the beginning of the twenty-first century has not deterred such investors. History suggests that such funds are likely to recover in the relatively short term against most other forms of managed investment.

Property unit trusts

A unit trust is a legal claim to a fractional part of the trust's portfolio. It allows the small investor to benefit from full portfolio diversification and specialist management, without requiring the individual expertise and financial resources that would be required if investing directly in the same portfolio. A property unit trust invests solely in property. Units in property unit trusts can only be acquired or redeemed through the unit trust, while shares can be traded on the stock market. See also Chapter 2 which describes the recently introduced real estate investment trusts (REITs).

Building societies

These are an important source of development finance, especially for domestic property, but they had no direct involvement in commercial property until after the Building Societies Act in 1986. They are mutual or friendly societies rather than public limited companies (PLCs). Whilst many changed their status and became banks during the 1990s, they still rely extensively on savings schemes and mortgages as their core business. The 1980s also saw a blurring of activities between the banks and the building societies. Prior to 1987, the only difference between the different societies was size. They effectively operated a cartel system with precisely the same savings and mortgage accounts. After the so-called Big Bang, in 1989, and the move towards the deregulation of financial institutions, building societies were given much greater freedom to use their funds in different ways and to apply them more widely than solely to the domestic property market. The potential can be measured by their large collective size and assets. Their assets were in excess of £300 billion in 1996, but due to conversions from mutual status their assets by 2001 were collectively worth £160 billion. Whilst their main source of business is likely to remain with home loans, they will need to use their assets in the best possible way to satisfy investors on the returns on their capital.

Financial deregulation meant that the small to medium-sized building societies were unable to compete in the market. The decline in the housing market, their main sector of activity in the early to mid-1990s, together with the larger societies being able to offer higher discounts, credit cards and savings accounts at more attractive interest rates have caused the further decline in the number of societies and this trend is set to continue.

The first building societies were established in 1775 in the Midlands and the North of England. It is interesting to note that a large number of the original building societies had their origins on the Lancashire and Yorkshire border. By 1825, there were 250 societies and by 1860, 750 societies in London and a further 2000 in the provinces. In 1910, there were 1723 societies with assets of £76 million. This number had reduced to 819 by the early 1950s. By 1982, the number was 227 and in 1992 the number had further reduced to 87, due to mergers or takeovers by banks. There are now just 60 building societies with assets worth £310 billion. In 1989, the Abbey National converted to a bank with many more following in short succession. Since 2004, Abbey has been part of the Santander Group, the seventh largest bank in the world by profit. Data from the Building Societies Association claims that it costs 35% more to run a bank than a building society, due primarily to the demands of shareholders.

Of the top ten societies in 1992 in terms of their assets, only one remains as a building society, the remainder having converted to banks awarding millions of pounds in the form of windfalls to their members. The second largest society in 1996, the Nationwide, is now by far the largest of the remaining building societies (see Table 8.10), but is constantly under threat to demutualise its status.

Table 8.10   The top Building Societies (2007)

Society	Assets £m
Nationwide	118 288
Britannia	28 953
Portman*	17 724
Yorkshire	16 178
Coventry	11 090
Chelsea	9602
Skipton	8821

Note: * The Portman merged with the Nationwide in 2007.
(Source: Building Societies Association)

Table 8.11 and Figure 8.4 provide an indication of the main sources of house purchase, showing the rapid increase between 1980 and 1990 of loans from banks, which

Table 8.11   Sources of finance for house purchase (%)

	1980	1990	2000	2004
Building societies	79	74	23	16
Insurance companies	4	1	0	0
Banks	7	23	70	69
Local authorities	6	0	0	0
Other	4	2	7	15
	100	100	100	100

(Source: British Banking Association)

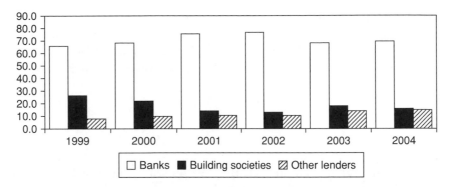

Figure 8.4   Net mortgage lending (£bn) (Source: BBA, Bank of England and ONS)

virtually absorbed all sources other than the building societies. The increase in bank lending was also due, in a large part, to building societies converting to banks.

Mortgage interest rates and their comparison with bank rates are shown in Figure 8.5. The mortgage rates mirror the pattern of base lending rates. During the 1990s mortgage rates were maintained at a constantly low level which had not been achieved during the volatile years of the 1970s and 1980s. During the 1970s and particularly the 1980s rates were much more erratic and at a higher level, peaking at 15.4% in 1990. The early 1990s provided good news for home buyers, with the rate falling almost to its 1950 level. This should have kick started the house building industry but a lack of confidence amongst prospective purchasers failed to start the upward spiral of the economy in general, until the end of the decade and into the early part of the twenty-first century.

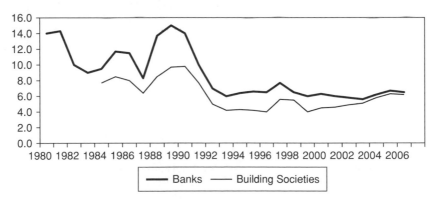

Figure 8.5   Mortgage lending rates: percentages (Source: ONS, Financial statistics)

In 1974 there were 450 000 mortgage loans worth in total £2984 m. The typical percentage advance was 68%. By 2005 the number of loans were in excess of one million and worth in total almost £130 000 m and the average percentage advance had increased to 78%. By far the largest mortgage lender is the HBOS with 20% of the market. The four largest lenders (HBOS, Nationwide, Abbey National and Lloyds TSB) have between them almost 50% of the market. Only one of them is now a building society.

Banks

The major British banking groups (MBBG) are shown in Table 8.12. The banking market after 1945 was eventually dominated by the big four, i.e. Barclays, National Westminster, Lloyds and Midland. However, since deregulation the distinction between banks and building societies has become less evident. The Abbey National was the first to convert to a bank in 1989. Table 8.13 indicates the respective size of these banks. These also figure quite strongly in the top ten list of British companies shown in Table 8.14, although this table frequently changes due to mergers, acquisition and performance of the companies concerned.

Table 8.12   Composition of the Major British Banking Groups

Group	Associated Companies
Alliance and Leicester	
Barclays Group	Barclays Bank plc
	Barclays Bank Trust Company Ltd
	Barclays Private Bank Ltd
HBOS	Halifax plc
	Bank of Scotland
	HBOS Treasury Services plc
	Capital Bank plc
HSBC	HSBC plc
	HSBC Trust Company (UK) Ltd
Lloyds TSB Group	Lloyds TSB Bank plc
	AMC Bank Ltd
	Cheltenham & Gloucester plc
	Lloyds TSB Private Banking Ltd
	Lloyds TSB Scotland plc
	Scottish Widows Bank plc
Northern Rock	
The Royal Bank of Scotland Group	The Royal Bank of Scotland plc
	Adam and Company plc
	Coutts & Co
	National Westminster Bank plc
	Tesco Personal Finance Ltd
	Ulster Bank plc
	Ulster Bank Ireland plc

(Source: British Banking Association)

Table 8.13   Banking and insurance groups within the top 200 companies (Major banking groups in bold)

Ranking		Company	Market capitalisation	
2001	2007		2001	2007
63	73	Alliance and Leicester	4224	4883
10	7	**Barclays Group**	37 067	50 527
14	10	**HBOS**	17 161	42 887
4	1	**HSBC**	77 457	106 584
8	13	**Lloyds TSB Group**	40 659	34 168
92	68	Northern Rock	2674	5269
7	5	**Royal Bank of Scotland**	50 492	65 074

(Source: FT Interactive Data, 2007)

Table 8.14   The top ten listed companies and their market capitalisation values

Ranking		Company	Market capitalisation	
2001	2007		2001	2007
4	1	HSBC	–	106 584
1	2	BP	136 937	104 484
2	3	Glaxo Smith Kline	116 276	83 344
3	4	Vodafone	90 979	78 817
7	5	Royal Bank of Scotland	50 492	65 074
6	6	Royal Dutch Shell A	57 005	62 881
10	7	Barclays	37 067	50 527
–	8	Royal Dutch Shell B	–	46 991
5	9	Astra Zeneca	57 748	44 777
–	10	HBOS	–	42 887

(Source: FT Interactive Data, 2007)

Bank lending to developers was often restricted to short-term bridging finance to cover the expenses of development prior to the property's eventual sale. At the start of the 1970s, bank lending to property companies was small at around £2 billion and represented only 1% of all commercial loans. Their lending then peaked in 1974 at about £14 billion, but by the mid-1980s had fallen back to around £5 billion. From then onwards, lending to property companies progressively increased, throughout the property boom years, and by 1991 was in excess of £40 billion. This was followed by the deep recession of the early 1990s, resulting in the collapse of some major property companies. Bank lending was then accordingly reduced and by 1993 was about £36 billion or 11% of all the banks' commercial lending. By the beginning of the twenty-first century, bank lending to property companies had reached in excess of £150 billion or 12% of commercial lending (see Figures 8.6 and 8.7). The total bank lending to British property companies includes that from foreign banks. The direct risk to the

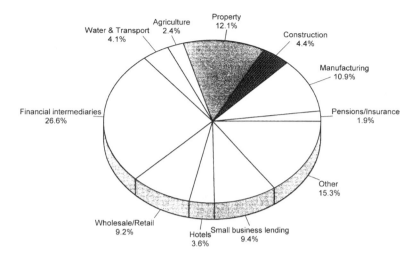

Figure 8.6 Bank lending to industry and commerce (Source: British Banking Association)

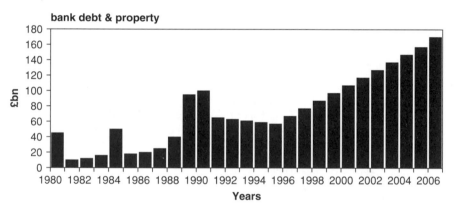

Figure 8.7 Bank debt outstanding to property companies (Source: DTZ)

banks is also limited since they have been operating tougher rules on the ratio of capital to outstanding loans.

Bank lending to property companies increased sharply in the latter half of the 1980s, and in 1992 stood at 8.5% of total bank lending. This was almost at the level that triggered the secondary banking crisis in the mid-1970s. However, the Bank of England is keen to stress the difference between the 1974–75 crisis and that of the early 1990s in two respects: the heavy exposure to property of the weakly capitalised secondary banks and the limited role of the foreign banks in the British property market then. In the 1990s, the leading British banks and foreign investment were much more involved. The total bank lending to British property companies was just over £39 billion. Of this just over £17 billion is owed to foreign banks, including £4.5 billion to the Japanese. Whilst the City and Docklands were troubled areas, bank lending on property in other parts of the country performed well. The direct risk to the banks is also limited since they have been operating tougher rules on the ratio of capital to outstanding loans.

Landowners

This section includes the large land and property owners such as the Crown Estates, the Church Commissioners, local authorities and other major private and public companies. Whilst many of these have considerable funds available for investment, they are controlled by their own trust deeds and articles of association. Large investment portfolios have been acquired through bequests, intestate deaths and government sanction. Many of the large aristocratic families have large estates, often in London and are major developers in their own right. They are often able to obtain borrowing facilities, contribute from their own available sources of funding and may be able to offer a site or land for the development project (see also Chapter 1).

Government funding

There are now a wide variety of different sources of assistance available in terms of funding for development. The requirements for support are often stringent in their nature and are often aimed at encouraging development in areas that would otherwise be unattractive to developers and investors. A number of major initiatives are discussed later, although these are constantly under review and change with the regulations for their implementation varied to suit political, social and economic circumstances. Assistance may be available as a grant or loan, through taxation relief or through incentives in respect of relief in terms of local taxation such as business rates. Charitable foundations may be prepared to offer their financial support to projects of many different types. There are different European funds that can also be tapped for grants and loans.

The growth in these areas of funding is now such as to require the publication of several guides and directories that soon become outdated. Computer online services have been introduced to take account of the frequent changes to sources, availability, amounts, regulations, etc. Whilst the application for and the procedures to be complied with obtaining relevant funding is often fraught with bureaucracy in all its guises for the sake of public accountability, there has been a real attempt to simplify the procedures employed. However, there still remains a time-lag delay with some of these grants and this can cause cash flow problems for the developer.

In many cases the refusal to offer financial support for a project is clearly understood, with little opportunity for the developer to redesign the works, an unwillingness to relocate to a designated area or an inability to comply with conditions that are stipulated as a requirement for grant aid. In some cases it is due to a misunderstanding either on the part of the developer regarding the grant provisions or on the content of the proposals. For example, a developer of a new industrial site found that the application was unsuccessful, because the project was inaccessible to a potential workforce that it was being created to serve. This would have required the construction of a new access road. This was not part of the proposal. However, through discussions with the funder, it was felt that such an access road was worthy of funding, since it increased the overall access to other amenities close by. The project then went ahead successfully.

Property shares

Investors who decide to purchase property shares have a wide choice of different companies in which to place their funds. Whilst most of these funds cover a wide spread of interests, those with holdings in prime areas, such as central shopping areas or central business districts are often regarded as the most attractive. The very best quality or prime sites will usually always be in demand and in this sense there may be little risk attached to such an investment. The quality of the management employed will largely determine the future prospects for the shares. Inactivity, particularly over long periods of time, is unlikely to commend shares to the investor. The larger companies which are able to provide a professional team approach are significant particularly as size and complexity increase. New development schemes, which in the longer term may be very profitable, are sometimes difficult to assess. They need to reach the commercial test of lettability, and until this is achieved may be unprofitable.

The development and trading activities of a company are much less secure than the incomes available from established property assets. A highly geared property investment company will produce a more proportionate increase in dividends when rents are increased and mortgages remain unchanged. The property companies that have financed their activities on a sale and leaseback arrangement, may have false share values, since the terms of the leaseback transactions are not always clear.

The most attractive companies are those whose income is mainly of a receiving nature. Whilst some companies have diversified into other activities such as contracting, the trading activities and associated future growth will require a more positive activity than that which is to be acquired solely from an increase in long-term agreed rents. Other companies have invested overseas and, whilst these may prove to be valuable in the long term they can present further uncertainty that will discourage investors.

Public sector developments

Although a substantial proportion of development and construction work is undertaken in the public sector, this continues to diminish for the reasons already stated. Public sector projects are undertaken by both central and local government and for a range of other agencies and quangos. Public projects are those that are constructed and maintained at public expense. The privatised utilities, for example, no longer fall within this category, although the health service trusts remain as public sector companies but with a greater freedom to direct their own activities.

The public sector currently remains responsible for the construction and maintenance of parts of infrastructure, such as roads and rail networks, but not now for sewers, which form a part of the privatised water companies. The transfer of the ownership of British Rail into the private sector has been fraught with difficulties to such an extent that Railtrack plc entered into administration in 2001 only five years after it was privatised. There are suggestions of private road programmes that would build toll roads if they were constructed. The former utility services of electricity, gas and telecommunications became PLCs during the 1980s, and these have generally been successful operations.

A local authority will often raise finance for development purposes in the same way as a private developer. Although it cannot own shares for sale it is able to create investment opportunities by offering bonds and other forms of securities to the public. It can also form a development company for the purpose of carrying out particular capital schemes. These may include urban regeneration and development or projects to attract industry and enterprise into its geographical area. In addition, it can also apply to central government for loans and grants for approved development projects.

Local government is restricted in its activities through administration and control by central government and by law since it can only act within the powers that have been conferred upon it.

Financial assistance for development

Although the various planning regulations are able to prevent undesirable development from taking place they cannot encourage socially desirable development to be undertaken. As a part of the planning process, the authorities are able to suggest to a developer that certain specified works are carried out as a part of the approval for development. These might include, for example, leaving part of a proposed housing development as a public open space. This is termed development gain. However, in order to encourage desirable developments to take place in unattractive locations, some form of financial assistance is often necessary. This assistance may be obtainable from several different sources, but especially from government agencies through grants, loans, taxation relief, etc.

The intention of such financial assistance is to support projects in areas where they otherwise might not take place, or in circumstances where there might be little obvious economic benefit to a developer. One of the biggest difficulties for developers is that government policy on these matters is constantly under review and change in order to meet new demands and to help generate improved economic conditions. Such assistance may be offered from a local or regional authority or central government department. In the case of Action for Cities launched in 1988, the promotion included three major government departments working jointly. A major aspect of this initiative was also the effective cooperation between the public sector, local and central government and private business and voluntary organisations. Government invested about £3500 million on urban regeneration in Britain in 1989–90, with an expectation of several times that amount from the private sector. Capital expenditure which is incurred on new buildings, adaptations to existing buildings, and new machinery, plant or equipment under the right circumstances are eligible for grant aid.

Financial assistance associated with construction projects arises in several instances, which include:

- Urban renewal programmes
- Regeneration of industrial areas
- Investing in jobs to benefit areas of high unemployment
- Land reclamation schemes
- Property improvement, such as housing improvements
- Slum clearances and derelict land clearance.

The aim is to encourage private companies or public managing agents to develop areas, either as a means of improving the standards and amenities or through investing in projects that will help wealth creation and at the same time reduce levels of unemployment in a region. Financial assistance is therefore targeted in those areas where it might be difficult to encourage companies to otherwise invest. Although financial assistance may be available for a variety of different purposes, it is in the designated areas where the amounts of such grants are the highest. Higher levels of grant are available for approved schemes in Urban Programme and Assisted Areas. Urban Policy Initiatives also include Enterprise Zones and Urban Priority Areas and their respective successors.

Investment grants are made by the government to manufacturing and extractive industries only, in respect of new buildings or adaptations and plant and machinery. The grants are treated as non-taxable capital receipts. Loans are treated in a similar way and may be free of interest. Loans are sometimes offered to companies who for a variety of reasons are unable to secure finance in the normal conventional ways through commercial banks. The financial assistance offered may be a combination of the following.

- *Taxation allowances* on capital expenditure for buildings and plant. The company in receipt of such an allowance must in the first instance make a profit to secure the benefit.
- *Low rents or business rates* which are offered by a local authority as inducement to locate a business in their area. This will be offered to a company for a limited number of years.
- *Grants for capital items* to assist the firm in new developments. These grants may be as high as 50% of the total capital costs of building, extending, converting and improving industrial and commercial property.

- *Amenity grants* which can represent 100% of the costs associated with providing access roads, car parking and other amenities.
- *Bridging finance* to close the gap between developing a building and its market value.
- *Interest relief grants* to help offset some of the costs of borrowing finance.
- *Building loans* at preferential interest rates which may be available to cover up to 90% of the market value of land and buildings. In addition government may act as a guarantor for bank finance for building.
- *Enterprise Zone* benefits, which include 100% tax allowances for money invested in commercial and industrial buildings, exemption from local property taxes. Enterprise Zone status also provides for simplified planning procedures for developments.
- *Subsidies* paid to companies who employ additional employees in specified occupations. These, like low rents, are an inducement for a limited period of time only.

Regional initiatives

Regional industrial policy operates within a general economic framework designed to encourage enterprise and economic growth in all areas of Britain. However, in some areas specific additional help is required under a regional initiative. Help is thus focused on the assisted areas, which are designated *intermediate* and *development* areas, and were based in the existing industrial conurbations. Also included under the assisted areas scheme are the north of Scotland and Cornwall. Northern Ireland has its own full range of incentives. The Scottish and Welsh Development Agencies also promote industrial development in their respective countries.

These initiatives are aimed at those areas of high unemployment caused by the demise of traditional industries or the loss of major employers. In order to obtain assistance the project envisaged must create or in exceptional circumstances safeguard jobs within a designated area. The project must also have a good chance of long-term viability. In addition, the greater part of the project costs must be financed by the applicant or from private sector sources. The applicant must also be able to show that without this assistance the project would not take place at all on the proposed basis. A further criterion is that improved economic efficiency and greater security of employment should result. Grants are based on the fixed capital costs of new buildings or adaptations of existing buildings, plant, equipment, machinery, vehicles, etc.

The European Union

The European Union (EU) seeks to increase the degree of economic cohesion and to ensure a more balanced distribution of economic activities within the Community. The principal responsibility for helping depressed areas remains with the national authorities, but the Community may complement schemes through aid from a number of sources. The European Regional Development Fund (ERDF) was established in 1975. Its purpose was to contribute to the correction of the main imbalances within the Community by participating in the development and structural adjustment of regions whose development is lagging behind and in the conversion of declining industrial regions. By 1988, £395 million had been allocated to Britain, mainly for assisted areas and Northern Ireland. In addition, the EU extended eligibility to certain non-assisted areas, with Britain receiving about £170 million for special programmes to improve the environment and to encourage employment initiatives in certain textiles, steel, coal, shipbuilding and fisheries closure areas.

Interactive financial assistance websites

There are a range of interactive financial assistance and business support websites (Table 8.15), providing details on all business funding opportunities available throughout the United Kingdom and Europe.

Investment from overseas

The levels of activity seen in the UK at the beginning of the twenty-first century are well ahead of other mature markets in Europe, such as France, Germany and Sweden. There are strong volumes of cross-border purchasing activity that indicate that non-domestic investors remain a key participant in the UK real estate investment market. As a percentage of total acquisitions, overseas purchases in 2005 accounted for about a quarter of the total. This was broadly in line with the long-term average. With the UK economy remaining strong, overseas investors are expected to continue to be an important source of capital for the UK real estate market.

Over 80% of overseas activity was from Ireland, the United States, the Middle East, Germany and the Netherlands. Irish investors continue to represent the largest single source of cross-border capital representing 22% of total, although marginally down from 2004. With interest rates expected to remain stable over the short term, Irish investors are expected to remain an important source of capital for the UK real estate market although, increasingly, they are turning their attention to a range of European real estate markets, which include Central and Eastern Europe.

The US appetite for UK real estate continues to be robust although this represents a significantly lower volume of purchasing activity compared with the past. This is largely due to the absence of any transactions on the scale of Canary Wharf.

Central London remained the most popular destination for non-domestic investors in 2005, accounting for £8.2 billion of all foreign transactions in the country. This was down from 2004 when £9.4 billion was recorded, which was skewed by the Canary Wharf transaction. If this transaction is discounted from the headline figure, overseas

Table 8.15   Interactive financial websites

Website	Description
Business link	This is an interactive financial assistance and business support website, providing details on all business unities available. (www.businesslink.gov.uk/finance)
Grant net	This allows a free access to use an interactive financial assistance site that includes a comprehensive and up-to-date database of grants, loans and other initiatives operating in the UK. It includes European Union grants, UK Government grants, Local Government grants and grants provided by corporate sponsors and charitable trusts. (www.grantnet.com)
Government funding	This is an interactive portal containing details on grants from the Department for Communities and Local Government, the Department of Health, the Department for Education and Skills, the Home Office and the government offices for the regions. (www.governmentfunding.org.uk)
J4B Grants	This is an interactive website containing comprehensive information on government and lottery grants for both business and voluntary groups (www.j4bgrants.co.uk)
Lottery funding	This a joint website run by all Lottery funders in the UK, the site allows individuals to search information on current funding programmes across the UK. (www.lotteryfunding.org.uk)
EU funding	This is an interactive website that has been designed for professional users who apply for EU funds, participate in tenders and develop European contacts. (www.myeucenter.org)
Local authorities	Local authorities can provide grant support to companies which are investing in their area. This can take the form of a grant to offset either some of the property tax associated with a building or assistance with rental costs and/or building alterations. This support is usually linked to the number of jobs created.
Regional Development Agencies	The Regional Development Agencies play a pivotal role in assisting companies and providers including business angels, private investors and venture capital funds. As well as the delivery of a range of financial products, the Agency evaluates investment proposals, offers guidance on sourcing finance and can groom SMEs and entrepreneurs in presentation techniques
Prince's Trust	The Trust offers grants for 14–30-year-olds for jobs and training and group projects. It provides grants and loans to those who have an idea for a business but can't raise all the cash needed to start a business from anywhere else. (www.princes-trust.org.uk)
Rural funding	There are several websites that contain details on rural and community funding schemes where entrepreneurs can gain access to grants and growth capital from a wide range of sources

acquisitions in Central London in 2005 actually reached the highest level recorded. The City market remained the most popular of the London sub-markets in terms of deal activity, followed by the West End market.

Investors continue to move into new markets across Europe. To some extent, this will be driven by continued strong levels of competition for products in the larger and more established real estate markets, although the growing range of indirect investment vehicles offering access to specific European locations and sectors is also likely to encourage investors to increase allocations to non-domestic real estate.

Chapter 9
Project Construction Costs

Early price forecasting

An estimate or forecast of costs of a construction project is done at different stages of the project. A budget sum for the client on behalf of the design team, or design and build contractor, will be prepared at the inception stage. This might be based upon a financial method or one of the single price methods listed in Table 9.1. If the amount is acceptable to the client or developer, then a more detailed method should be used to provide a framework for cost planning. Once the project has been fully designed, contractors will estimate their costs as a basis for their tender sums. This is normally based upon a bill of quantities or work measurement packages using an agreed method of measurement. The work is priced in sufficient detail to allow for cost reconciliation to take place at a later stage. The successful contractor's bid is then incorporated in the contract documents as the contract sum.

Table 9.1 Methods of early price estimating

Method	Description
Conference	Based on a consensus view
Financial methods	Cost limits determined by the client
Unit	Used on projects having standard units of accommodation
Superficial	Based upon floor area
Superficial–perimeter	Based upon a combination of floor area and the building's perimeter
Cube	Based upon the project's volume
Storey–enclosure	Based upon a combination of weighted floor, wall and roof areas
Approximate quantities	Major items measured
Elemental estimating	Used in conjunction with cost planning
Resource analysis	Used mainly for contractor estimating
Cost models	Mathematical computerised modeling
Expert systems	Computer based systems developed using practitioner's expertise

(Source: Ashworth, 1999)

Early price estimating, during the design stage, uses a variety of techniques as shown in Table 9.1. These have become known as single price methods, even though in some cases they use a limited number of cost descriptors. These methods are also described as approximate estimating, since they provide only an indication of the expected costs of construction. They rely upon design information, cost data and the skills and expertise of the person using one of the methods. Since the design brief is still being developed at this stage, this increases the uncertainty and accuracy. More detailed information of these different methods can be found in *Cost Studies of Buildings* (Ashworth, 2004).

In most cases during the early part of the design stage, the drawings and specifications are uncertain and imprecise. However, a forecast of costs will still be required. As the design progresses the forecast can be refined, but on average this may vary by at least 10% from the eventual contract sum. This in itself is still an estimate of the actual cost. It is therefore more usual at the design stage to offer a range of possible estimates or confidence limits rather than a single lump sum forecast. There is a large amount of data available which measures pretender estimates against final costs. The relationship between the two is often poor but this is because of factors which are outside the control of the cost adviser. The future is always difficult to forecast and errors and inconsistencies in pricing will occur. Such estimates may also have been made a number of years ahead of the actual construction of the project, when conditions may have been very different.

While the forecasting of construction costs remains imprecise, like any forecasting technique, this is a universal problem, rather than one restricted to the British construction industry alone. It should also be understood that other industries concerned with capital budgeting fare no better in this respect. An estimate, by definition, will never be a precise forecast.

Some of the above differences in costs can be accounted for by changes in design and specification, changes in client's requirements, the introduction of new technologies and for reasons of inflation. Also some projects represent innovatory solutions, in terms of design and technology, which often require considerable modifications during construction. The usual levels of cost estimating accuracy that are achieved in practice are shown in Table 9.2. These can be applied to the full range of different types of construction projects.

Table 9.2   Estimate classification and accuracy

Estimate	Purpose	Accuracy
Order of magnitude	Feasibility studies	±25–40%
Factor estimate	Early stage assessment	±15–25%
Office estimate	Preliminary budget	±10–15%
Definitive estimate	Final budget	±5–10%
Final estimate	Prior to tender	±5%

(Source: Ashworth, 1999)

Contractor's estimating is based upon measuring a large number of work items and analysing their unit costs, based upon previously recorded site performance data. The measured items should aim to consider only the cost important items. Contractor's estimating is considered further in Chapter 12. In theory, the site performance data or labour outputs and material and plant constants are retrieved from work done on site. Research has shown that this theoretical concept is flawed in practice due to poor and inappropriate recording systems used by contractors and the lack of confidence that estimators have in individual site feedback. The time taken to undertake the different construction operations is also highly variable. The difficulty of capturing this data in a meaningful form that can then be reused is a complex task beyond the profitable occupation of most contractors. A comparison of similar items priced by different contractors also reveals differences or discrepancies as high as 200%. Even published data on guide prices can vary by as much as 50% (Ashworth, 1980).

Cost planning

Cost planning is a term that can be used to describe any system of bringing cost advice to bear upon the design process. In the context of the development of construction economics the term has developed its own significant meaning and describes a process that has evolved since its first inception. To be effective, cost planning requires a close working relationship and cooperation between engineer, architect and quantity surveyor. It also requires an appreciation of each other's aims and objectives.

Principles of cost planning

The objectives of cost planning are aimed at:

- Achieving value for money for a client
- The sensible expenditure between the different parts of a building
- Keeping the total expenditure within the amount agreed by the client.

The underlying principles involved have been defined as follows (RICS, 1976):

- There is a standard reference point for each definable part of a building.
- It allows performance characteristics to be related to each reference.
- It allocates costs in an apportioned and balanced way throughout the building.
- Previous projects can be classified in a standard manner.
- The same process and procedures with design methods are adopted.
- Costs can be checked and amended as the design develops.
- Designers can take the necessary action before committing themselves to any one design solution.

- Design risks and contingencies can be taken into account.
- Costs can be presented in a logical and orderly way to the client as the design develops.

In projects that are cost planned effectively, the standard procedure will often be activated by a proposal to consider several different options before making a design decision. Thus a designer may wish to know the relative or absolute costs of different types or configurations of structural frame. Equally, the cost implications of alternative procurement arrangements will need to be considered. The outcome of such exercises will be a statement of the relative costs of the options together with an assessment of their acceptability within the budget. It is important that all cost advice is related to the agreed budget, since cost planning assumes that the clients want to control their financial commitments (Bennett, 1981).

Cost planning systems

Two separate systems of cost planning evolved. These were known as *elemental* and *comparative* cost planning.

Elemental cost planning

This method is sometimes referred to as target cost planning since a cost limit is fixed by the financial method of approximate estimating (Ashworth, 2004), and the architect must then design the project to fit within this cost. The cost plan is the architect's design in financial terms. The process is therefore described as designing to cost. The method has been used extensively throughout the public sector, where financial limits were placed on proposed expenditure.

Comparative cost planning

This method of cost planning is often described as costing a design. The main difference between this and elemental cost planning is that no fixed element budgets are necessarily considered. Different design solutions for each element are considered on their respective merits, with the design team and the client making the appropriate selection based upon as much information as possible. The most expensive solution may be selected, but this is taken in the full knowledge and awareness of the cost consequences.

In practice, the cost planning process is usually a combination of the two theories. Most projects recognise the need to set element cost limits, but not necessarily at the strict expense of eliminating the advantages of an overall good design solution. This overall aspect will include the full consideration of life cycle costs, which could as far as building costs are concerned increase initial costs.

Advantages of cost planning

The general advantages claimed for the use of cost planning include:

- The tender sum is more likely to equate with the approximate estimate
- There is less possibility of abortive redesign being necessary at the tender stage owing to tender sums being higher than expected
- Some element of cost effectiveness through a balanced design is more likely to be achieved
- Cost considerations are more likely to be considered as a full and integral part of the process
- The amount of pre-tender analysis and evaluation by the design team should result in a smoother running of the project on site
- Cost planning provides a useful means of comparing the costs of individual projects.

The cost planning process

Whilst there are two systems that are used for cost planning the processes involved are similar. These include the following stages shown in Table 9.3.

Table 9.3   The cost planning process

Early cost advice	A discussion with the client on the broad aims and objectives of the project
Preliminary estimate	This is usually prepared using a single price method of approximate estimating
Preliminary cost plan	The preliminary allocation of the budget among the various elements of the project
The cost plan	This is based upon an elemental analysis involving the setting of cost targets for the whole project
Cost checking	The evaluation of changes in the design of elements during the design process
Tender reconciliation	The comparison of the final cost plan with the accepted tender sum
Post-contract cost control	The control of costs during the construction period

Early cost advice

This is usually done to provide a client with some indications of the likely costs involved based upon the client's broad aims and objectives regarding the project. The methods used for financing the project and the determination of its general financial viability would also be provided at this stage.

Preliminary estimate

This is usually prepared at the outset of a project. A quantity surveyor is asked to provide some ideas on the likely project cost. If drawings are available these are likely to be sketches that may provide an indication of the size (floor area), possibly the materials that might be considered and ideas of how the client's brief may be developed. If the client undertakes a considerable amount of building work, then the information and the brief are likely to be more precise. In this case the client may also have some idea of the expenditure that is involved. The preliminary estimate will attempt to confirm, or otherwise, the client's expectations. If the project is a one-off scheme for the client, then the information is likely to be vague with aims and objectives that will need to be agreed as the project develops.

The methods used for the preparation of early price estimates are listed in Table 9.1. Many of these are applied in a deterministic way, providing clients with a single estimate of total cost that is used for budget purposes. The assessment of design and price risks in preliminary estimates are also considered in Ashworth (2004). This indicates that the risks associated with each of these factors are greater at inception, where many decisions still have to be taken, than later when the project is more fully developed and where costs have become more explicit. It is now generally considered to be insufficient to provide clients with only a single estimate of cost and therefore more appropriate to offer a range of values of a forecasted tender sum. The technique of *multiple estimating using risk analysis* (MERA) attempts to provide such a range of estimates. This procedure was devised by the former Property Services Agency within the Department of the Environment.

Traditionally, early price or approximate estimates are prepared to provide clients with a budget of their expected costs. This is to avoid the expensive design fees on aborted schemes that have to be abandoned because they are too expensive. Historically no measure of accuracy or confidence limits was given, other than the estimate having been prepared in accordance with the appropriate skills and expertise that were expected. Estimate deviation, between preliminary estimates and contractors' tenders was usually explained away with explanations for any discrepancy being provided.

It is considered to be good practice today to provide preliminary estimates of cost within a range of values and to offer confidence limits on these values. Such information is now automatically requested by informed clients and this provides them with additional information on which to make more sound judgements.

Providing the preliminary or initial cost estimate is therefore crucial to the whole scheme. It is usually concerned with the seven factors listed in Table 9.4. This requires considerable skill, experience and judgement. Prior to cost planning being introduced this is where predesign estimating ceased.

Preliminary cost plan

The preliminary cost plan is really the first phase of the cost planning process. Its main purpose is in determining the cost targets for the different elements involved. The preliminary estimate will have been accepted by the client and the architect. This estimate may be subject to revision based upon a preliminary investigation of the site and further consideration of the proposals outlined in the client's brief. A sum will eventually be

Table 9.4   Seven factors associated with early price estimates

Market and contract conditions	Factors to consider: workloads, labour availability, type of client, contractual conditions, etc.
Design economics	The effects on cost of the design, shape, height, size, constructional details, etc.
Quality considerations	The quality of materials and the standards of workmanship, including compliance with government regulations
Engineering services	The type and amount of engineering services
External works	The size of the site and the features that are to be provided
Exclusions	Items excluded from the estimate include: land costs, VAT, professional fees, interest charges, loose furniture, but also items such as fitting out on some projects
Price and design risk	Allowances to cover design and its impact on construction methods and the volatility or otherwise of the construction market.

accepted that satisfies the brief and the proposed design, and this will form the basis of the cost target. Alternatively, there will be circumstances where a target expenditure is imposed, as in the case of government sponsored projects. The evidence in these cases suggests that for defined standards of specification and the amount of accommodation required it is possible to design within a cost limit.

The cost plan

Once the sketch plans have been completed and agreed by the client, the task of allocating sums of money to the various elements can take place. The individual element unit target costs must of course equate with the total target cost, otherwise the cost plan will be out of balance. It may also be desirable to present these within a range of probable costs, although it is essential that an overall agreed total cost is also identified. The methods described later for cost analysis will be employed to arrive at element unit totals. The following information will be required:

- *Drawings*: at least plans and elevations
- *Specification*: an indication of the quality of materials and standards of workmanship
- *Contracts*: the likely method of appointing a contractor
- *Cost analysis*: available from other comparable projects
- *Other analyses*: to be used as a second opinion.

The target costs that are calculated should be adhered to by everyone involved, or at least any variation from these costs must be noted for information and appropriate action. A separate sum should be set aside for price and design risks. This amount will vary depending upon the experience and skills of the designer and quantity surveyor. The percentage adjustment can vary considerably depending upon the complexity and innovation in the design. The adjustment will reduce the closer the project gets towards the tender date.

Cost checking

During the design stage of the project the different components or elements of the design evolve. Throughout this process it is necessary to compare the costs in the evolution of design with the amounts allocated in the cost plan. Designers usually design in this way and cost planning that adopts the elemental approach allows the changes in costs to be easily calculated and compared. Where the architect does not diverge from the information and ideas incorporated in the cost plan then time and effort will be minimised. This is frequently achieved in those circumstances where similar projects are being designed for the same client. In the case of an individual design to meet a one-off solution, great care will be required if the desired results of cost planning are to be achieved.

Where cost checking reveals compatibility with the original cost plan, then only limited action will be required. It must be appreciated that an element may be designed and redesigned several times in order to achieve the correct solution. In a similar way cost checking will also replicate this process. Where a cost check reveals differences from the target cost, then different courses of action are required as follows:

- Redesign the element so that the target cost for the whole scheme is not affected.
- Approve the change in the element unit total and amend the overall cost plan amount.
- Approve the element unit total but examine ways in which other element unit targets can be amended so as to leave the cost plan total unchanged.

It is always dangerous to assume that reductions in the costs future elements, yet to be designed, can be achieved. The element unit totals should not be adjusted unless there is good evidence that this is the case.

The cost checking should be carried out as soon as the design details are received, in order to attempt to reduce any future design work, should the element total be too costly. The advantage to clients of this process is to provide them with the full implication of design decisions that they approve. It provides a reasonable assurance that the budget estimate will not be exceeded. The greatest amount of attention by the quantity surveyor should be given towards the cost-sensitive elements, but this does not imply that other less important elements should be ignored. The pricing of cost checks will be carried out using current market prices. During the later stages of the process, the costs of specialist works will be replaced with firm quotations from suppliers and subcontractors.

Tender reconciliation

Where the process of cost checking has been carried out thoroughly then the receipt of tenders should provide few surprises. It is nevertheless appropriate to carry out certain checks on the tender to be accepted, to highlight any differences between this and the final cost check. This will provide some insight into the cost planning of any future projects. Any discrepancies between the two should be easily explained.

Errors on the part of the quantity surveyor and the builder's estimator do occur. The deliberate distortion of prices by the contractor may be made for possible future gain. The quantity surveyor will report on the sufficiency of the contractor's prices which will include a technical and arithmetic check of the tender.

Post-contract cost control

The cost planning process does not cease when the contractor starts work on site. The design is likely to have stayed within budget, as evidenced by the accepted tender. Whilst the design may change little, there are provisions for variations authorised by the architect or other designers. Some of these changes will be due partly to site conditions and revisions instigated by the client. The client's main financial interest will now be to ensure that the budget and final cost remain in agreement. Whilst the budget and tender cost will have broadly achieved this, changes and unforeseen items of expenditure can result in unacceptable differences.

Cost planning during construction is normally achieved through the regular preparation of financial statements. These will advise the client of any probable changes between the agreed tender sum and the probable final cost of the project. The size and complexity of the project will determine how frequently these statements are prepared. The financial report is in two stages. The first considers the current financial position of the project and the second the expected final cost based upon adjustments to the contract sum. These may include the costs of any variations, adjustments to prime cost and provisional sums, daywork accounts, increased costs and contractual claims.

Whole life costing

Whole life costing involves the application of established economic techniques to the decision-making process associated with the design and commissioning of capital works projects. The combination of initial capital expenditure and future costs-in-use may be fitted to the constraints of the client and the project under consideration. It is a trivially obvious idea, in that all costs arising from an investment are relevant to that decision. The image of a life cycle is one of progression through a number of different phases.

The pursuit of economic whole life costs is the central theme of the whole evaluation. The method of application incorporates the combination of managerial, functional and technical skills in all phases of the life cycle. The proper consideration of the costs-in-use of a project is likely to achieve improved value for money and improved client satisfaction. Different objectives may be set at different times of the project's life.

Whole life phases

The sequence of the seven phases of a building's life is described appropriately in British Standard 3811, and whilst this adopts engineering terminology, that definition can also include the physical assets of buildings as shown in Table 9.5.

Table 9.5 Whole life phases

Phase	Description	Cost implications
Specification	The formulation of the client's requirements at inception and briefing. Feasibility and viability of different proposals	Initial costs associated with land purchase, professional fees and construction.
Design	Translating ideas into working drawings from outline proposals scheme and detail Design	Cost planning including whole life costing of alternative design solutions Associated contract procurement documentation
Installation	The construction process	Interim payments and financial statements
Commissioning	Handover of the project to the client	Final accounts
Maintenance	The project in use	Recurring costs associated with repairs, running and replacement items
Modification	Alterations and modifications necessary to keep the project to a good standard	Costs associated with major refurbishment items
Replacement	Evaluation of the project for major changes or the site for redevelopment	Redevelopment costs

(Source: Ashworth, 2004)

The major issues associated with whole life costing

During life cycle costing there are many difficulties that need to be resolved. Since it is a forecasting technique and forecasts are invariably not precise, some added judgement needs to be applied. The quality of the life cycle cost forecast is determined by the availability and reliability of the data and information that are used in the calculations and the skills that are employed by the practitioner when making judgements.

Building life

The life cycles of buildings are diverse during their inception, construction, use, renewal and demolition. There also lies a varied pattern of existence, where buildings are subject to periods of occupancy, vacancy, modification and extension. A building structure may be designed using materials, components and technology that may last for about 100 years or more depending upon the quality and standards expected from users. There are numerous examples of buildings that exceed this time span. However, the engineering services components in buildings have a much shorter life with an expectancy at the most of about 15 years and the life expectancies of finishes and fittings are now frequently less than 10 years. By comparison, information technology hardware and software systems are becoming outdated even after a period of only 3 years.

The determination of building life expectancy is of fundamental importance in a life cycle cost calculation. However, in practice only limited consideration is given towards the assessment of building life expectancy at the time of its inception and design. By contrast engineering systems are more carefully designed to meet expected and determined life cycle predictions. Materials and methods of manufacture, assembly and construction are selected on this basis to coincide with predicted life spans. Many industrial processes are reliant on life expectancy, often assuming that rapid changes in technology will make some processes obsolete and there is little point in attempting to design processes beyond limited life spans.

Component life

The life span of the individual materials and components has a contributory effect upon the life span of the building. However, data from practice suggests widely varying life expectancies, even for common building components. It is also not so much a question of how long a component will last, but of how long a component will be retained. The particular circumstances of each case will have a significant influence upon component longevity. These will include the original specification of the component, its appropriate installation within the building, interaction with adjacent materials, its use and abuse, frequency and standards of maintenance, local conditions and the acceptable level of actual performance required by the user. The management policies adopted by owners or occupiers are perhaps the most crucial factors in determining the length of component lives. There is a general lack of such characteristics in retrieved maintenance data.

The design must recognise the difference between those parts of the building with a long, stable life and those parts where constant change, wide variation in aesthetic character and short life are the principal characteristics. There seems to be little merit in including building components with long lives in situations where rapid change and modernisation are to be expected. All components have widely different life expectancies depending upon whether the physical, economic, functional, technological or social and legal obsolescence is the paramount factor influencing their life.

Figure 9.1 shows the variation in building component life for softwood windows. It is based upon a sample of building surveyors' findings. The life expectancy of softwood windows and doors, from this survey, varied between 1 and 150 years. Typically it represents a life expectancy of about 30 years. Furthermore, it would be foolish, for example, to prepare a life cycle cost based upon 150 years, even where guaranteed maintenance is promised, owing to the possibility of advancing obsolescence in buildings as identified

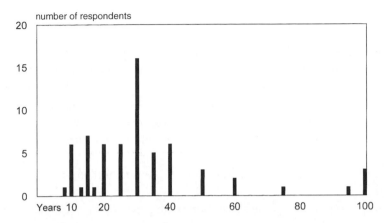

Figure 9.1 Life expectancies of softwood windows. Responses were based on a sample of experienced building surveyors. Mean life expectancy 32 years (Source: RICS/ BRE, 1992)

above. Changes in use, the implications of fashion and the development of new technologies will also have some impact on life expectancies.

The survey does not provide an indication of the possible reasons for the expected different life expectancies. The replacement of the windows may be due to general decay, vandalism, fashion, the installation of double glazing, in order to reduce long-term maintenance, development of new technologies, etc. These and other data characteristics are not provided. If this information were included, then the range of values under a particular set of circumstances would be reduced. This would then allow its reuse in new situations to be made with greater confidence. On the basis of this and other information alone, it is not possible to select a precise life expectancy for a particular building component. Different techniques, such as sensitivity analysis or simulation can be used to test the effects of best, worst, typical and any other scenario in terms of assessing the life expectancy.

Sources of life span data

A number of organisations including research groups, professional bodies and manufacturers provide information on building component life. However, an important point is that the prediction of component lives for life cycle costing purposes is not so much concerned with how long a component will physically last, but how long it will be retained. The scientific data is almost solely concerned with component longevity and not with obsolescence. Whilst manufacturers and trade associations offer a valuable source of information it needs to be remembered that the component life of a product may be described under ideal or perfect circumstances that rarely occur in practice.

Inflation

Throughout almost the whole of the twentieth century the United Kingdom experienced erosion in the purchasing power of the pound. Much has been written on the causes and its possible cure. The effects of inflation and the problem that it causes to capital investment decisions need to be taken into account in a life cycle costing comparison. Even with relatively low levels of inflation, prices will be substantially affected over long periods of time. The following are some of the characteristics of inflation.

- Inflation refers to the way that the price of goods and services tends to change over time.
- Inflation causes money to lose its purchasing power because the same amount buys less.
- The most commonly used measure of inflation in the UK is the *Retail Prices Index* (RPI). This is supposed to measure the costs of goods and services of a typical family's spending.
- The *nominal* rate of return on an asset or investment is the amount you get back; the *real* rate of return is the return after inflation has been taken into account.
- Cash deposits such as savings accounts, although secure, do not keep pace with inflation.
- Interest rates are used to control inflation. By raising interest rates, governments can dampen consumer spending which results in reducing economic activity.
- Low inflation is believed to be good because it leads to price stability.
- The opposite threat of deflation is considered to be just as much a threat as inflation.
- Zero inflation is rarely desirable. The level of interest rates needed to achieve this would probably discourage economic activity.
- Europe measures inflation using a harmonised index. If this were adopted in the UK, Britain's inflation record would look much better.

The principal problem facing the decision-maker is whether to forecast future cash flows associated with an investment project in real terms or in money terms. Real terms here means in terms of today's (the date of decision) price levels. Money terms refers to the actual price levels which are forecast to obtain at the date of the future cash flow.

Two different approaches may therefore be used to deal with the problem of inflation. First, inflation could be ignored on the assumption that it is impossible to forecast future inflation levels with any reasonable degree of accuracy. The argument is reinforced in that there is often only a small change in the relative values of the various items in a life cycle cost plan. Thus, a future increase in the values of the cost of building components is likely to be matched by a similar increase in terms of other goods and commodities. There is therefore some argument for working with today's costs and values. Also, since we are attempting to measure comparative values real costs can perhaps be ignored.

However, changes in costs and prices and their interaction with each other are not uniform over time. Also, property values tend to move in booms or surges whereas changes in building costs are much more gradual. Costs do not necessarily increase in line with inflation. Reference to a range of different material or component costs over a period of time will show that these do not follow a uniform trend or pattern. Even similar materials, such as plumbing goods, can show wide differences even over a ten-year cycle of comparisons. To ignore such differences will at least create minor discrepancies in the calculations.

The alternative approach in life cycle costing is to attempt to make some allowance for inflation within the calculations. This may be done using evidence of market expectations, published short and long-term forecasts and intuitive judgements relating to the prevailing economic conditions. It is worth remembering, that in common with all forecasting there will be a degree of error. The forecasting of inflation is both a science and an art in its own right. Mathematical models are constructed using a wide range of data to assist in their predictions. The models can only consider future events that may occur. In reality events occur that could not be predicted even a few years earlier.

Discount rate

The selection of an appropriate discount rate to be used in the life cycle costs calculations will depend upon the financial status of each individual client. The discount rate to be selected will be influenced by the sources of capital that are available. The client may intend to use retained profits or to borrow from one of a number of commercial lenders.

It can be argued that the choice of a discount rate is one of the more crucial variables to be used in the analysis. The decision to build or to proceed with an investment may be influenced by the discount rate that is chosen. The selection of a suitable discount rate is generally inferred to mean the opportunity cost of capital. This is defined as meaning the real rate of return available on the best alternative use of the funds to be devoted to the proposed project. In practice the discount rate that is selected often represents the costs of borrowing, whether from the firm's own funds (loss of interest) or at a higher rate through borrowing. The discount rate that is proposed should then be adjusted by the expected rate of inflation or the time that the project remains live.

It is very important that for each option that is being considered, the respective cash flows are calculated on exactly the same basis. If cash flows are to be estimated in nominal terms, i.e. include an estimate for inflation, they should be discounted at a nominal discount rate. This should then be applied to all of the options under consideration. It is difficult to be definitive regarding which approach to adopt. Where cost estimates are assumed to inflate at the same rate then it is preferable to perform all calculations in current prices and to apply a real discount rate. However, where inflation is expected to operate differentially then the calculations should be done in nominal terms with explicit account then being taken of the differential rates of inflation.

To select too low a discount rate will favour or bias decisions towards short-term, low-capital cost options. A discount rate selected that is too high will give an undue bias towards future cost savings at the expense of higher initial outlays. The most 'correct' discount rate should reflect the particular circumstances of the project, the client and prevailing market conditions. It is all too easy to tamper with the discount rate to make the calculation reflect the desired outcome. It is a matter of judgement but one that is done within the context of best professional practice and ethics.

Taxation

Cash flows associated with taxation must be included in the calculation for practical reasons. Most projects will have different corporation tax effects. This may be due to capital expenditure attracting relief through capital allowances, profits from the project resulting in additional taxes or losses attracting tax relief. Tax is not assessed by the Inland Revenue project by project but for the company as a whole. Cash flows must therefore be considered in this context and calculated on whether the project is carried out, delayed or abandoned. The matter is further complicated since the project may be spread over one or more tax years. Careful accounting may result in beneficial effects through tax avoidance measures.

Capital allowances are set against taxable profits in order to relieve the expenditure on fixed assets. Statute provides for several categories of asset among the various types of business fixed assets. Each of these has its own rules and basis for granting the allowance. In practice, they are a combination of writing down allowances and balancing charges.

Sensitivity analysis

During a life cycle cost analysis a large number of different assumptions need to be made, such as building life expectancy and the longevity of the building components. While historic data may be used, the variability of such makes calculations difficult to perform. It is necessary therefore to test the judgements or assumptions in order to reduce as far as possible any distortions or misleading information and data that may have been introduced. One way of testing whether the results of the life cycle cost analysis remain stable under varying conditions is to repeat the calculations by changing the values that have been attributed to the individual variables. This is a technique that is known as *sensitivity analysis*. In practice changing the values of the different variables used will result in changes to the overall outcome.

There are two different scenarios that can be employed. The first is to make changes to the life cycle cost model resulting from variations in the design, the materials used or the construction techniques to be employed. This will result in alternative life cycle costs being calculated. It may be obvious from these calculations which is the preferred alternative design or construction solution. However, the results may be so similar that expert judgement needs to be applied in making the final design decision. The alternative approach is to provide life cycle cost models that test the stability of the model under varying circumstances. For example, the comparison of two alternative designs may reveal that design A is always preferred to design B in terms of the life cycle cost under normal circumstances. But the future might not necessarily be normal as we understand it today. The alternative approach is therefore to test the model at or even beyond the extremes of possibility to ensure that it remains stable under all conditions. This will rarely be the case in the real world and under different circumstances the alternative solution will be preferable.

It is therefore possible to test the effect on the life cycle cost of any variable used in the calculation. It needs to be remembered that life cycle costing is a technique to assist in the selection of the best or most economic of the alternatives that are available. Sensitivity analysis alone will not do this for us, but it will provide us with a range of different values that can be used to help formulate an overall judgement. Sensitivity can also only be applied under known or expected circumstances. There are many examples where the future cannot yet be imagined or described.

Least cost solutions

Traditionally the focus on construction costs was to ensure the lowest tender from a construction firm. This was the main consideration. The introduction of cost planning refocused some of our ideas towards adding value or value for money as it was then termed. This recognised that the construction solutions could be improved by examining more carefully the spatial design and the construction methods to be used to help achieve the same objectives. It also soon became recognised that to ignore the recurring costs of ownership was foolish. If it was possible to spend a little more at the design stage that would result in year-on-year cost savings that outweighed the extra initial construction costs, then this would be a good policy and practice. While cost planning has demonstrated conclusively that more could be obtained for less, this was often set within the parameters of a cost target. Once this was achieved, then no further cost savings were sought. The development of value management and the lean construction methods (see Chapter 13) saw no boundaries but sought to remove waste and hence costs from the construction process. Facilities management further refined this by adding in the costs associated with operations of the project which were often very diverse from *construction* alone. Further practices are now refocusing on least cost solutions since often in the above scenarios the costs of professional fees were sometimes excluded as were financing and land costs.

Chapter 10
Budgeting, Costing and Cash Flows

Introduction: business plans

The majority of business failures occur in the initial stages of operation. It is then that there is the greatest potential for unaccounted factors to occur and produce problems. It is the normal practice therefore for a bank manager to request a business plan together with a forecast of the expected cash flows. The business plan should include the following:

- Information on the product and an assessment of the likely market with supporting information.
- The potential size of the market and the competition expected from other firms.
- An appraisal of the prices being charged for the product or service and the extent to which these fluctuate at different times of the year.
- The likely costs of production, including the need for plant, machinery and equipment, materials, labour and overheads. It should also anticipate the potential for any changes in prices or the availability of these resources.
- The projected level of profit that the business seeks to achieve, and how satisfactory this will be when compared with other forms of investment that might be available.
- The proposals regarding the management of the business, the experience of the people involved and their responsibilities for the different aspects of the business.
- The business strategy will also outline the possible future developments of the firm in the short and medium term as well as possible expansion and growth together with contingency plans should these original ideas not come to fruition.

Budgetary control

Many businesses do not make a profit or have enough cash available at the right time because the management has not planned ahead. Too often, they do not know how much profit or loss has been made until months after the end of an accounting period. Often profits are not properly thought out until there is a crisis.

Budgetary control is the process of setting and monitoring the short-term objectives of different aspects of the organisation's operations. It involves the day-to-day financial management of the different departments of the business and is based upon individual targets, goals and objectives. It is essential to achievement of the overall plan, whether for the company as a whole or the individual building site and its various operations. Essentially a budget has three distinct aims:

- To allow the organisation to meet its objectives through the coordination of a range of activities
- To allocate the appropriate level of finance to allow the achievement of these objectives
- To permit the efficient management of the organisation's financial resources and ensure that it is aware of the extent and timing requirements for finance.

Without detailed adherence to individual budgets, the organisation's wider objectives may be hindered. As an integral part of the overall plan, it must progress through a number of stages.

Budgeting is a useful planning tool that helps to make an organisation run more smoothly and profitably. Budgets are used for planning and controlling the income and expenditure in many different organisations. It is through the budget that a company's plans and objectives can be converted into quantitative and monetary terms. Without these a company has little control. The overall budget is known as the master budget. This comprises the different departmental budgets. Anyone who influences cost should be provided with a budget against which to measure their actual expenditure. Budget headings must exist within a well-defined structure to avoid the possible overlap of virement (transfer of items from one account to another) where this is not desirable. Budget targets may be set by management, through negotiation with the budget holder or on the basis of the past performance of the budget heading. However, if budgets are always based solely upon the previous year's expenditure, inefficiencies and over-spending may be carried forward from one year to the next. Wherever possible, the budget should be re-examined and new targets set on the basis of the proposed expenditure targets set in the strategic plan.

Successful budgeting is essentially people-centred and depends upon a number of different factors, which include:

- cooperation and communication between budget holders
- targets that are perceived to be realistic, achievable and fair
- managers' objectives consistent with the overall objectives of the master budget
- feedback should be constructive and supportive rather than critical.

The budget reports should be:

- In a format that is easily understood
- Produced on a timely basis to allow as much corrective action as possible to take place
- Accurate in order to maintain the budget holder's confidence in the system
- Increasingly detailed for the lower aspects of the organisation
- Supported by frequent meetings with management.

Advantages of budgeting

Budgets can be time consuming to prepare and review. Where they are determined solely from management's perspective they may also constrain and discourage other individuals in the organisation. However, they do have significant advantages, some of which are as follows:

- The organisation's aims and objectives are clearly set out in financial terms.
- The budget provides a basis on which to compare financial performance.
- Early warning of possible problems are highlighted so that corrective action can be taken.
- Interdepartmental conflict can be reduced.
- The budget can provide a positive motivating stimulus.

It is important that the various subheadings include a time scale since the expenditure under a budget heading may run for a few months or extend over a few years. This is especially the case in the construction industry where the contract period for the project is of at least equal importance to the calendar year. While the contractor will have a work programme for the project and this can be costed, the procedure may be disrupted by delays on the part of all those involved and through changes (variations) to the original scheme and these will need to be taken into account.

The budgetary information will give a rate of expenditure and a rate of income throughout the project, and by deducting income from expenditure the amount of capital required at the different times can then be calculated. This information can be compared with the actual performance, and differences can then be easily spotted in order that action can then be taken that is appropriate to the company. This is discussed later in the sections on Contractor's income and expenditure curves and Project cash flow. The contractor will also need to aggregate this information from all projects in order to determine the company's overall position. Budgetary control is a continuous process undertaken throughout the contract duration. When variances occur from the budget, the contractor will need to assess the reasons for such differences.

The building client's budgetary control procedures are somewhat similar, although the control mechanism is different. The client's main concern is with the total project expenditure which has been forecast. The ability to control this depends upon the sufficiency of pre-contract design, the number of subsequent variations, the steps taken to avoid unforeseen circumstances and matters which fall outside of this control such as strikes, etc. The client will also be concerned about the timing of expenditure for funding purposes, but in this case although control will be influenced by the above factors, the contractor's method of construction will also be influential. In the case of a large building client such as a county council, the interaction of the many different projects that are under construction at any one time will need to be aggregated together, in a similar way to all of the contractor's projects, to establish the total periodic payments which are required. For a large client or contractor, research has indicated that when such projects are grouped together this often produces similar cash outflows for each month of the year. The importance of this information for the building client will influence scheme approvals, starts and completions.

Costing

Costing is a control procedure used by the building contractor and involves analysing what was planned with the work that actually has taken place. Without an adequate ongoing costing system, it will be necessary to await the preparation and agreement of the final account in order to establish the profitability or otherwise of the company. The main purpose therefore of this procedure is to reveal the efficiency or otherwise of specific identifiable activities, usually by comparison against a preset budget, in order to allow the appropriate action to be taken.

Site cost control is concerned with the control of those variables on site which can be analysed in order that action can be taken to either improve productivity or reduce costs. The construction industry is unlike manufacturing industry, essentially because it deals with one-off projects usually on the client's own premises. Even where projects might be described as of a repetitive nature, the different site conditions and especially weather conditions require each project to be examined independently. The cost control systems used in manufacturing industry might not therefore always be appropriate, since work that is undertaken in factories is done under more controlled situations. Any system of site cost control will have the following three facets:

(1)   The selection of desirable and achievable targets for the work to be carried out
(2)   Comparisons with these targets of the actual performance
(3)   A means of corrective action should performance fail to meet the expected criteria.

The system used should seek to identify the problems that arise, provide indicators of these and allow the corrective action that is necessary to be taken. Each individual site will have its own peculiarities but the cost control system should be sufficiently detailed to allow the project to be analysed in the form of work packages. Perhaps one of the biggest difficulties to overcome is that of the time–lag delay between the time the work was carried out, recorded and eventually submitted to the site management. A good cost control system will thus be designed with an early warning system to allow the appropriate corrective action to be taken before it becomes too late to do so. Site management will not be able to take the necessary corrective action if the performance data only reaches them some time after the site operation was completed. In this case the information is only of value for record purposes or perhaps in support of a claim for additional payments.

The costs of a construction project can be considered in several different ways. They may be identified in the manner outlined in Figure 10.1. This shows the diagrammatic representation of the unit rate. Alternatively, they may be considered in the context of a hierarchical structure or levels of different costs shown in Figure 10.2.

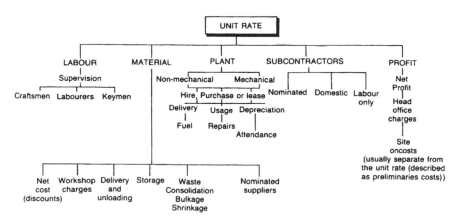

Figure 10.1   Diagrammatic representation of a unit rate (level 6 Figure 10.2)

Types of costs

Direct costs

These can be directly related to actual units of production on a construction site. They may, for example, include items of labour, materials or plant and their respective costs. In theory they should be able to be identified easily and precisely. In practice, owing to the complex numbers of items against which costs can be recorded and the poor recording and clerking procedures employed they inevitably and always will include errors which relate to an incorrect allocation.

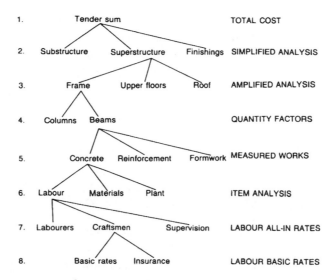

Figure 10.2   Hierarchical structure of cost data

Indirect costs

These cannot be allocated to individual items of work since they represent costs of a more general nature. Their costs nevertheless count towards the items of production. They must therefore be allowed for within the overall costs of production of the project. Furthermore, these items are much less precise in terms of their costing and it is difficult to even allocate some of these items realistically to individual projects. They include some of the items described as preliminary costs or site on-costs, the contractor's general overheads (head office charges) and expenditure costs that are incurred away from and separate from the project. Many of these items can therefore only be allocated to the projects on some form of rough proportion basis or by means of a calculated percentage.

Cost centres

In attempting to classify construction costs the main objective is to identify a particular activity or operation of work to which such costs can be allocated. It is desirable that these activities are easily defined and, wherever possible a clear line of demarcation can be drawn between adjacent or similar activities. These activities are often referred to as cost centres. Within the construction industry it is possible to attempt to identify a whole range of different cost centres for different activity purposes. Individual contractors have devised their own classification systems to suit their own individual needs. The building industry, for example, has sought to identify cost centres as building trades, elements or site operations. It is also important to be careful when using this terminology, since different industries may use these words in a different context or place a particular emphasis upon their meaning. The costs which are allocated to the cost centres can be further analysed into the following five sub-categories:

- Labour costs
- Material costs
- Subcontractors
- Plant
- Overheads.

Labour costs

These costs include all of the labour charges that might normally be allocated to building operations. They include the labour costs associated with manufacture and assembly on site, but exclude off site operations particularly in respect of the costs associated with standard stock production items such as windows and doors. The cost control of these items of production is more akin to manufacturing industry and is dealt with in that way separately from contracting. Labour costs on site include foremen, craftsmen, gangers, labourers, machine operators, etc. They will also include labour costs expended in the contractor's own workshops. They will also include labour only subcontractors since these will contribute towards the contractor's overall site expenditure. The costs of labour should include all of the items associated with the employment such as national insurance, sick pay, bonus and productivity schemes, holiday pay, severance payments, any training levies, etc. They include any payments that are required under the National Working Rule Agreement.

The items included within these costs are those that are normally considered in the analysis of the contractor's all-in labour rates used for estimating purposes. The actual costs to the building contractor of employing labour are frequently about twice the cost of the basic rates of pay. This will vary between the conditions applied by individual contractors, relevant legislation and the availability of labour. The reliable prediction of the labour costs on a construction project is the most difficult part of the estimator's job. Salaried site staff and the costs of head office employees are normally excluded from this cost centre heading but are recovered under a building contract as either head office overheads or as a lump sum within the site on-costs.

Material costs

The costs of materials include their net costs delivered to the site or to the contractor's workshops. Added to this are the costs of storage and an amount to cover the difference between the quantities of materials purchased and those eventually paid for by the client. This latter factor of wastage of materials is in practice frequently underestimated at the time of tender. It must also be remembered that the measured quantities that are included in bills of quantities, for example, are net – exclusive of any bulking or shrinkage factors. The building contractor will normally pay for these materials a month in arrears with the settlement of the supplier's account resulting in a cash discount off the trade prices of at least 2.5%. Value added tax will need to be allowed for where this cannot be reclaimed from Customs and Excise. The estimator will also need to have made some allowance for buying margins, where, for example, the price allowed for at the time of tender is different from the price actually paid for the materials when delivered to the site. Vandalism, materials deterioration and misuse of materials will also need to be considered and allowed for under this heading.

Subcontractors

The control of subcontractors' costs, where proper documentation has been prepared in the first place presents the main contractor with fewer problems of a financial nature. The contractor will relinquish some control and profit, but there are immense benefits in employing subcontractors to carry out the work on the main contractor's behalf. Hence subcontracting has become a norm in the industry with main contractors using relatively little directly employed labour.

The main contractor's costs for organising, supervising and accepting the subcontractor's work on completion are partially recouped through the cash discount of 2.5–5%, the allowance for the contractor's profit on this work and the costs allocated to general and other attendance items. In addition, the main contractor may be able to use the subcontractor's due payments for a short period of time and the main contractor does not need any general outlay of cash in order to finance the work.

Plant

This can be classified as either non-mechanical, mechanical or small tools. The latter is typically regarded as a general overhead expense, although in some instances it can also be included as a site on-cost. Large items of plant, such as tower cranes, may be charged for in the preliminaries section of a bill of quantities or where they are directly related to a specific item of work, as in the case of earth moving they can be included in full against the measured quantity of work. Medium to large contractors will hire the plant from either a separate plant hire firm or through one of their own subsidiary plant companies. The choice between purchase, hire or lease needs to be correctly evaluated for each project. Often, a combination of these alternatives is used on large projects. Unless the building contractor can foresee a large plant utilisation then it is usual not to purchase. In any event it may not be desirable economically to lock up huge amounts of capital in mechanical plant. The capital may be better utilised elsewhere.

Overheads

These items represent the costs associated with managing a company or of facilitating the construction project. They include the costs of maintaining a head office, workshop and off site storage compound for plant and materials. They are recovered from a project by means of a percentage addition to the costs directly associated with the construction project together with the site on-costs. The percentage applied can only be approximate and relates to the turnover expectations of the company. While the real percentage, if this could be measured, will vary between the different projects that are under construction at any one time, it is not really feasible to attempt to disaggregate these costs and to allocate them independently to the individual projects in turn.

The purpose of costing

The contractor's site costing system is a part of the whole process of financial control on site. The process commences with the analysis of costs that are required during the estimating process. After the contractor has submitted a successful bid for the work, then budgets for the various parts of the project are established by which to help plan and execute the work and to use as a yardstick against the actual costs incurred.

Costing is used as the means of this site measurement. The costing process consists of recording all those expenditures which are related to specific items of work in such a way that the cost of the work can be readily identified. The final part of the process is that of accounting, which is a process that centres around the business activities of the firm, by ensuring that it receives and makes the right kind of payments at the appropriate time. This latter part is largely an historical process. Costing has a number of objectives.

- To provide cost data or feedback for future management and estimating purposes.
- To enable appropriate budgets and targets to be set for future work that is to be undertaken.
- To highlight uneconomic activities of working either:
  - for the purpose of reappraising how the work might be carried out more effectively and economically;
  - to support the basis for a future claim for an additional payment, where the work being executed is different to that which was originally envisaged.
- To identify any excessive wastage of materials, the inefficient use of plant or the appropriate application of labour.

• To provide information which will enable the firm to undertake work that it is best equipped to carry out.

Variable and fixed costs

It is very easy to attempt to simplify construction costs and to misinterpret their possible cause and effects. A bill of quantities or measured work package is criticised for many reasons, although it does attempt to measure and describe construction work in the context of the costs involved. Often criticism arises when attempts are made to use the bill of quantities for purposes for which it was never designed. In essence, measured quantities of construction work are only really useful on those projects that have been properly designed. Rightful criticism has been expressed in the past by building contractors where the bill of quantities had measured work that had not yet been designed. Costs can be broadly described as follows, with quantity related and time related costs being identified as variable costs.

Quantity related costs

The costs of a particular site operation are related to the quantity of work performed. The majority of work on building sites falls broadly into this category. Cost over a given range of quantity values is assumed to be linearly related. For example, if 100 m^2 of a particular item costs £100 then 200 m^2 will cost £400. The assumption of this linear relationship can only be made over a limited range of values, since the relationship of 10 000 m^2 at a cost of £40 000 or 1 m^2 at £2 are both likely to be invalid. Many different factors affect the cost–quantity relationship, with quantity itself being an important variable to consider. Minor changes in quantity between that which was forecasted and that which actually takes place will however, generally easily fit within the quantity-cost relationship. Figure 10.3 indicates the typical relationship between cost and quantity over a range of values.

Time-related costs

Some costs associated with building operations are more related to time than quantity. Scaffolding, for example, once erected on site will be charged for largely on the basis of time. The hire of most equipment is based often on a time related charge rather than in relationship to any quantity of work that might be performed. However, similar principles exist as with those for quantity–cost related items in that the relationship is linear only over a given range of values. For example, plant required for long periods of time will be hired at rates far below those where the plant might be required for just a few weeks.

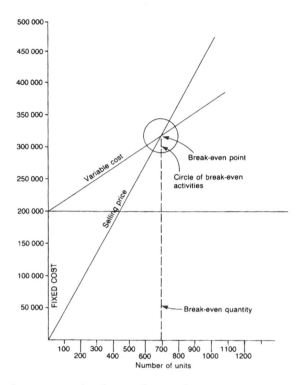

Figure 10.3   Break–even graph of manufactured joinery components

Fixed costs

Some costs do not relate to either time or quantity. They often exist even when no work is being undertaken. These are often referred to as fixed cost items. They include, for example, rent, rates, heating, lighting, insurances, etc. They also occur on the building contractor's site. For example, bringing items of plant to site constitutes a fixed cost. This cost is expended and does not directly relate to the quantity of work to be performed nor to the time that the plant might be on site. It does not matter what quantity is excavated by a dragline, nor how long the process of excavating might take, the fixed costs of bringing the item of plant to site remains the same. Insurances that are often priced as a lump sum are, however, not strictly fixed costs since their calculation will have been based upon the quantity of work being performed and the amount of time that is necessary to complete the work.

The contractor can, of course, choose to allocate the costs involved in any way that is considered suitable. Essentially, however, construction site operations, either of a temporary or permanent nature, can have their costs analysed in a combination of the above three ways. In analysing bill rates, it is necessary to be able to identify the costs concerned separately in the above ways.

Costing methods

Unit costing

The basis of unit costing for a particular site operation is the calculation of expenditure either in cash terms or on the basis of the hours involved. For example, if it takes 5 men 7 hours to place and consolidate concrete in a wall 10 m long × 4 m high × 300 mm thick, then the unit time for placing and consolidating concrete in this situation can be calculated as:

$$\frac{5 \text{ men} \times 7 \text{ hours}}{10 \text{ m long} \times 4 \text{ m high} \times 300 \text{ mm thick}} = 2.92 \text{ hours/m}^3$$

This sort of data is useful for future estimating purposes. If the all-in costs of labour are £7.00 per hour, then the hourly cost of labour in money terms will be £20.44.

Where similar operations are being carried out, the unit costing approach will allow site management staff to compare these costs from one week to another, examining any differences that occur. Factors other than quantity will affect the unit costs, so that it would be unusual to find the figures constant, even though the same operation was being carried out. The men, for example, would be on a learning curve and one would expect the unit costs to decrease in successive weeks. However, there are other factors involved and these might cause the unit costs to increase in some of the weeks. It needs to be emphasised that the building site is not an indoor factory where these other factors could be kept to a minimum. Building sites have unique characteristics and even though projects may appear to be repetitive they seldom are what they seem. Where the unit cost shows a considerable variation then this will need to be investigated to establish whether the variations were due to factors beyond the control of the building contractor.

Standard costing

The definition used by the Chartered Institute of Cost Management Accountants states that 'Standard costing is the preparation and use of standard costs, their comparison with actual costs and the analysis of variances to their causes and point of incidence'. This definition helps to highlight the similarity with budgetary control mechanisms.

The variances calculated are then analysed and where possible, action is taken to counter the reoccurrence of excess costs in the future. Standard costing has been used more successfully in manufacturing industry, where controlled conditions exist and in those companies where there is a considerable amount of repetition in their production.

A standard cost is calculated through normal work study procedures or alternatively on the basis of normal distribution data from records retained from previously similar site operations. These standards are then set out on a cost record sheet against which the actual values are entered and verified. A simplified example is shown below.

> The total number of facing bricks required to build a speculative house of consistent design is estimated to be 3500. These bricks cost the building contractor £150 per thousand, delivered to site. However, the actual number per house is calculated to be 3750, on the basis of purchased quantities, and these cost the building contractor £155 per thousand. The material cost variances can therefore be calculated as follows:

Standard cost $3500 \times £150.00$ per $1000 = 525.00$
Actual cost $3750 \times £155.00$ per $1000 = 581.25$
Total variance $= 56.25$

This can be further analysed to show

(1)   Quantity cost variance (cost variance)
        $3750 - 3500 = 250 \times £150$ per $1000 = 37.50$
(2)   Material cost variance (usage variance)
        $3750 \times (155 - 150$ per $1000)$ $= \underline{18.75}$
        Total variance $= 56.25$

The cost variance can be used by the purchasing department in order to assess their ability to obtain the lowest possible prices. The usage variance has either resulted from inaccurate estimating at the time of tender, because of poor control by site management staff or because changes have occurred to the design resulting in a change in quantities or a higher wastage of materials.

Standard costing is therefore likely to be of most use in the construction industry where repetitive tasks occur, for example:

- Manufacture of off-site joinery components
- Installation of services in standard houses
- Erection of shuttering to repetitive structures
- Assembly of prefabricated components.

Marginal costing

This technique seeks to differentiate between marginal costs or variable costs which occur in connection with production and therefore tend to vary directly with any variations in production. Fixed costs tend to remain static, almost irrespective of production levels. Marginal costing therefore seeks to identify and to distinguish between direct and indirect costs. Even where no active production is being carried out, fixed charges are still likely to occur. It will then be realised that until a sufficient level of productive work can be achieved then no profit can be made. The higher the number of production units manufactured the greater the profit, or the lower the price. Volume manufacturing therefore results in the fixed costs being distributed on a lower proportionate basis.

Example

A manufacturer of standard joinery components, such as doors and windows, has fixed annual costs of £200 000 in respect of premises, machinery and other overhead costs. The variable manufacturing costs in respect of labour and materials represent £180 per unit. It is anticipated that 1000 units per annum will be manufactured, each selling at £475. On the basis of these assumptions break even will occur after manufacturing 678 units (see Fig. 10.3). Assuming, however, that all of the joinery units are sold then an overall profit on cost of 25% would be expected.

Costs	Fixed	= 200 000
	Variable 1000 units × £180	= 180 000
	Total	= 380 000
Price	Variable 1000 × £475	= 475 000
	Difference (profit)	= 95 000

If more of these joinery units can be manufactured and sold, then the price will consequently fall should the same profit percentage be applied. For example, 1200 units could be sold at £433.33 each based on the following.

		£
Costs	Fixed	= 200 000
	Variable 1200 units × £180	= 216 000
	Total	= 416 000
	Plus 25% profit	= 104 000
	Total	= 520 000
Price	Then 1200 units will each be	
	sold for 520 000/1200	= 433.33

Where more units can be manufactured and sold then the price will reduce accordingly, where fixed costs remain unaltered.

Absorption costing

This is a method of costing which charges a share of all the overheads (variable and fixed) to the items produced. The Chartered Institute of Management Accountants defines it as 'A principle whereby fixed as well as variable costs are allotted to cost units and total overheads are absorbed according to activity level'. The distinguishing feature of absorption costing is that the total cost of an item is calculated by including or absorbing a share of the fixed overhead cost. The allocation of company overheads to the individual measured or work items on a construction project is typical of absorption costing.

Break-even analysis

This technique is concerned with calculating the break-even point between two or more operations or activities. It generally involves the preparation and interpretation of a break even chart. The points of intersection on this chart are described as the break even point.

Figure 10.3 illustrates a typical example of a break-even chart. In the construction industry a large number of site operations comprise a fixed cost, a variable cost and a price. The fixed cost may, for example, represent site set up charges which are largely unaffected by the actual quantity of work that is involved, once they have been established. Variable costs, on the other hand, are believed to have some direct relationship with quantity (other factors being excluded). Price is often established in direct relationship to quantity alone over a given range of values. In order to determine this price or rate for the work involved, some form of break even analysis is desirable.

There are many different applications of this technique. For example, it can be used to identify the economic choice amongst a number of competing alternatives. Figure 10.4 shows the cost in relationship to quantity for three products that all meet the same specification. Where relatively small quantities are required then system A should be selected. If large quantities are required, then the break even chart indicates that system C is the most appropriate on the basis of cost selection criteria.

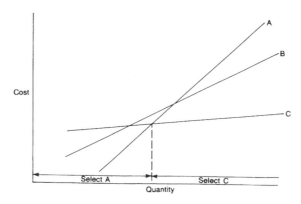

Figure 10.4   Optimisation chart

Figure 10.5 illustrates a further use for this technique in the determination of the economics of quality. This implies that there is a limit, in economic terms, regarding the possible improvement in the standards and quality of construction work. The break even point on this chart indicates that methods of quality control at a high level are costing more than the failure costs themselves. The separation of costs into different categories is in practice not as straightforward as is suggested in this example. The construction industry until more recently has also not been accustomed to thinking in these terms. The linear presentation of these charts may also represent an oversimplification of practice. However, research has been able to show that some relationship between costs and quantity does occur within a range of values. Too much emphasis should not be placed upon the actual break even point. It is desirable to interpret this as the centre of a range of activities rather than a single point value.

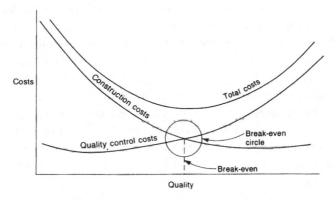

Figure 10.5   Ascending and descending costs

The recording of costs

The contractor's future decisions relating to the price to be charged for the various items of work will be dependent upon an analysis of past performance coupled with future market intelligence. The past performance of the company will be contained in the cost records for the different jobs that have been undertaken. If these data are to be of any future use they must have been recorded and collected in as systematic a way as possible to ensure reliability. A costs code classification system must have been developed against which the appropriate costs can then be recorded. It is first necessary to identify the type of costing system that might be appropriate and to as clearly as possible define which costs are to be allocated under which classification. The latter point requires the system developed to be clear and unambiguous, and that the cost surveyor is fully conversant with allocation of costs and has been properly trained in its use. It is not difficult to misallocate costs in practice. In some circumstances, an inexperienced surveyor can misallocate costs to such an extent that a profitable operation of work might be described as loss making and vice versa.

Time might then be wasted in examining information, with a consequence that part of the control is then lost. The distribution of preliminary items to cost codes must also be clear. It is not unusual in respect of these items, that the costing system is not properly established at the start of a project and so the costs are not properly or fully recorded and that when the project is 'completed', cost surveyors are moved on to a new project and the remnant costs are not recorded. In both cases these items of cost may be significant. All costs attributable to a particular cost code must be recorded, otherwise the costing system remains incomplete and of only limited value to the managers of the company.

There are several different ways in which the costs can be recorded and allocated and different firms will use different methods. In the case of some preliminary items, these may be either costed independently from the individual site operations, or where they contribute specifically to one or more site operations then their costs can be distributed accordingly. For example, it is difficult to allocate the costs and upkeep of a temporary haul road to the permanent site works. In this case its actual costs are better analysed separately and compared with the budgeted costs. The costs of a crane, however, may be better allocated to the site operations to which it contributes. Projects must also contribute towards the costs associated with running the head office, and the allocation of these costs to the individual projects must also be clear. In many respects a straight-forward percentage addition to the cost code is perhaps the simplest and most straight-forward and perhaps the fairest method to use. Different company policy might choose to ignore this item so that it is clearly understood that all site costs are exclusive of company overheads.

The total costs associated with the construction of a project can be easily recorded and can be precisely calculated. However, to be of any use whatsoever, this single sum must be further analysed into its constituent parts. It is recognised that the more detailed a costing system, the more prone to misallocation and error are the costs attributed to the individual items of work. Whilst it may be possible to calculate exactly the costs of a single trade, this precision may not even be correct since some of the tradesmen involved might do small tasks beyond their normal remit to increase their own progress of work. The total trade costs might also include some remedial or rectification work that will distort what really should have taken place. However, further more detailed subdivisions of the site operations will be necessary and it is important to recognise that deficiencies are likely to be present in any system. Sampling at various stages throughout the project will also highlight discrepancies and differences that cannot be explained.

Contractor's cost comparison

The contractor, having priced the project successfully enough to win the contract through tendering, must now ensure that the work can be completed at the most for the estimated costs. One of the duties of the contractor's quantity surveyor is to monitor the expenditure, and advise site management of action that should be taken. This process also includes the costs of subcontractors since these form a part of the main contractor's total expenditure. The contractor's surveyor will also comment on the profitability of different site operations. Where losses are encountered decisions need to be taken to reverse this position if at all possible. Wherever an instruction suggests a different construction process to that originally envisaged then details of the costs of the site operations are recorded. The contractor's surveyor will also advise on the cost implications of the alternative construction methods which could be selected by site management. For example, the following items were included as a part of the drainage bill. The items were not identified separately from the remainder of the work.

Trench excavation, average 750 m deep		
for drainpipe	80 LM @ £12.00 =	960.00
Granular filling around 150 mm pipe	80 LM @ £ 8.00 =	640.00
150 mm drain pipe	80 LM @ £ 9.00 =	720.00
Total price of these activities		= 2320.00

This drain to be laid was 2 m away from, and ran parallel to an existing building within an industrial complex of structures. The contractor had intended to use a machine (JCB or similar) for excavating. It was quickly realised that this would not be possible due to the number of drains and services which crossed the proposed line of the drain. A cost record was kept of the operation, which included mainly hand excavation with the machine working at a reduced efficiency due to the nature of the work. These actual costs are shown below.

Excavation at Goose Mill		
JCB excavator 8 hours	@ £40.00 =	320.00
Granular filling 18 m^3	@ £15.50 =	279.00
150 mm drain pipe 80 m	@ £ 3.25 =	260.00
Labourer 362 hours	@ £ 5.75 =	2081.50
Dumper 6 hours	@ £ 9.25 =	55.50
		2996.00
Head office overheads 5%	=	149.80
Cost of operation	= £	3145.80

In order to make a direct comparison with the price charged, the contractor would need to include a sum for profit. On this project, according to the contractor's pricing notes, this represented 5%.

Contractor's costs	3145.80
Profit 5%	157.29
Total	£3303.09

This represents an under-recovery against the work included in the documents of £983.09 or 42%.

Discounting the fact that estimators can sometimes be wide of the mark when estimating, even with common items, the contractor would seek reasons for such a wide variation between costs and prices. This will be done for two reasons. First, in an attempt to recoup some of the loss. Second, to remedy such errors in future work.

The above situation may have arisen for one of the following reasons:

- The character of the work is different to that envisaged at the time of tender.
- The conditions for executing the work have changed.
- Adverse weather conditions severely disrupted the work (but clearly not in the above case).
- There was an inefficient use of resources.
- There was an excessive wastage of materials.
- Plant had to stand idle for long periods of time.
- Plant had been incorrectly selected.
- Delays had occurred because of a lack of accurate design information.

This list is, of course, not exhaustive, and often when the project is disturbed by the client or designer this can have a knock on-effect on the efficiency and outputs of the contractor's resources. Contractors may also suggest that they always work to a high level of efficiency. This is not always the case, and the loss is sometimes due to their own inefficiency. However, in the above case the contractor appears to have a good case for the recovery of some of the loss. The client's quantity surveyor in accepting the argument would then try to attempt to see just how reasonable the quantified data was and whether it had been independently checked or recorded by the clerk of works. A prudent contractor would always draw such work to the attention of the clerk of works to ensure that at least the recording was fair and reasonable.

There will be circumstances arising where an inefficient use of resources is entirely at the fault of the contractor. Costing which shows that a project has lost money is of limited use where the contractor cannot remedy it. The contractor needs to be able to ascertain which part of the job is in deficit and to know as soon as it starts to lose money. The objectives therefore of a cost control system are:

- To carry out the works so that the planned profits are achieved
- To provide feedback for use in future estimating
- To cost each stage or building operation, with information being available in sufficient time so that possible corrective action can be taken
- To achieve the benefits suggested within a reasonable level of administration charges.

Ideally, therefore, a cost control system should provide for reports on a daily basis. Since this could become an extremely expensive procedure, the cost of work done is consequently checked weekly in an attempt to allow for some corrective action to be taken. If the information is appropriately recorded, it can then be used as a basis for bonusing and valuations for interim certificates. The measurement of the work done should be undertaken by someone who correctly understands the demarcation between the various operations, since misrecording of costs can easily occur. The costs can then be properly compared with the value of work concerned.

Estimated, target and actual costs

Estimated costs

These are the costs calculated by the estimator for the preparation of the contract's tender. Each company uses its own tables of standard work outputs for the different operations of work, and these are used by the contractor when preparing pricing for new projects. A prudent building contractor will review and revise these regularly, but owing to the variability of construction work generally, the outputs that might be derived from site feedback are only rarely revised. This then often has more to do with new methods of construction working or to changes in techniques or terminology. The process of estimating therefore becomes more subjective, based upon experience, rather than necessarily on the use of performance data. Research (Skitmore, 1999) has also indicated that the more experience estimators have, the more accurate they are likely to be in their estimating of costs.

Target costs

The outputs that the estimator has used are then adopted as a basis for the construction operations on site. The target or budget costs attempt to relate costs to specific tasks that are undertaken on site. Whilst the estimator may choose an average output for the whole of an item in a bill of quantities, the cost surveyor will need to analyse these average outputs separately depending upon the various levels of difficulty experienced on site. The location of the work on site is known to have an influence upon the outputs of the operatives' work. For example, estimating the brickwork costs in a multi-storey block of flats will require the estimator to use a single output representing each of the floors. The various target costs are, however, likely to increase as each brickwork lift is undertaken. The cost surveyor will use the average estimator's rate as a basis but will vary this in relation to actual achievements expected. Target costs are also often lower than estimated costs, since they are used as a basis for the contractor's incentive scheme. They are also generally lower to allow for the possibility of unforeseen circumstances.

Actual costs

These represent the actual recorded data retrieved from site in the 'as-done' mode. Actual labour outputs when compared to budget outputs may show a considerable difference due largely to the fact that incentive schemes are in operation. The differences between the target and actual costs would, however, not be as large since the latter would be priced at enhanced bonus rates.

Choosing a cost control system

The following points should be borne in mind when choosing a cost control system.

- The costs involved in setting up and managing the system. There is little or no point in devising a system to save the company £10 000 if it costs £20 000 in cost surveyors' salaries to administer.
- The largest savings are likely to be achieved immediately after such a system has been introduced. This is due to the awareness of the site personnel and the fact that it is always easier to make initial savings in wastage, but much more difficult to continue to show improvements.
- The costing system may take some time to implement fully. It is likely to detect anomalies that will need to be rectified.
- There will come a point when no further savings or refinements are practicable.
- A simplified system is to be preferred. Where a large number of cost codes are used then deterioration in recording accuracy will be expected.
- The system should be as simple as possible. Complex systems that behave like a black box are looked upon with mistrust.
- The system should be forward looking to allow for future corrective action to be incorporated.
- Effective budgetary and cost control procedures are unlikely to be cheap to instal or administer. In the construction industry, because of the variety and differences of work anticipated, they will by necessity need to be complex.
- It will be essential to invest in training for all staff concerned.

Cash flow

The cash flow forecast is an attempt to show the anticipated inflow and outflow of money for a business. Inflow represents the revenue received from work done. In practice, most businesses have to wait some time before payments are received. Also, many customers expect to receive some trade credit and others are simply slow payers. Most industries have established custom and practice which should be followed. Outflows or expenditure pay for wages, materials and other bills that appear with a regular frequency. There will be difficult times when the business is short of available funds to make payments. The money can of course be borrowed, but at a cost, and the lender will require some form of security that repayment in full can be expected at the appropriate time. Creditors, too may lose patience, and refuse to supply any more goods or materials until payments of past invoices are made in at least part. The unsympathetic creditor may eventually institute bankruptcy proceedings.

Cash flow is not simply a means of impressing the bank manager in order to secure a loan. It is an important management tool that allows the financial performance to be planned and then compared with what actually takes place. The cash flow forecast will:

- Indicate when there might be insufficient funds available and do so before it happens
- Indicate when there is a surplus of funds so that this can be used more effectively
- Ensure that there is sufficient funds available for any necessary capital expenditures
- Establish how to use resources more efficiently and thereby reduce costs.

Table 10.1 shows an example of a simple cash flow statement. This would need to be supported with explanations of income and expenditure and be read in conjunction with the business plan. This cash flow is typical of a business start up, when substantial sums are spent in advance of income being received. Provided the business plan is effective and the company run profitably then a reversal in the overall trend should be expected.

Table 10.1   Cash flow statement

	January	February	March	April	May
Opening cash balance	0	−35 000	−55 000	−75 000	−20 000
Capital introduced	100 000	0	0	0	0
Income: work done	0	80 000	100 000	200 000	150 000
Receipts	**100 000**	**80 000**	**100 000**	**200 000**	**150 000**
Wages/Salaries	60 000	60 000	60 000	60 000	60 000
Materials	40 000	30 000	40 000	50 000	60 000
Equipment	10 000	0	5000	10 000	20 000
Premises costs	5000	5000	5000	5000	5000
Others	20 000	5000	10 000	20 000	20 000
Payments	**135 000**	**100 000**	**120 000**	**145 000**	**165 000**
Movement in cash	−35 000	−20 000	−20 000	55 000	−15 000
Closing cash	−35 000	−55 000	−75 000	−20 000	−35 000
Borrowing facility	60 000	60 000	60 000	60 000	60 000
Additional needs	0	0	15 000	0	0

There is also the added danger of possible overtrading, where apparently profitable work could be undertaken, but there is a lack of capital to cover the periods before any payments are received. The cash flow forecast will allow a clear indication of this possible situation occurring in advance.

Client's cash flow

In addition to the client's prime concern with the total project costs, the timing of cash flows is also important. Equal monthly instalments cannot be assumed, indeed as the project proceeds a peak in activity is achieved about two-thirds of the way through the contract period. The client's advisers will therefore need to prepare an expenditure cash flow which is linked to the contractor's programme of activities. On large and complex projects and in periods of high inflation, the timing of payments might result in higher tender sums being a better economic choice for the project as a whole. Table 10.2 represents the projected cash flows for two contractors based upon their own programmes for executing the works. Contractor A intends to set up a highly mechanised system on site that will produce cost benefits and savings later in the contract. Contractor B intends a more traditional approach, typifying much of the scenario in the United Kingdom. The client's opportunity cost of capital is 10%. On the basis of submitting the lowest tender sum, contractor A is the logical choice, since both contractors meet the requirements in terms of quality and time constraints. However, when cash flows and the costs of finance are taken into account, contractor B is the better alternative. If the interest rates were higher then this would make contractor B an even better choice. In practice such cash flows would be calculated on a monthly basis.

Table 10.2   Client's cash flow

Contract period (years)	Cash flow	Discount factor @10%	Time value of payment (£)
Contractor A: High mechanisation			
1	850 000 ×	0.90909	772 727
2	610 000 ×	0.82644	504 128
3	800 000 ×	0.75131	601 048
4	970 000 ×	0.68301	662 520
5	510 000 ×	0.62092	316 669
6	450 000 ×	0.56447	254 011
Tender sum	4 190 000	Net present value	3 111 103
Contractor B: Traditional			
1	300 000 ×	0.90909	272 727
2	500 000 ×	0.82644	413 220
3	750 000 ×	0.75131	563 482
4	1 200 000 ×	0.68301	819 612
5	1 200 000 ×	0.62092	745 104
6	350 000 ×	0.56447	197 565
Tender sum	4 300 000	Net present value	3 011 710

The Department of Health (Hudson, 1978) has developed a method of expenditure forecasting for clients using a formula. This has been developed on the basis of plotting the totals of interim valuations for a number of contracts. The curves of best fit are drawn which results in an S curve of approximately the same shape for each contract.

Therefore within a range of cost and over a limited contract period a standard S curve can be used to help predict the expenditure flow for future contracts. This can then be used as a comparison against actual expenditure from valuations. The former Property Services Agency also developed a similar system.

Contractor's cash flow

Contractors are not, as is sometimes supposedly believed, singularly concerned with profit or turnover. Other factors also need to be considered in assessing the worth of a company or the viability of a new project. The shareholders, for example, will be primarily concerned with the measurement of their return on the capital invested. Contractors have become more acutely aware of the need to maintain a flow of cash through the company as shown in Table 10.1. Cash is important for day-to-day existence, and some contractors have suffered a downfall not because their work was not profitable but due to an insufficiency of cash in the short term. In periods of high inflation, poor cash flows have resulted in reduced profits which in their turn have produced an adverse effect for the shareholders' return. It is necessary therefore to strike the correct balance between these objectives.

Contractor's income and expenditure curves

Table 10.3 represents the costs and payments for a project with a contract period of 16 months and a defects liability period of six months. The S curve for the project is shown in Figure 10.6.

Table 10.3   Contractor's income and expenditure

Month	Expenditure	Cumulative Expenditure	Cumulative Valuation	Net Income
1	15 100	15 100	21 100	20 045
2	21 200	36 300	47 200	44 840
3	28 700	65 000	80 000	76 000
4	47 700	102 700	125 000	118 750
5	40 500	143 200	177 400	168 530
6	52 800	196 000	240 000	228 000
7	61 500	257 500	309 100	296 212
8	57 500	315 500	366 500	353 612
9	47 500	363 500	408 000	395 112
10	31 500	395 000	428 000	421 556
11–15	11 000	406 000	428 000	421 556
16	2500	408 500	436 200	436 200

Retention is 5%, with a limit of 3% (similar to ICE Conditions of Contract). Contract sum was agreed at £429 600 on a fixed price basis, therefore the retention fund maximum equals £12 888.

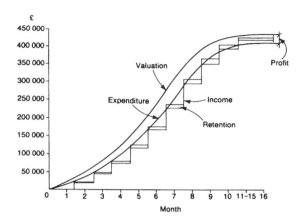

Figure 10.6   S–curve

Expenditure

This is a combination of all of the contractor's costs and will include wages, materials, plant, subcontractors and overheads (head office charges). Expenditure is an ongoing item which will be made at irregular intervals other than for payment of wages. The expenditure S curve would therefore show as a continuous curve.

The expenditure on this contract peaks at month 7 and then begins to decline up to the completion of the contract. Any expenditure beyond month 10 is likely to represent minor items of work which had still to be carried out after the certificate of practical completion and the making good of defects. The contractor's income and expenditure data provides a good indication of the progress of the works.

Valuation

This is also shown as a continuous curve, since although it may be impracticable to value the work in this way, this is how the value of the works changes. For convenience and clarity it is shown as a smooth curve even though in practice this will not strictly be the case. The comparison between expenditure and valuation will show profitability for that point in time. Some care of course needs to be exercised here since this may not be a true representation due to unevenness in the way that the works have been priced.

Income

The third line on the graph in Fig. 10.6 represents the amount and timing of receipts paid to the contractor. Payment under the majority of forms is now almost standard practice at within 14 days of the date of certification to the client. For this reason income is shown as the mid point of the month. The unbroken line represents the income received, with the amount of retention shown as a dotted line. Retention is released to the contractor normally in two equal parts, at completion and with the issue of the final certificate.

Cost comparison

In practice it is always difficult to make comparisons between costs and valuations, since either the full items of expenditure are unavailable or the valuation has only been prepared approximately. However, the contractor does need to determine which contracts are profitable and which are not, and also to determine which operations either gain or lose money. The information which is then generated may be used to form the basis of contractual claims or to assist in future tendering and the contractor's selection of projects for which to tender.

From Table 10.4 both the actual profit and percentage on cost can be calculated. This offers the contractor an indication of the financial trend, although in order to measure this realistically these figures need to be compared with their respective budgets. It is unclear from this data alone whether the contract was successful. By inference it can be suggested that the project reached its most successful stage in terms of profit alone at month 7. The profit–expenditure ratio had, however, been decreasing since the commencement of the contract, indicating a possible front loading of the contract. This is also a typical feature of fixed price contracts if the anticipated profit has been distributed evenly throughout the project. Towards the end of the contract the project was probably losing money. For example, compare the expenditure with the valuation for month 9. This may suggest that the work has been deliberately over valued during its early stages.

Table 10.4   Contractor's profitability

Month	Actual profit (Expenditure – Valuation)	Profit % on Expenditure
1	6000	39.73
2	10 900	30.02
3	15 000	23.07
4	22 300	21.07
5	34 200	23.88
6	44 000	22.44
7	51 600	20.03
8	51 500	16.34
9	44 500	12.24
10	33 000	8.35
11–15	22 000	5.42
16	27 700	6.78

Project cashflow

Using the data shown in Table 10.4, the contractor can also establish the cash flow and borrowing requirement. Although in this case the information is retrospective and therefore of historical use only, the forecast amounts could be used to build a model of these requirements. The contractor expends £15 100 in month 1, and this is then indicated as a negative cash flow. A valuation is then prepared and agreed but 14 days may elapse before the contractor receives any payment. One half of month 2's expenditure is therefore spent before this payment is received. Payment is indicated as a positive cash flow. This same procedure is then repeated throughout the contract. The expenditure is shown progressively from month to month since expenditure on wages, materials, etc. will be made throughout. Income is a single payment made once per month and is represented by the vertical line. The diagram is often referred to as a saw tooth diagram.

The effects of inflation and interest charges will therefore influence the actual profits received from a contract. Although the actual profit measured as a ratio between income and expenditure is shown to be 6.78%, this percentage will be reduced by the size of each cash outflow and the interest rates applied by the banks for borrowing short-term finance. The adverse effects of a negative cash flow can be to some extent mitigated by the following.

- *Profit margins*: the amount of this and whether it has been unevenly applied will ensure that the project is self financing more quickly. In addition it will also bring into benefit the secondary factor of reducing bank charges.
- *Interim payments*: the period is traditionally one month. If this can be reduced to provisional amounts paid weekly, as occurs on some of the large projects, then this too will improve the cash flow.
- *Retentions*: The more of this that is retained and the greater the delay in its release the worse will be the cash flow.
- *Delay in receipt of payments*: Where the client does not honour certificates promptly this will cause the cash flow to deteriorate.
- *Delays in making payments*: the greater the time between receiving goods or services and paying for them, the better will be the cash flow, even if it results in loss of discounts for normal trade terms. This is especially so in times of high inflation.
- *Over valuation*: this has the effect of improving cash flows, but caution is more likely to result in the contractor's valuations being undervalued creating the opposite effects.

In the past, many contractors have faced bankruptcy or liquidation because they over-stretched their commitments which resulted in cash flows that then became unsustainable. It is always tempting to take on every project available and expand too quickly in an unplanned fashion, but an apparently lucrative contract, unless based upon sound financial analysis and planning, often results in disaster.

Chapter 11
Procurement

Introduction

Traditionally, clients who wished to have projects constructed would invariably commission a designer, normally an architect for building projects or an engineer for civil engineering projects. The designer would prepare drawings for the proposed scheme and, where the project was of a sufficient size, a quantity surveyor would prepare estimates and documentation on which contractors could prepare their prices. This was the procedure used throughout the last century and is known as *single stage tendering*. Even up to the mid–1960s there was a limited choice available of methods used for contract procurement purposes. Procurement was a word that only entered this process during the 1980s. However, since the early 1960s there have been several catalysts for change in the way that projects are procured. These are listed in Table 11.1.

Table 11.1   Catalysts for change in procurement methods

Government intervention through committees, such as the Banwell Reports of the 1960s and more recently through the Department of the Environment (now DEFRA) and the Latham and Egan Reports
Pressure groups have been formed to encourage change for their members, most notably the British Property Federation
International comparisons, particularly with the USA and Japan and the influence of the Single European Market in 1992
The apparent failure of the construction industry to satisfy the perceived needs of its customers, particularly in the way in which the industry organises and executes its projects
Influence of educational developments, research and innovation
Trends towards greater efficiency, effectiveness and economy throughout society in general
Rapid changes in information technology, both in respect of office practice and manufacturing processes
Attitudes amongst the professions
Clients' desire for single point responsibility

There remains no panacea and procedures will continue to evolve in order to meet new circumstances and situations. Procurement methods of a hybrid nature are taking place in an attempt to utilise best practice from the various competing methods. Each of the different methods has been used at some time in the industry; some more than others, due largely to user familiarity, ease of application, recognition and reliability. New procurement systems will continue to be developed to meet new requirements and demands from clients, contractors and the professions.

Reports

Several different reports, many of which have been government sponsored, have been issued since the early 1950s. Their overall themes have been aimed at improving the way that the industry is organised and the way that construction work is procured. These various reports are listed in Table 11.2.

Table 11.2   Reports influencing procurement

- *The Placing and Management of Building Contracts*, 1944. (The Simon Report)
- *A Code of Procedure for Selective Tendering*, 1959
- *Survey of Problems before the Construction Industry*, 1962. (The Emmerson Report)
- *The Placing and Management of Contracts for Building and Civil Engineering Works*, 1967. (The Banwell Report)
- *Action on the Banwell Report*, 1967
- *Faster Building for Industry*, 1983
- *The Manual of the British Property Federation System*, 1983
- *Construction Contract Arrangements in EU Countries*, 1983
- *Thinking about Building*, 1985
- *Faster Building for Commerce*, 1988
- *Building Towards 2001*, 1991
- *Constructing the Team*, 1994. (The Latham Report)
- *Rethinking Construction*, 1998. (The Egan Report)
- *Rethinking Construction Innovation and Research*, 2002. (The Fairclough Report)
- *Modernising Construction*, 2001 (National Audit Office)

Major considerations

Table 11.3 identifies the major variables to be considered in the selection of an appropriate procurement system. The different options are not mutually exclusive.

Table 11.3   Major considerations in procurement

Consultants	or	Constructors
Competition	or	Negotiation
Measurement	or	Reimbursement
Traditional	or	Alternatives

Consultants or constructors?

The arguments for engaging either a consultant or a constructor as the client's (promoter's) main adviser or representative are to a large extent linked with tradition, fashion, loyalty and the satisfaction or disappointment with a previous project. There is also the (sometimes mistaken) belief that had an alternative approach to procurement been used then some of the difficulties would not have occurred and some of the problems would have been more easily solved. However, like any process, procurement must continually evolve to enhance itself. The emphasis on a single point responsibility for the client is an attractive proposition and in itself may outweigh other important factors. However, this should not automatically be assumed to be design and build by the constructor, but a reevaluation of existing procurement and contractual arrangements. Table 11.4 lists some of the factors to be considered.

Table 11.4   Consultants v constructors

- Single point responsibility
- Integration of design and construction
- Premier client interest
- Impartial advice
- Needs for inspection, payments, warranties
- Overall costs of design and construction

Competition or negotiation?

Businesses, be they designers or constructors, are able to secure their work or commissions in a variety of different ways. These can include invitation, recommendation, reputation, speculation, etc. However, irrespective of the final contractual arrangements that are chosen by the client, designers and constructors need to be appointed to the project in order to carry out their respective tasks. Either a client can choose to appoint a single firm or organisation through some system of negotiation, or a number of suitable firms can be invited to compete for the project against some specified and clearly defined criteria. Most of the evidence which is available suggests that the client, under normal circumstances of contract procurement, will be able to obtain a better deal or arrangement if some form of competition on price, quality or time exists. However, there are circumstances under which a negotiated approach with a single firm or organisation may provide particular benefits to the client. Table 11.5 lists some of these considerations.

The list in Table 11.5 is not exhaustive. It should also not necessarily be assumed that negotiation will always be preferable where the above conditions exist. Each individual project needs to be examined on its own merits and a decision taken, based upon the particular circumstances concerned and with the interests of the client remaining paramount.

Where competition is envisaged then a choice between selective or open competition has to be made. Selective competition is the traditional and most popular way of

Table 11.5   Reasons for negotiation in procurement

- Business relationship
- Early start on site
- Continuation contract
- State of the construction market
- Contractor specialisation
- Financial arrangements
- Geographical area

appointing a contractor. In essence, a number of firms of known reputation are selected by the client's advisers to submit a price. The firm that submits the lowest price is then usually awarded the contract. In the case of open competition, the details of the proposed project are advertised, and any firm that feels able to complete the project within the stipulated conditions is then able to submit a tender. The use of open tendering may remove from the client the moral obligation of accepting the lowest price since firms are not normally vetted before tenders are received, and factors other than price must also be considered. With selective competition the number of tenders is restricted to about six firms. With open tendering there is, in theory, no limit to the number of tenders and exceptionally there are records of almost 100 firms submitting a tender for the same project. It should be noted that the preparation of tenders is both expensive and time consuming and that these costs must be borne by the industry. It should also be remembered that there is no such thing as *a free* estimate.

Measurement or reimbursement?

There are essentially only two ways of calculating the costs of construction work. One is on the basis of paying for the work against some predefined criteria or rules of measurement. The alternative is to reimburse the contractor the actual costs involved in construction and to use a system of reimbursement.

Measurement contracts distribute both more risk and incentive to the contractor to complete the works efficiently. Reimbursement contracts result in the contractor receiving only what is spent plus an agreed amount to cover profits. The rates or prices quoted by a contractor on a measurement contract must allow for everything which is contained in the contract. Table 11.6 lists some of the main points to be considered in making this choice.

Table 11.6   Measurement or reimbursement

- *Cost forecast:* other than in the broadest of terms this is not possible with cost reimbursement contracts
- *Contract sum:* reimbursement contracts are unable to provide a contract sum
- *Efficiency:* cost reimbursement contracts may encourage the contractor to be inefficient
- *Price risk:* measurement contracts allow for this (clients may therefore pay for possible non-events)
- *Cost control:* there is little scope for controlling construction costs on a reimbursement contract
- *Administration:* reimbursement contracts require a large amount of clerical work and record keeping

Traditional or alternatives?

Until recent times the majority of the major building and civil engineering projects were constructed using a procurement system known as single stage selective tendering. The British construction industry had developed this system within the parameters of good practice and procedures. Whilst there remains real criticism of this method the general lack of impartial research makes real comparison and evaluation of it difficult. However, in more recent years the traditional method of procurement has been under reappraisal for many different reasons because of the consideration of new procurement procedures and a better knowledge of other practices around the world. These reasons are listed in Table 11.7.

Table 11.7 Difficulties with traditional procurement systems

- Appropriateness of the service provided
- Length of time from inception through to completion
- Projects over-running their contract periods
- Final costs being higher than expected
- Problems of quality control
- Mismatch between design and construction
- Line of legal responsibility
- Limitation on the procurement advice available

The various alternative procurement procedures have therefore been devised and developed in an attempt to address some or all of the issues identified in Table 11.7. However, it is difficult to select one of these procurement methods to offer a solution to all of these criticisms. A review of the past shows that an almost rigid and uniform set of procedures were applied to all projects, almost irrespective of the needs of individual clients. It is now accepted that none of the newer contractual procedures singularly address all of these criticisms. Procurement practice is really a case of trading off different methods against client objectives, needs and wants, in an attempt to achieve a best possible solution.

Procurement strategy

The selection of appropriate contractual arrangements for any but the simplest type of project is difficult owing to the diverse range of options and professional advice which is available. Much of this advice is in conflict and lacking in a sound research basis for analysis and evaluation. Some of the advice is tainted with fashion. Often a failure in procurement practice has more to do with the individuals concerned than the refinements of the particular methods selected. Identical projects using identical procurement methods can often achieve very different outcomes. Procurement practices are a learning process.

Design and build contractors are unlikely to recommend the use of an independent designer. Such organisations believe that the full integration of design with construction is likely to achieve the best long-term solution both for the project and the client. The professions which provide only a design service will generally take the opposite viewpoint. Individual experiences, prejudices, vested interests, familiarity, the need and desire for improvements are all factors that have helped to reshape procurement in the construction industry. The proliferation of differing procurement arrangements have resulted in an increasing demand for systematic methods of selecting the most appropriate arrangement for a particular project. Intelligent knowledge based systems (expert systems) have been used with some effect in this respect. The reality of this situation is that a defined project with clear objectives should always result in the selection of the same and appropriate procurement options. Table 11.8 identifies factors which should be considered when choosing the procurement path.

Table 11.8   Factors to consider in the choice of procurement options

- *Size*: small projects are not suited to complex arrangements
- *Design*: aesthetics, function, maintenance, buildability, contractor integration, need for a bespoke design, design before build and the use of prototypes
- *Cost*: price competition/negotiation, fixed price arrangements, price certainty, price forecasting, contract sum, bulk-purchase agreements, payments and cash flows, life cycle costs, cost penalties, variations and final cost
- *Time*: inception to handover, start and completion dates, early start on site, contract period, optimum time, phased completions, fast tracks, delays and extension of time
- *Quality*: quality control, defined standards, independent inspection, design and detailing, single and multiple contractors, coordination, buildability, constructor reputation, long term reliability and maintenance
- *Accountability*: contractor selection, ad-hoc arrangements, contractual procedures, auditing, simplicity, value for money
- *Organisation*: complexity of arrangements, standard procedures, responsibility, subcontracting and lines of management
- *Risk*: evaluation, sharing, transfer and control
- *Market*: workloads, effects of procurement advice
- *Finance*: collateral, payment systems, remedies for default and funding charges

Price determination

Building and civil engineering contractors are paid for the work which they carry out on the basis of either measurement or reimbursement of costs.

Measurement

Using this method the construction work is measured on the basis of the actual quantities which are incorporated into the finished project. The contractor is paid for this work on the basis of quantity multiplied by rate. In addition, the contractor is also paid hire charges for plant, temporary materials, etc. The work may be pre-measured for tendering purposes in which case it is known as a *lump sum contract*. Alternatively it may be measured after being executed for payment purposes, in which case it is known as a *remeasurement* contract. On most projects, however, some form of remeasurement is always likely to occur to take account of variations, unexpected or otherwise. Building contracts are more often lump sum contracts, whereas civil engineering contracts are typically of the remeasurement type. Table 11.8 shows the main types of measurement contracts which are used.

Drawings and specification

This is the simplest type of measurement contract and is really only suitable for small or straightforward projects. Each contractor has to measure the quantities from the drawings, interpret the specification and prepare a tender price.

Performance specification

With this method the contractor is required to prepare a price on the basis of the client's brief and user requirements alone. The contractor is left to decide upon the type of materials and construction which meet the performance criteria.

Schedules

These form two categories. A *schedule of rates* is similar to a bill of quantities but with an absence of any quantities. A *schedule of prices* is an already priced document, where the contractor's tender is simply a percentage adjustment to these prices. Neither of these methods is able to calculate a contract sum.

Bills of quantities

The contractors prepare their tenders on detailed quantitative and qualitative data. The information is prepared in accordance with some agreed rules of measurement, e.g. Standard Method of Measurement of Building Works, Civil Engineering Standard Method of Measurement, etc. The contractors tendering for the project are all able to use the same detailed information for pricing purposes. In circumstances where the work is unable to be properly defined then bills of approximate quantities may be prepared.

Cost reimbursement

On these types of contract the contractor is able to recoup the actual costs of materials which have been used and the time spent on the project by the operatives. An amount is agreed to cover the contractor's profit. Day works, under a measured contract, are valued on a similar sort of basis. Because these types of contractual arrangement are unable to provide a clear indication of expected costs they offer little incentive for the contractor to control costs. They should, therefore, only be used in special circumstances such as those indicated in Table 11.9.

Table 11.9   Situations where cost reimbursement might be used

- Emergency works projects
- Where the character and scope of the works cannot be easily determined
- Where new technology is being used
- Where a special relationship exists between the client and the contractor
- Where it would be unfair to ask a contractor to price the works, for example, in cases of exceptional risk

Cost reimbursement contracts can take several different forms: the *cost plus percentage* arrangement, where the contractor's profits are in a direct relationship to costs, i.e. the higher the costs the higher the profits; the *cost plus fixed fee*, where the fee is fixed beforehand and does not alter with a rise or fall in costs and the *cost plus variable fee* which supposedly provides some incentive on the part of the contractor to control costs. The difficulty with these latter two methods is related to a general inability in the first instance to be able to forecast construction costs accurately in advance, on which the fixed and variable fees are then to be calculated.

Procurement options

There are a wide variety of procurement options available in the construction industry, which are aimed at addressing criticisms of poor quality, long construction periods and high costs. A summary and description of these can be found in *Contractual Procedures in the Construction Industry* (Ashworth, 2006). The methods vary from traditional single stage selective tendering, where a client uses a designer to prepare drawings and documentation on which contractors are invited to submit competitive prices, to schemes where a single construction firm will provide the truly all-in service, the 'turnkey' project. This last method might even include land acquisition, furnishings and the long-term maintenance of the project. There are procurement methods that have been devised to get the contractor on site as quickly as possible, such as two-stage tendering and fast tracking, with of course the anticipation that the contractor will also complete the works sooner than by using the traditional arrangement. Other methods have recognised contractors' improved management skills and their influence on the whole construction process. Some methods have evolved to recognise changes which have occurred in the industry, such as the proliferation of subcontracting, the increase in litigation and the need for one organisation to accept total responsibility for the whole process. Another method of procurement attempts to recognise that if a project is fully designed prior to tendering, then the time spent on site should be able to be shortened with a consequent reduction in construction costs. The difficulty with this approach is to combat the client's desire for work to start on site as soon as possible. Table 11.10 lists the alternative procurement arrangements that are available.

Table 11.10 Alternative procurement arrangements

Contractor selection
- Selective competition
- Open competition
- Negotiated contract

Contractual options
- Traditional
- Design and build
- Package deal
- Turnkey
- Management contracting
- Construction management
- Project management
- Management fee
- Design and manage
- Develop and construct
- British Property Federation
- Measured term contract

Contract variables
- Early selection
- Fast tracking
- Serial tender
- Continuation contract

The principle involved in the options shown in Table 11.10 is that the contractor selection, contractor option and variable arrangements are amalgamated to form a procurement system. In practice it may be necessary to amalgamate the different possibilities or even to use a combination of systems on a single project. This latter suggestion needs to have established clear procedures and arrangements for its operation.

Contractor selection

There are only two ways of selecting a contractor: negotiation or competition. Competition may be restricted to a few selected firms or may be open to any firm that wishes to submit a tender for the project. The contractual options described later are used in conjunction with one of these methods. Contractors are now sometimes asked to provide a refundable deposit that is returned upon the submission of the tender to client. This is more common with open tendering, in an attempt to discourage the long tender lists that have sometimes occurred. There have been instances recorded where firms have had to pay sums of money to the client just to ensure that their names appear on the tender list. Such a practice has received criticism from all quarters of the industry. A Code of Procedure that exemplifies good tendering practice was developed by the National Joint Consultative Committee (NJCC) and has subsequently been modified by the Construction Industry Board (CIB). Whilst this code is not mandatory, its use is to be recommended. Its main points are listed on p. 310.

Selective competition

This is referred to as the traditional method and in still the most common method of awarding construction contracts. In essence, a number of firms of known reputation are selected by the project team as possible firms who might tender for the works. Since there are prerequisites for adding the names to the list, the firm that eventually submits the lowest tender should then be awarded the contract. These prerequisits refer to the contractors' reputations, as identified in the CIB Code of Practice described later in this chapter on p. 310.

Open competition

Using this method, allows those firms who might not fulfil all of the above criteria to submit a tender. The details of the proposed project are often advertised in local and trade publications, or through the local branch of the Building Employers' Confederation. In selective tendering, the number of firms is often limited to six, but in open tendering any number of firms might submit a price. The only restriction is on the number of documents that may be available. All of the firms who request the documentation may not necessarily receive it, particularly where the consultants concerned do not feel that such a firm would be a suitable contractor. It should also be remembered that the costs of tendering can be high, and are not free, as some might suggest. Such costs must be borne by the industry and absorbed into the prices of successful tenders. The use of open tendering removes the moral obligation of the client to accept the lowest tender, although most tendering documents state that the client is not bound to accept the lowest or any tender. With open tendering, factors other than price may need to be considered when assessing the different bids. It is generally accepted that this method will provide the lowest price for the client.

Negotiated contract

This method of contractor selection is used where a price is to be agreed with a single contractor in the absence of competition. Some of the reasons for this are provided in Table 11.5. The contractor will price the documentation in the usual way and this will then be reviewed by the client's quantity surveyor. Any rates for which there is a disagreement are then the subject of negotiation. Since there is an absence of competition, this type of arrangement may add about 5% to the costs of the contract. However, a negotiated contract should result in fewer errors in pricing. This will result in fewer claims to compensate for such potential losses. It might also involve the contractor in greater participation in the project and result in more cooperation during the construction period. The smoother running of the project may achieve financial savings in other ways. However, because of the higher tender sums involved, public accountability and the suggestion of possible favouritism, public works contracts are rarely of this type.

Contractual options

The following options have been developed in an attempt to address some of the failures and criticisms of traditional tendering arrangements. They are not mutually exclusive. It is possible, for example, to award a serial contract using in the first instance competitive tendering. Fast tracking may be incorporated as characteristic with any of the options suggested. Competition or negotiation will be a constituent part of the arrangement.

Traditional

With this option, the client commissions an architect to prepare a brief, a scheme outline and working drawings, and to invite tenders and administer the project through issuing instructions, inspecting the work under construction and preparing certificates for payment. The architect may also, with the client's approval, appoint other consultants such as quantity surveyors and structural engineers. The contractor, who has no responsibility for the design, will usually be selected by competitive tendering. The design team work independently from the contractor, who is responsible for executing the construction work in accordance with the terms of the contract. There are many variations to this approach. For example, on some occasions the client may appoint a different consultant as a main point of contact.

Design and build

It has long been argued that the separation of design and construction is one of the reasons for difficulties in the construction of buildings in the United Kingdom. Whilst this statement is apparently correct, there is little real evidence, other than hearsay, to say that it is so. The construction industry has frequently been likened to manufacturing, where design and production are generally undertaken in the same firm. Although the construction industry has many similarities to manufacturing, there are also many differences. For example, these include construction of the works on the client's own premises, the effects of the weather and the frequently individual nature of a construction project.

With a design and build arrangement, instead of using an architect or engineer for a separate design service, the client chooses to employ a contractor directly for an all-in design and construction service. It may also be necessary and desirable to employ independent professional advisers to monitor the progress and quality of the contractor's work and to agree the value of interim certificates for payment purposes.

While there are clear advantages for appointing a single firm to accept all of the responsibilities, there are also disadvantages that the client will need to consider. A project commences with a design and it is important that this is correctly formulated. It could be argued that a contractor's design might be more suited to the contractor's own organisation and construction capability rather than the interests of the client. Conversely, however, the project might result in lower production costs on site, a shorter overall design and construction period, consequent overall savings in price and perhaps even an implied warranty of suitability, since the single organisation has provided an all-in service.

Design and build firms have made positive responses in recent years to the criticisms of unimaginative designs. Many design and build projects, in respect of their aesthetic attributes, are now almost indistinguishable from projects that use independent designers. This is often because a contractor may employ independent designers as a part of the contractor's all-in service to a client.

The advantages claimed for the use of design and build arrangements include:

- The contractor's familiarity with the project from inception
- The option for contractors to include their constructing capability in the proposed design
- A possible reduction in the overall timescale of the project through the possible telescoping of aspects of the design with construction
- No claims for possible delays due to a lack of drawn information
- Direct contact between the client and the contractor.

Where it is necessary to provide some basis of competition between contracting firms, design and build has been found to be unwieldy. It is frequently difficult to evaluate a project across a wide range of attributes. Clients do not usually prioritise their requirements in a manner that assists design and build contractors to prepare the successful tender. This sort of arrangement also discourages, through the costs involved, changes to the project once the design has been accepted. For projects that may take several years to complete, this is not an acceptable condition.

Package deal

In practice the terms 'package deal' and 'design and build', when applied to construction projects, are interchangeable. However, a package deal is really a special type of design and build arrangement, where the client effectively chooses a building from a manufacturer's catalogue. The package deal is off the peg, whereas design and build is bespoke. Under a package deal arrangement clients are often able to view similar types of buildings that have been completed elsewhere. This type of contractual arrangement has been used extensively for the closed systems of industrial buildings of timber or pre-cast concrete, such as multi-storey offices and flats, low-rise housing, light industrial units and farm buildings. The client will typically provide a contractor with a site, supplying relevant design information and performance specifications and the package deal contractor prepares a tender on this information.

Whilst these buildings may not provide the best economic solution, either initially or in the longer term, they do have the advantage that they can be erected very quickly. The client may wish to retain an independent adviser, such as an architect or surveyor to advise on the different systems that are available, on time, cost and quality and during the construction of the project.

Turnkey

This form of contractual arrangement is relatively uncommon in the United Kingdom, but it has been applied in both the Middle East and the Far East with apparent success. The true turnkey contract includes everything from inception to completion, often including furnishings, being undertaken by a single contractor. On some occasions this might also include finding a suitable site for the development and long-term routine maintenance. It is interesting to note that whilst automobiles now frequently include a warranty for future defects, for up to five years or more, the quality guarantee on building projects still lasts only the six-twelve months. Many clients are now becoming more interested in the long-term aspects of their projects, through the application of whole life costing philosophy. On industrial projects the contract might also allow for the design and installation of manufacturing process equipment. The method gets its name from the concept of turning the key. The client therefore expects that on completion of the project, it will not only be ready for occupation but also ready for use.

Management contracting

The term *management contracting* is used to describe a method of organising the project team and operating the construction process. The management contractor acts in a professional capacity, providing the management expertise and buildability requirement to the overheads and profits involved in return for a fee. The contractor does not therefore participate in the profitability of the construction work itself and does not directly employ any of the labour and plant, except possibly for those items involved in setting up of the site and the costs normally associated with preliminary works.

Because the contractor is employed on a fee basis, the appointment can take place early during the design stage. The contractor is therefore able to provide a substantial input into the practical aspects of the building technology process. Each trade required for the project is tendered for independently by subcontractors, either upon the basis of measured work packages or a lump sum. This should therefore result in the lowest cost for each trade and thus for the construction works as a whole. The management contractor assumes full responsibility for the control of the work on site.

Construction management

This procurement route offers an alternative to management contracting and has been adopted on a number of large projects since the mid-1980s. The main difference is that the individual trade contractors are in a direct contract with the client. The client appoints a construction manager, either a consultant or a contractor, with the relevant management expertise. The construction manager, if appointed first, would then take the responsibility for appointing the design team, who are usually in direct contact with the client.

The construction manager is responsible for the overall control of the design team and the various trade contractors, throughout both the design phase and the construction phase of the project.

Project management

Although the definition of the term 'project management' is not universally agreed, the following description generally conforms to what is understood in the United Kingdom. Project management requires the client to appoint a professional adviser to the position of project manager, who will then in turn appoint the appropriate design consultants and select a contractor to carry out the construction works. The method is more appropriate for large building projects or engineering structures. The function of the project manager is therefore to organise and coordinate the design and construction programmes and to ensure that the project does not over-run in either cost or time. It is essentially conceived as a client function.

Management fee contract

Management fee contracting is a system whereby a contractor agrees to carry out construction works at cost. In addition, the contractor is paid a fee by the client for the management input and for profit. Some clients also like to include contractor incentive arrangements and use as a basis a target cost. This type of arrangement is similar to that used on cost reimbursable contracts, and therefore has the same advantages and disadvantages.

Design and manage

This method of procurement is really the counterpart to design and build. In this case the architect, engineer or surveyor undertakes the role of the contractor in addition to their design activities. Much of general contracting is now almost restricted to the management, administration and coordination of subcontractors, a function that the design firm can develop through the employment of suitable personnel with contracting skills. The contract is agreed between the client and the design and manage firm, offering the client a single point responsibility arrangement. The method of subcontractor selection has all the advantages of traditional tendering arrangements coupled with those of design and build.

This arrangement is suitable for all types of construction project. Clients undertaking large projects might however, because of past experience, prefer to use a more traditional form of procurement involving one of the larger proven contractors. A further difficulty for the design and manage firm is the need to acquire the necessary plant and equipment that a contractor may already own.

This type of arrangement should be able to offer competitive completion times compared with the other methods that are available. Since there is independent control of the subcontractor's construction work, quality should be at least as good as that offered by the other contracting methods. Regarding overall costs, these should at least be in line with design and build costs.

Develop and construct

This is a speculative arrangement that is common for projects such as private housing. It is also used on office and industrial developments. The firms, some of whom are contractors, purchase a site, obtain the necessary planning approvals and then design and construct buildings for either sale, rent or purchase. The developers will have undertaken extensive market research prior to these activities and may undertake the work in conjunction with financial institutions. Speculative developments are undertaken on the basis that the demand for a particular type of building will be available on its completion.

British Property Federation

The British Property Federation (BPF) represents a substantial commercial property interest in the United Kingdom. It is thus able to exert some influence on the construction industry, its professions, contractors and thus the procedures employed for the awarding of contracts. The BPF system unashamedly makes no apologies that the interests of the client are paramount. Its aim therefore has been to devise a more efficient and cooperative system of organising the construction process, by attempting to make a genuine use of the different professions involved. It was developed largely in response to the dissatisfaction that its members had experienced with existing contractual arrangements. It was claimed that buildings cost too much, were frequently delayed and took longer to construct than similar buildings in other parts of the world and often did not achieve the standards of quality that its members were now requiring.

The BPF manual sets out the operation of the system in detail, and while this might appear to be a rigid set of rules and procedures, its originators claim that it has the potential to be used flexibly and in conjunction with the best of the other methods of procurement. However, since the procedures and recommendations have been devised by only one of the parties involved in the construction process, it lacks the compromises that have been jointly agreed by the different bodies representing the client and contractor interests and are inherent in the other forms of contract. Ideas not normally found in the traditional methods include for example, single point responsibility for the client, the utilisation of the contractor's buildability skills, reductions in the pre-tender period, re-definition of risks and the preference for specifications rather than measured quantities.

Measured term contract

This type of contractual variable is used on major maintenance projects. It may be awarded to a contractor to cover a range of different buildings in different locations. The contract will usually apply loosely to predefined work, the nature of which may make the application very variable. It is often limited to a fixed time scale, although the needs of the client may often require this to be extended. The contractor will initially be offered the work to cover a number of different trades usually on the basis of prices contained in a schedule prepared by the client. The schedule may be particular to the contract concerned or be of a standard type such as that of the former Property Services Agency Schedule of Rates.

Where the client supplies the rates for the work, the contractors are given the option of quoting a percentage addition or deduction from these rates depending upon their desire to win the project. The contractor offering the most advantageous percentage to the client will then be awarded the contract. An indication of the amount of work expected over the specified period of time will assist the contractors in the consideration of the percentage to be quoted. The JCT publishes a Measured Term Form of Contract that is suitable for use with any schedule of rates.

Contract variables

Early selection

This method is sometimes referred to as *two-stage tendering*. Its main aim is to involve the selected contractor as soon as possible with the project. This method tends to succeed in getting the person who knows what to build (the designer) in touch with the firm that knows how to build (the constructor) before the design is completed. The contractor's expertise in construction methods and processes can therefore be used to benefit the design. A further advantage is that the selected contractor will also be able to start work on site sooner than might otherwise be expected using other methods of procurement.

It is first necessary to select a contractor for the project. This is frequently done on a competitive basis, unless there are special circumstances that would warrant negotiation. This can be achieved by inviting suitable firms to price major items of work for the project. Some form of simplified bill of quantities is therefore required that will include items such as preliminaries, major items of work and specialist items, allowing the contractor the opportunity to price the profit and attendance items prepared. The contractors will also be asked to state the percentage additions to be applied as overheads and profit. This information will then be used as a basis for negotiation of rates on the contract documentation once it has been prepared.

The NJCC have also prepared a Code of Procedure for Two-stage Selective Tendering. Thus includes recommendations and good practice that should be observed when dealing with procurement arrangements that include this consideration.

Fast tracking

This approach to contract procurement results in the letting and administration of multiple construction arrangements for the same project. It is applicable generally to large projects. The process results in the overlapping of the various design and construction operations, so that whilst some of the construction work is being completed on site, some of the design work has yet to begin. The various stages may be let as work packages, resulting in the creation of separate contracts or a series of phased starts and completions. When the design for the complete section or element, of work, such as foundations, is completed then a contract is awarded to a contractor for this part alone. This staggered letting of the work has the objective of shortening the construction time from inception through to completion for the project overall.

This type of arrangement will require considerably more skills in its organisation and planning. In practice, therefore, some form of project management must be envisaged. Although the handover and completion dates may be earlier than with more conventional arrangements, there is a stronger likelihood that this might be at the expense of cost and quality.

Serial tender

Serial tendering is a development of the system of negotiating further contracts, where a firm has successfully completed a contract for a project of a similar type. Initially contractors tender against each other for a single project. This is on the legal understanding that a number of other similar projects will be automatically awarded using the same prices. The only adjustment would be to allow for the possible fluctuations in price to cover inflation.

The contractors therefore know at the initial tender stage that they can expect to receive a number of further projects to provide them with some continuity in their workload. The arrangement has been used for school building projects, petrol filling stations and restaurant chains. Conditions are generally incorporated to prohibit the awarding of further work in the case of a contractor's poor performance. Serial contracts should result in lower prices for the client, since the contractors are better able to organise their work more effectively, by for example, investing in mechanical plant.

Continuation contract

This arrangement occurs where the client decides that on completion of the project, further works of a major nature will be carried out on the same site. This situation may occur, for example, in major highway construction, where it is usual to award the contracts for the different lengths or sections of road once they have been designed. Sometimes it occurs on industrial projects because of the underestimation of demand for a product, and further buildings are required to meet a projected product capacity.

A continuation contract can be awarded as an add-on to the original contractual arrangements. Continuing with the same construction team, where this has been satisfactory, is a sensible arrangement under such circumstances. Each of the parties will be used to the methods of working and there will be cost savings, because the contractor is already established on the site. It may be desirable to review some of the contractual arrangements, particularly where market forces have changed since the original contract was awarded. It is also not unreasonable to expect that the contractor will be prepared to share in the monetary savings that are likely to be achieved under such an arrangement.

Construction Industry Board (CIB) Code of Procedure

Good procurement practices that relate fairness and reward are important in the construction industry. The former National Joint Consultative Council for the Building Industry has developed codes of procedure for tendering purposes to identify good practices in the awarding of construction contracts. Many of its ideas were developed by the Construction Industry Board, following recommendations form the Latham Committee and its report, *Constructing the Team* (1994). They represent the principles of good procurement practice, many of which can be applied with the different methods which are currently available. The following are some of the more important recommendations:

- Use of a standard form of contract with which the various parties in the construction industry are familiar.
- Restriction on the numbers of potential tenderers. It should be remembered that the costs of unsuccessful tenders will be borne by the construction industry.
- When preparing a shortlist, the following should be considered:
    - firm's financial standing and record
    - recent experience of building over similar contract periods
    - general experience and reputation of the firm for the type of project envisaged

- – adequacy of the contractor's management
- – adequacy of capacity.
- Each firm on the list should be sent a preliminary enquiry to determine if it is willing to tender. The enquiry should contain the following information:
  - – job title
  - – names of employer and consultants
  - – site location together with a description of the works
  - – principal nominated subcontractors
  - – approximate cost range
  - – form of contract
  - – dates for possession and completion
  - – anticipated date for dispatch of tender documents.
- If it is necessary for a contractor to withdraw after the decision to tender the client's adviser should be notified as soon as possible.
- Contractors who have shown a willingness to tender but are not chosen for the shortlist must be informed immediately.
- The tender period should not be less than four weeks but this will depend upon the size and complexity of the job.
- Where a tenderer seeks some clarification of the tender documents then all other tenderers should be informed of the decision.
- If a tenderer submits a qualified tender then this should be withdrawn or the tender rejected.
- Under English law a tender may be withdrawn at any time before its acceptance. Under Scottish law, it cannot be withdrawn unless the words 'unless previously withdrawn' are inserted in the tender after the stated period of time it is to remain open for acceptance.
- All tenderers should be informed as soon as possible of the result.
- All tenderers should be provided with a list of tender prices once the contract has been signed.
- The tender prices should remain confidential at all times.
- Where errors in pricing occur, the alternative ways in which these should be dealt with should be specified. Any corrections should be confirmed in writing or initialled.
- The employer is not bound to accept the lowest or any tender, or take responsibility for the costs involved.
- If the tender under consideration exceeds the specified costs then addendum bills can be prepared in association with the lowest contractor's prices.

Coordinating Committee for Project Information (CCPI)

Research at the Building Research Establishment (BRE) on 50 representative building sites has shown that the biggest single cause of events which restrict working together by site managers, designers or tradesmen is unclear or missing project information. Another significant cause is uncoordinated designs, and at times much wasted effort is directed towards searching for missing information or in reconciling the inconsistencies which arise. In an attempt to overcome these weaknesses the Coordinating Committee for Project Information (CCPI) was formed with the task of developing a common arrangement for work sections of building works that could be used throughout the various forms of documentation. The CCPI consulted widely to ensure that the proposals were practicable and helpful, and its work has been undertaken by those working in practice who are fully aware of the different pressures involved. It is not sensible to produce incomplete, inconsistent and contradictory project information during the design process and then to expend wasted time trying to rectify the arising difficulties on site with the additional time dealing with claims for disruption, delays, extra costs, etc. It is preferable if complete and coordinated information can be prepared in order to avoid such problems. Regardless of the procurement methods used, the principles of the CCPI should be applied and adopted in practice. Procurement methods which cannot incorporate such principles are probably in themselves severely flawed in practice.

Procurement management

Clients of the construction industry rely extensively upon the advice given in respect of the most suitable method of procuring their project from inception through to completion. The advice provided must therefore be both relevant and reliable and based upon the appropriate levels of skills and expertise which are available. The need to match the client's requirements, which are sometimes vague and generally imprecise, with the capability of industry is of vital importance if customer satisfaction is to be achieved and the image of the industry improved. The procurement management role is shown in Table 11.11.

Table 11.11   Procurement management role

- Determine the client's requirements and objectives
- Discover what is really important and what is of secondary need
- Assess the viability of the project and provide advice in respect of funding, taxation and residues
- Advise on the organisational structure for the project as a whole
- Advise on the appointment of consultants and contractors
- Manage the information and coordinate the whole process from inception through to completion

Conclusion

The traditional approach to construction has been to appoint a team of consultants to prepare a design and estimate, and to select an independent constructor. The latter would calculate the actual projects costs, develop a programme to fit within the period laid down in the contract, organise the workforce and materials deliveries and construct to the standards specified in the contract documentation. In practice, design details were sometimes inappropriate, the quality of workmanship below acceptable standards, costs above what were forecast and even in those circumstances where projects did not over-run, the completion date was later than expected in other countries. In addition, too little attention has been given towards future maintenance aspects and life cycle costs. The separation of the design role from construction and the different procurement practices have been the target in dealing with these obvious deficiencies. However, many clients have wished to retain the services of independent designers, believing that they will serve their needs better. They also believe in competition as a form of efficiency and have been reluctant to commit themselves to a contractor while costs remain imprecise. Ideally, however, the client would prefer a single point responsibility and a truly fixed price and for projects to be completed as required.

Procurement procedures remain a dynamic activity. They will continue to evolve to meet the changing and challenging needs of society and the circumstances under which the industry will find itself working. There are no standard procurement solutions, but each individual project needs to be considered independently and analysed accordingly. There is, however, a need to evaluate more carefully the procedures being recommended in order to develop good practice in procurement and to improve the image of the industry.

Chapter 12
Estimating and Tendering

Introduction

The Chartered Institute of Building's *Code of Estimating Practice* (2008) provides the authoritative guide to the practice and procedure of contractors' estimating. It is of use to students and practitioners and others who are involved in either the production or use of estimates. The Code distinguishes between *estimating* and *tendering*:

- *Estimating* is described as the technical process of calculating construction costs by building up rates for each of the items in the tender
- *Tendering* is the separate commercially based function that uses this estimate of cost as a basis upon which the final price is determined and submitted to the client for approval.

Organisation and administration of estimating

The principal function of the estimating department within a contractor's organisation is to calculate as accurately as possible the costs of proposed construction projects. It is not the function of the estimator to calculate the right price for the job. This is the role of management, although because of the skills and knowledge that they possess, estimators will nevertheless play an important part in the adjudication that follows the preparation of the estimate.

The estimating department must have ready access to current pricing information for all aspects that are likely to affect the costs of construction. Estimators must be familiar with the contents of the forms of contract and methods of measurement and have a good understanding of construction methods. They must be familiar with the different working rule agreements, particularly if they are to price work in civil engineering as well as building projects. During the preparation of estimates they need to visit sites in order to familiarise themselves with problems that might be encountered during the construction of the project.

Estimators will have working relationships with the majority of the other departments in the contracting firm, working closely with site management and quantity surveying. There will be close and important links with those who are responsible for buying materials and plant. The establishment of the estimating department varies widely between companies, depending upon the type of work that they carry out and the nature and size of the business. In the smaller firms, the estimator is also likely to undertake other activities, such as surveying. As a rough guide the workload of a typical contractor is frequently twice the value of the annual turnover. Contractors might on average expect to win about one job for every eight tenders submitted. This ratio will depend upon the state of the market, the niche area in which that contractor might be working and the policy regarding the submission of competitive tenders. The use of negotiated tenders will save a lot of abortive time in tendering process although these arrangements are more costly for the contractor to prepare.

Tendering

Decision to tender

The invitation for a contractor to submit a tender may arrive at the contractor's office in a variety of different ways. Usually, however, the client (or consultant who may be dealing with the project), would first enquire whether the firm was interested in submitting a price before the documents are sent to the contractor. This is generally done whilst the documents are still being prepared in order to draw up the list of tenderers. It provides the consultants with a list of interested tenderers and also allows the contractor time to consider the project and to know that a possible tender is to be prepared. The preliminary enquiry identifies the main characteristics of the project. On other occasions the tender documents will arrive at the contractor's office unexpectedly. This has disadvantages in that the work of the tender office may need to be rescheduled to cope with its tender preparation.

Unfortunately, the steady flow of enquiries and the winning of suitable contracts is in line with the ups and downs in the economy. If a regular workload could be sustained, then contractors might be more easily able to plan ahead, charge more consistent prices and undertake the work more efficiently, effectively and economically. This would be good news for all concerned with the construction industry. However, even in circumstances of industry recession, some contractors will sometimes refuse to join a tender list and even return the documents unopened to the client or consultant. The time allowed for the preparation of tenders must be sufficient without requiring the work to be undertaken in haste. The contractor, prior to any work on the documents will need to satisfy themselves on the following points:

- Are they capable of carrying out the work?
- Do they want to carry out this work?
- Do they want to work with this client and the consultants?

If adopting good tendering practices and procedures a client and its advisers should have already established the various contracting firms':

- financial record and standing
- recent experience of similar types of project
- general experience and reputation
- management capability
- current workload.

In seeking answers to the question of whether or not to submit a tender for a particular project, a contractor will need to establish the following points:

- The approximate cost, duration and personnel and plant requirements for the project.
- Can the firm's resources adequately meet these requirements?
- Is the project suitably located or is it in a location posing special problems for the delivery of materials or the availability of labour?
- Is it a type of project with which the firm is familiar? If not will this impose, for example, higher than usual insurance terms and premiums?
- What competition is there for this project? The time and costs involved in unproductive tender preparation can be considerable, and estimators must use their time carefully.
- What other possible projects are likely to be on the market?

The answers to these questions depend upon a variety of different considerations. Most importantly, will the contractor be able to complete the work satisfactorily and at a profit; even under some circumstances this may be meagre. Having decided to submit a tender then the estimator will need to take a closer inspection of the documents.

The contract documents

On a building contract these normally comprise at least a form of contract, drawings and either a bill of quantities or a specification. On civil engineering projects it is more usual to include both a specification and a bill of quantities as a contract document. The use of the Joint Contracts Tribunal (JCT2005) form of contract allows only one of these alternatives as a contract document. The contract documents that will be signed by the client and the contractor will include as a minimum the following information:

Provided by the consultants:

- The work to be performed
- The quality and standards of work required
- The contractual conditions to be applied.

Provided by the contractor:

- The cost of the finished works
- The construction programme.

Project appreciation

Once a contractor has decided to submit a tender the following will need to be established:

- What type of contract conditions are being used and have they been altered in any way?
- Are there restrictions on the way that the work is to be performed, such as access to the site, sectional completion, evening or weekend restrictions?
- Are there financial disincentives concerned with insurances, period of payments, retention, liquidated damages or a guarantee bond?
- How much of the work can the firm undertake and how much will need to be done by subcontractors?
- What proportion of the project are prime cost sums, perhaps indicating more of a management role for the contractor and little work for the contractor's own employees?
- Are there any apparent inconsistencies in the documents?

Method statement and tender programme

It is essential that a meeting is arranged as soon as possible between the estimator and those who will be responsible for the programming of the construction works. This will help to establish the methods to be used for the construction of the project. Construction methods can have a huge impact upon costs. The amount of detail provided at this stage will vary from one firm to another and the familiarity with the type of project concerned. The method statement will assist in the preparation of the estimate by providing guidelines on how the work is to be undertaken. Where time permits, alternative ways of working and sequencing of the operations can be evaluated, since these can have an impact upon the overall estimated construction costs of the project.

The method statement thus has two main functions: to assist the estimator during the preparation of the estimate and to describe the method of working that will be adopted on site should the contractor be successful in winning the project.

The tender programme is a vital document for the contractor. Eventually this will be incorporated within the contract where the contractor's tender is accepted. For example, in clause 5.3 of the JCT2005 Form of Contract, this specifies this requirement, although it does not identify the amount of detail that should be provided. The length of time that the contractor is on site will have an impact upon the overall costs and hence on the contractor's estimate. The preliminary programme will identify the project's main requirements and the resources that will be necessary. It will seek to identify key dates and provide basic information that can be used for obtaining quotations for materials and subcontract work.

Site visit

Once it has been decided to proceed with the tender preparation, and the estimator has become familiar with the documentation, then a visit to the site should be undertaken. This may be done in conjunction with a contracts manager, who will be able to advise on the practical working conditions that might be encountered. This may also be required in order to price the works on site or spot items, normally associated with alterations and extensions projects. A comprehensive report should be produced on a standard format. This will seek to establish more clearly, the following aspects:

- Topography of the site and ground conditions
- Availability of existing service pipes
- Need for temporary roads and any site access difficulties
- Security needs such as, fencing the site perimeter, hoardings, security guards, etc.
- Nature and use of any adjacent buildings
- Any demolition work requirements
- Site accommodation and material storage locations
- General availability of labour and materials
- Special difficulties, such as restrictions that might be imposed, for example, on the use of tower cranes and other similar mechanical plant.

Visit to the consultants' offices

It may also be necessary to visit the offices of the different consultants involved with the project to discuss the project in more detail and to gain some insight into the client's main objectives for the development. These visits will allow the examination of other detail drawings, and will help to identify the advancement in terms of the finished design. Whilst it is assumed that the project has been fully designed prior to the invitation to the contractors to tender for the work, this is not invariably so. This may give rise to problems throughout the construction of the project and this can be costly to the contractor. These visits should also be used to clarify any outstanding queries that have arisen during any of the preliminary investigations outlined above.

Any adverse reports after these preliminary enquiries have been carried out may result in a reconsideration of whether or not to submit a tender for the proposed project. While this remains a possibility throughout the tendering period and even at the adjudication stage, it becomes more of an unlikely possibility as the tendering process progresses.

Enquiries and quotations

Subcontractors

The estimator, or one of the assistants, will separate the bills of quantities (or work packages) into convenient trades or sections. These will then be separated into those that the estimator will price and the remainder will be sent off to several (normally at least three) subcontractors for their pricing. Where the contractor has entered partnering arrangements and supply chain management with subcontractors this will need to be considered. These are the contractor's own subcontractors, sometimes referred to as domestic subcontractors. The main contractor will also need to obtain the architect's approval prior to awarding domestic subcontracts to these firms should the contractor be successful in winning the contract. The main contractor should ensure that all of the conditions of the main contract are included within the terms of the domestic subcontract.

Labour-only subcontractors are dealt with in much the same way. The main contractors in all cases need to know what attendance items they will require to be provided, such as assistance with the unloading of goods and materials, storage space and how much supervision of their work will be necessitated.

Goods and materials

A similar process is adopted for obtaining quotations for the different materials to be used, although the information from the bills is often transferred to schedules, since the suppliers are only concerned with the materials element and not the labour, plant, etc. The bills need to be checked very carefully since some materials may be restricted to a limited number of suppliers. The bills should be carefully used for the future ordering of materials, since the quantities may be incorrect or be subject to future variations. For the more common materials the contractor will use a regular local supplier. The information sent to the suppliers will include the following:

- *The specification of the materials*: it may be necessary to provide photocopies of the relevant preamble clauses from the bills of quantities or other contract documents.
- *The quantity of materials required*: quantity can have a significant effect upon price. It may be necessary to collate the various quantities from the different sections of the bills.
- *The address of the site*: for delivery purposes, including the nature of the site, access and unloading.
- *An indication of the expected delivery dates*: this will need to be a balance between too early a delivery resulting in the need for extra site storage costs and late deliveries which may cause a delay to the progress of construction. The programme may also require delivery at several different dates.
- *The terms of the quotation*: these should ideally be on the same conditions as the main contract. They should identify discounts required, typically 2.5%, although this might be different, the length of time that the prices will remain fixed and the account settlement period. This is typically 30 days.

Plant

Whilst the contractor may own some items of plant, it will be necessary to obtain quotations for the hire or purchase of other items of plant and equipment. In the case of hiring this often comes to site with a plant operator, but this will need to be carefully checked in the plant hire quotation and conditions. The choice of whether to purchase plant is normally a financial calculation based upon expected use compared with the costs to hire or lease. The advantages of ownership in addition include convenience and flexibility. However, there are also disadvantages such as plant becoming obsolete, tying up of capital, need for its maintenance and repair and for the charges concerned with transporting it to and from site. The main factors to consider when calculating the costs of mechanical plant include:

- Purchase cost
- Interest charges and taxation
- Insurances
- Depreciation
- Usage per annum
- Outputs in differing conditions
- Breakdown and repair costs
- Plant operators and banksmen
- Delivery to and from sites
- Consumable costs such as fuel, oil, etc.

Speed is essential in despatching all of these enquiries, but it is also of equal importance to ensure that the correct information is sent, and that pages of the tender documents are not missing. It is also desirable to give the subcontractors as much time as is possible to price the work. Whilst these firms may be concerned with only a single trade or section, the process involved is similar to that adopted by the main contractor. In order to facilitate the comparison of the different quotations they should all be on the same basis as directed by the main contractor.

Calculation of unit rates

The most common method of estimating the costs of construction work applies unit rates to the measured quantities in bills of quantities. The unit rates comprise all or some of the items shown in Table 12.1 (see also Figure 10.1).

Table 12.1   Components of a unit rate

Labour:	Craftsmen
	Operatives
	Keymen
	Trade supervision
Materials:	Manufacture or extraction
	Delivery or haulage to site
	Storage facilities
	Waste, consolidation, shrinkage or bulkage
Plant:	Mechanical
	Purchase or hire charges
	Depreciation
	Interest
	Fuel and consumables
	Repairs and maintenance
	Non-mechanical
Subcontractors	
Overheads	

The calculation of the unit rates for the individual measured items in a bill of quantities requires the consideration of the following procedure.

The collation of current cost information

This involves the collection or calculation of current rates for:

- Labour
- Plant
- Materials and subcontractors.

Labour

This includes the calculation of the all-in (costs to employ) hourly rates for the different kinds of labour that are directly employed by the contractor. Although the labour costs are defined in the Working Rule Agreement, contractors may in addition choose to supplement the amounts involved. Some of the labour costs might also be separately identified more appropriately as a part of the project's site on-costs. The *Code of Estimating Practice*, referred to earlier, sets out a typical calculation of an all-in labour rate. Needless to say that these calculations will change at least annually. The calculation provided may also not exactly mirror the way that every contractor assesses costs or calculates hourly labour rates. Items to be included are the basic rate and guaranteed bonus, allowances for inclement weather, non-productive overtime and sick pay, trade supervision and any extra payments to be made for example, for skill, responsibility or risk as a working rule allowance. Amounts to cover training levies, National Insurances, paid holidays, severance and employer's liability insurance should also be added. These additional costs that have to be borne by the contractor typically increase the basic rate by about 50%.

These costs need to be calculated on the basis of a year's employment. The annual costs are then divided by the total hours worked. The hours worked will vary with the types of project envisaged, and also the times of the year when operatives are employed. Due to better weather conditions and increased hours of daylight, the working hours during the summer weeks will be greater than those during the winter. The Working Rule Agreement states that at least a 39-hour week should be worked throughout the year. Typically, in a year an operative would probably work about 1850 hours, after adjustments for overtime, holidays, sickness and for inclement weather. It should also be remembered that this figure is an estimate. Like all estimates it is based upon an analysis of previous performance and future expectations. The actual hours worked will therefore be subject to some variation. The estimated figure may also vary for the different trades, since they are all, for example, unlikely to be affected by the same inclement weather conditions.

In addition, there are some labour costs that are better allocated as lump sums within the site on-costs or project overheads. These include pension payments, if made, although these sums relate directly to wages and they could therefore be added to the all-in rate described above. Daily travel allowances, lodging and extra payments to attract scarce labour are better identified separately from the all-in rates as a lump sum site on-cost. Where overtime is specially identified in the contract documents, either because of out of hours working or the need for speed in construction, this is also better shown as a lump sum. During the tender adjudication it is also less likely to be forgotten during the examination of the cost estimate. It is important to have a clear understanding of the Working Rule Agreement prior to estimating, since this may also identify other items of possible expenditure that will need to be included in the estimate.

Plant

The single most important question to be determined before anything else regarding plant costs is whether to purchase, hire or lease the plant for the particular project concerned. The only proper way to determine this choice is to consider the financial implications of each. No other criteria should be used, because if the financial considerations are properly evaluated, they will express in monetary terms all the factors that should be taken into account. For example, one factor to be considered might be the amount of use that a particular item of plant may have on the project. If this is only required for part of the time, then would it be better to hire? This can be expressed within the financial calculation. Using a less appropriate item of plant, that is already owned by the contractor, can also be evaluated as a financial calculation.

The items to be considered as part of the financial assessment are shown in Table 12.1. In addition it is necessary to convert the hourly rates calculated or supplied by the hire company into plant outputs costs per unit of production. In addition to these items, it is necessary to determine the hours the plant will be employed, after making adjustments for standing time, breakdowns and inclement weather, and the performance of the plant in carrying out the different tasks that need to be performed.

Materials and subcontractors

The costs of the different materials, including their delivery to site, involves the comparison of the various quotations received from the suppliers. It is essential when selecting quotations to ensure that they comply with all of the requirements requested. The quotations should specify the minimum quantities that will be delivered, the trade discounts offered and any requirements for unloading at the site which the contractor may need to provide. The estimator will also need to assess the costs of any storage or workshop charges that will need to be expended prior to the incorporation of the materials in the works.

In some circumstances storage will require secure accommodation from theft, vandalism and the weather. This will also necessitate the double handling of materials, the costs of which should not be assumed to be covered in other costs. It will also be necessary to assess the expected waste allowances for which contractors will not be paid, since the work is measured on the basis of finished quantities. Estimators are often too optimistic where these factors are concerned, with actual amounts being frequently higher than those estimated. The amount of wastage of materials should be of concern to all employed in the industry.

The assessment of labour and plant outputs

The establishment of realistic production standards is a major consideration of the estimating process. The traditional principle of gathering this information from previous

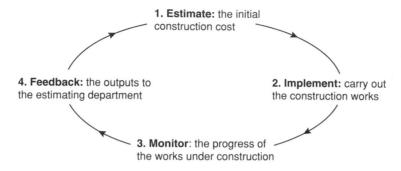

Figure 12.1   The traditional view of estimating feedback

projects, known as *feedback*, is illustrated in Figure 12.1. The estimator estimates the costs of the work to be performed using standard outputs, influenced by size, complexity, quality, etc. If the contractor is successful in submitting the tender then the work is put into practice and during its construction it is monitored by site management staff. The monitoring may be done for bonus or other purposes, but quantities of work and the time expended on these operations are recorded. These records are frequently copied to the estimating department in the form of feedback. However, such information is not routinely used by estimators in calculating or revising their outputs for the following reasons.

- It is very variable in terms of the outputs it generates
- There is insufficient confidence in the site recording systems
- The information is often not compatible for future estimating needs
- There is a difficulty in reusing the data because of the unique circumstances under which the work has been carried out.

The traditional method of assessment has been to develop a classification system and to record costs against it. The outputs achieved on similar work from previous projects should be the major source of information used in estimating. However, construction work requires a complex system against which to record this information. Research has shown that the reliability of any cost recording system substantially deteriorates when the number of cost codes exceeds 50. The cost code system used in the construction industry is a four digit system.

The complexity of construction work and the fact that most projects are bespoke one-off designs makes the process difficult to achieve in practice. Even projects that are considered to be virtually identical record different actual costs outputs. The use of new techniques or improved methods of working will, of course, encourage estimators to review their own tables of standard outputs. Amending these figures on the basis of feedback from one site will not, nor will the hope that things will work out better on the next project have much of an influence on the estimator's outputs.

Production standards, for both labour and plant, are likely to be influenced by a whole range of project characteristics. When adapting a standard output the estimator will need to assimilate these different factors, in order to arrive at a best estimate for the work. Some of these characteristics that need to be considered include:

- Location and accessibility of the work
- Amount of repetition
- Intricacy of the design
- Need for special skills
- Quantity of work involved
- Quality of materials and standards of workmanship
- Working environment such as safety, temperature, etc.

One of the most important aspects that affects the value of labour outputs is the incentive scheme operated by the contractor. This should be designed to improve the operatives' overall performance. In the past, it has to be said that some incentive schemes preferred improvements in speed of completion at the expense of the quality of work.

The calculation of the project overheads

Project overheads are sometimes described as site on–costs and are collected together in the bills of quantities under a section known as Preliminaries. The items concerned may be priced and shown in different ways in the bills of quantities. The preferred way is to calculate the costs of the individual items concerned and to indicate which items have been priced. Typically only about 20 items are priced. Alternatively, the preliminaries may be included as a lump sum, either on the basis of an analysis of the individual items or a percentage of the total costs of measured work items. Where the preliminaries appear not to have been priced, their costs will have been added to the individual measured rates. The calculation of project overheads involves an amalgamation of fixed and time related costs for items such as site accommodation, management and supervisory staff, temporary works, plant such as tower cranes that are difficult to allocate to individual measured items, site security, contract bonds, etc.

Since many of the preliminaries items are time-related, the length of the contract period has a particular influence upon their costs. These items are some of the most difficult to price. The examination of priced bills also indicates a wider variation in the prices of preliminary items, than amongst the major measured works items. The reason for this is that these items may be vague and each estimator therefore has to interpret from the information available the scale and extent of the costs involved and what should be allocated to these items.

Prime cost and provisional sums

Prime cost and provisional sums are included in bills of quantities for work that is to be the basis of nomination by the architect. It is sometimes referred to as specialist work, that is of a nature that most building contractors would not envisage carrying out. They might include, for example, lift installations, structural steelwork or electrical installations. There is provision in the bills of quantities to allow the contractor to include profit and attendance amounts. General attendance includes the use of the contractor's established facilities that are already on site, such as the use of standing scaffolding, sanitary accommodation, etc. Special attendance includes those items specifically requested by subcontractors such as special scaffolding, hardstandings, storage, power, etc. Rather than simply inserting a percentage on the PC amount, the contractor should analyse the costs that are likely to be involved.

Operational estimating

Operational estimating is a method that is based essentially on an analysis of the work content of a project on the basis of how costs are incurred. It is claimed that each identifiable site operation can be performed by a gang of men and materials without interruption from other operations. For example, the costs of a reinforced concrete suspended floor are an amalgamation of the formwork, reinforcement and concrete items. An *in situ* concrete staircase that is to be cast at the same time can also be part of this same operation. Instead of attempting to separate the costs and allocating these to the individual measured items, the cost of the entire operation only is calculated.

Operational estimating, to be effective, really requires the use of operational bills, since the layout of traditional bills is unsuitable for this form of estimating. However, in the 1960s, the possible use of operational bills was piloted by the Building Research Establishment, and rejected by all sides of industry as poorly orientated documentation. Also in practice the apparent clear and theoretical demarcation of activities is frequently blurred, with its relevant feedback data being misallocated. The development of operational estimating largely arose out of a criticism of the traditional unit rate estimating applied to bills of quantities as described above, for the following reasons:

- The unit rate provided in a bill of quantities does not segregate labour, materials and plant. This is an essential requirement for future cost control purposes.
- The unit rates are not reliable, principally because the individual locations, e.g. second lift of brickwork, are not separately classified.
- The costs, for example, of excavation are likely to be as much related to overall plant usage as the depth of excavation.
- The representation of unit quantities does not allow the opportunity to estimate in an organised manner.

- The calculation of the unit rates is time-consuming.
- Since the unit quantities are priced independently, the estimator may not realise the full potential of the work involved.
- Since the process of estimating is time consuming and expensive, it is important that as much as possible of this information can be reused throughout the project. Unit rates do not allow this.
- The application of unit rates to variations can give misleading results to both the contractor and the client.

However, whilst the above points are valid, the difficulty still remains of capturing the site feedback data, both economically and in a manner that allows their useful reuse in the future estimating process.

Estimator's report

The estimator's report will gather together all the relevant items that have been considered during the preparation of the estimate. The main objective of the report is to bring those items that were unusual to management's attention either about the project itself or during the preparation of the estimate, and which may have some influence on the tender to be submitted to the client. These might, for example, include unusual contract conditions that impose more risk than usual on the contractor, an identified difficulty regarding the recruitment of certain craftsmen or specified higher levels of quality than is usually the case.

The estimator's report and analysis should cover the following items:

- A brief description of the project and the methods chosen for its proposed construction (supported by the method statement).
- The conditions of contract to be used together with the identification of any risks, contractual difficulties and an assessment of the relevant completeness of the design.
- Assumptions made during the preparation of the estimate, its expected profitability, market and industrial circumstances at the time of tender and during construction.
- The need for any explanatory qualifications with the submitted tender, although these may invalidate its possible acceptance by the client.
- The terms used in the quotations for the labour and materials.
- The details of the project overheads and site on-costs allowed.
- Any other information regarding the client, consultants or other contractors who are also tendering for the project.

The estimator should also provide supporting documentation or pricing notes in case these need to be examined. This will include the site visit report, details of prime costs sums, the tender programme and method statement and cash flow projections.

Tender adjudication

The conversion of the estimate to the tender is the responsibility of management. This is a separate commercial function using the estimate and the supporting documentation as a basis. The main purpose now is to consider the financial and other implications that any business would need to assess. Whilst the estimator may work within a narrow project framework, it is management's function to look at the submission of the tender more broadly. For example, the contractor may have developed particular expertise over the years for the type of development or construction envisaged. The loss of this contract may necessitate the break up of this team of expertise through the need for redundancies and the selling off of some of the specialised plant and equipment. If the contract under discussion can be won, then there may be more projects of this type to be envisaged in the near future. But the contract may have to be won at the cost estimate value or even less. Can the company afford to do this? The various matters to be considered include:

- *Project type*: consultants, client
- *Management*: method statement, contract programme, manpower
- *Contractual*: contract conditions, contract period, damages, insurances
- *Financial*: fixed price, payments, retention, bond, cash flow
- *Mark-up*: profit, return, risk, overheads, discounts
- *Cost estimate*: analysis
- *Competition*: work load, other work, past performance
- *Allowance*: to be added in the case of firm price tenders.

The meeting may recommend that the all-in labour rates used in the cost estimate should be changed. This should not be done without realistic expectations of changes in performance. The use of computer aided estimating has now allowed the different computations to be easily performed and their effects on cost quickly calculated.

Bidding strategy

In a competitive tendering situation, contractors constantly face the dilemma of submitting a high price to maximise profits with the possibility of failing to win the contract and thereby suffering a shortage of work. Alternatively the contractor may submit low prices which win contracts, but show little profit margin. A bidding strategy may be evolved for determining the optimum bid, which will be the relationship between maximum profit and the possibility of being the lowest tenderer. As a basis for this it is necessary to analyse the bidding pattern of competitors and to compare these results with the firm's own estimated production costs over a number of contracts. A competitor's bid can be obtained from the list of tender results over a range of recent contracts.

In practice, due to contractors' marketing policies in a locality, a contractor is likely to find that the competition is confined to a limited number of firms. Knowing the competitors will then allow an optimum bid to be prepared for a project by using probability curves for the known competitors and developing a bidding curve for the particular contract.

Information and communication technology

The application of information and communication technology to the estimating process has allowed much of the repetition and routine processes to be removed. It has also allowed many of the previously discrete processes to be linked together, avoiding the need to rekey information and data. Electronic data interchange (EDI) is the electronic transfer of business data from one independent computer system to another using agreed standards to format the data. The receiving computer is then able to understand and react to the information, extracting and processing the data that is needed for the company's use. Such systems are now available which allow, for example, bills of quantities to be issued to contractors in an electronic form which can then be used by contractors to price the work based upon the estimator's analysis of outputs and other factors which will affect the eventual price which is charged. Even where contract information is not presented to the contractor in electronic format, computer software has now made it possible to use scanning devices to allow the rapid transfer of information from bills of quantities, without the need for rekeying of information and avoiding the tedium and possible errors in transformation that might occur.

The entire cost database systems of contractors are now also stored on computer, so that outputs and current prices can easily and routinely be updated and then applied to the individual items in the bills of quantities, allowing the estimator more time for analysis and evaluation of the project.

Traditionally many of the functions undertaken by contractors have been separate from each other. If the submitted tender was successful, then a separate process was started for the production of the contract information. Initially when computer systems were first developed they were stand alone packages. However, in more recent times computer-based systems have been introduced to allow each of the different processes used to be a part of an integrated system of data management.

Reporting on tenders

When the client and its advisers are considering the tenders received, factors other than price may be of importance. The time required to carry out the work, if stated as a requirement on the form of tender, may be compared. Time is frequently an important matter to the client. Although there may be reasonable excuses for failing to keep to the time agreed, and even justification for avoiding the liquidated damages provided for by the contract, the time stated by a reputable contractor may be taken as a reasonable estimate, having regard to the prevailing circumstances.

If the contract is subject to adjustment of the price of materials, the schedule of basic rates of materials must also be considered. The question should be asked 'Has the tenderer assumed reasonable basic prices for materials?' If they are too low there can be an excessive increased cost on a rising market or too little in reduced costs on a falling market. Where tenders are very close, the schedules of basic rates may be compared, since the lower tenderer may have less favourable prices. Only a preliminary examination will be made at this stage to ascertain which tender or tenders should be considered for acceptance.

The quantity surveyor will write a report for the client or committee concerned, setting out clearly the arguments in favour of acceptance of one tender or another.

Examination of priced bills or schedules

Before acceptance of a tender, the tenderer whose offer is under consideration is required to submit a copy of the priced bills to the quantity surveyor for examination. If this has not been delivered with the tender, an additional copy should be sent to the contractor for this purpose. Sometimes, to save time, the original bills may be requested. However, the original bills are often marked with the estimator's pricing notes which it would be injudicious to disclose. There is no justification for the certificate that is sometimes required that the copy of the bill has been compared and checked with the original. The tender is a lump sum tender and the sole purpose of obtaining the pricing is to provide a fair schedule for the adjustment of future variations.

The first check is an arithmetical check. Clerical errors are still common, even in an age of information technology. It is important that the contract bills are as correct as they possibly can be prior to the signing of the contract. If, for example, an item has been priced at £0.50 per metre and extended at £0.05, it is not fair that either the additional quantity or omission of the item should be priced out at the incorrect rate in adjusting accounts. All clerical errors should be corrected in the contract copy of the bills. The amount of the tender will of course not normally be altered. Any difference will be shown as a rebate or addition as an addendum to the summary in the priced documents.

In addition to the arithmetical check, a technical check is also made of the pricing by examining the contractor's rates and prices. Deliberate or accidental errors may be found. Items may accidentally have been left unpriced. Items billed in square metres may have been priced at what is obviously a linear metre rate, or vice versa. An obvious misunderstanding of a description may be noted. Corrections should be made so that a reasonable schedule of rates for pricing variations results. *Practice and Procedure for the Quantity Surveyor* (Ashworth and Hogg, 2007) includes details of how errors in priced bills should be dealt with.

A secondary reason for examination of the priced bill is to ensure that the tenderer has not made such a serious mistake that they would prefer to withdraw the tender. Under English law contractors may do this at any time prior to the acceptance of a contract. When such a serious error is detected, it is always advisable to bring this to the attention of the tenderers. The error will sooner or later be discovered, resulting in a risk that constant attempts will then be made to recover the loss, to the detriment of the client's interest. The contractor should, however, be recommended to stand by the tender that has been submitted.

E-tendering

E-tendering is an electronic tendering solution that facilitates the complete tendering process from the advertising of the requirements through to the placing of the contract. This includes the exchange of all relevant documents in electronic format. Some e-tender solutions are packaged with evaluation tools, which assist in comparing tenders from contractors in order to select the successful tender.

E-tendering is a relatively simple technical solution based on secure email and electronic document management. It involves uploading tender documents on to a secure website with secure login, authentication and viewing rules. E-tendering must comply with the relevant EU requirements. Compliance with a company's standing orders also needs to be considered and changes to these made if necessary. Tools available in the current market offer varying levels of sophistication. A simple e-tendering solution may be a space on a web server where electronic documents are posted with basic viewing rules. This type of solution is unlikely to provide automated evaluation tools, instead users are able to download tenders to spreadsheet and compare in an electronic format. Such solutions can offer valuable improvements to paper-based tendering. More sophisticated e-tendering systems may include more complex collaboration functionality, allowing a number of users in different locations to view and edit electronic documents. They may also include an email trigger process control which alerts users, for example, of a colleague having made changes to a collaborative tender or a contractor having posted a tender.

The most sophisticated systems may use evaluation functionality to streamline the tender process from start to finish, so that initial tender documents are very specific and require responses from vendors to be in a particular format. These tools then enable evaluation on strict criteria which can be completely automated. The market sector must be considered since there may be a loss of bidders due to a perceived complexity in the process.

The manual tender processes can be long and cumbersome, often taking three months or longer to complete. This is costly for both client and contractor organisations. E-tendering replaces the manual paper-based tender processes with electronically facilitated processes based on best tendering practices that save both time and money. Contractors are able to manage the tenders inquiries, with all tenders stored in one place. Contractors can select data from the electronic tender documents for easy comparison in a spread-sheet format if this is required. Evaluation tools can provide automation of this comparison process. Contractors' costs in responding to invitations to tender are also reduced as the tender process cycle is significantly shortened. E-tendering offers an opportunity for automating most of the tendering process. This includes the preparation of the tender specification, advertising, tender aggregation and the evaluation and placing of the contract.

The business benefits of using e-tendering include:

- Reduced tender cycle-time
- Fast and accurate pre-qualification and evaluation, which enables the automatic rejection of suppliers that fail to meet the tender specification
- Faster response to questions and points of clarification during the tender period
- Reduction in the labour-intensive tasks of receipt, recording and distribution of tender submissions
- Reduction of the paper trail on tendering exercises, reducing costs to the design team, contracting organisations and their subcontractors
- Improved audit trail, increasing integrity and transparency of the tendering process
- Improved quality of tender specification and supplier response
- Provision of quality management information
- Compliance with EU requirements is easier.

The typical features of an e-tendering system include secure electronic communication that allows contractors to submit their tenders to an e-address. All of the relevant documentation is stored centrally and this might also be a repository of previous tenders, standard terms and conditions and other contractual information. It also acts as an audit trail that assists management information and reporting. Other features include workflow functionality which routes documents to appropriate people or alerts individuals of actions required in the system. It also provides collaborative facilities enabling a number of people to work on creating a tender document or to evaluate tenders. It can also provide an automated evaluation of tenders and award notification and tools to assist any prequalification of tenderers.

E-tendering solutions are usually offered as web solutions, which are either hosted on the buyer's own servers through the internet or intranet or by third party service providers. E-tendering systems may also be provided as part of a suite of e-procurement solutions, bundled with e-ordering, contract management or e-auctioning systems, for example. The effort and cost involved in implementing an e-tendering solution are relatively low, especially in comparison to some complex e-ordering systems, mainly due to the low-tech nature of many systems and limited necessity for change management and training.

Many end users and suppliers regard e-tendering as a positive development eliminating many mundane tasks and speeding up the procurement cycle. The financial benefits of using e-tendering are based on process efficiencies rather than reduced tender prices.

Chapter 13
Developments and Trends

Change

Building and civil engineering contractors and property companies make a number of key assumptions about their clients and customers, about their own organisations and about their competitors. They also need to consider the future implications of changes in practices, which are occurring more frequently today than ever before. Practices and procedures that have no real rationale for their existence will be short-lived.

Change takes several different forms, such as industry reappraisal and consolidation, expansion, increased customer requirements and sophistication and deregulation and globalisation. This latter aspect, coupled with the advances both real and expected, in e-commerce and e-business present important challenges in every industry including both property and construction. These changes are sometimes termed discontinuities.

Before the onset of discontinuity, there are established leaders in every industry which others tend to follow. Many of these are household names and changes and innovations which are not adopted by these leaders have in the past often failed to succeed. However, after discontinuity has struck an industry, new leaders and followers emerge, with new rules for winning.

The Swiss watch industry is a good example of this. It was once dominant in the time telling industry. It was the first to discover the technological break through allowing quartz precision to replace mainsprings. However, it was Seiko in Japan and Texas Instruments in the United States who were the ones to capitalise on the Swiss break-though and make quartz watches universally available. Intellectual capital that is not properly deployed is like any other asset that is under-deployed.

Discontinuities change industry patterns and norms and create new industry orders. The Cooperative Society movement, which started in Rochdale in 1844 had an unprecedented lead on its competitors. Today, at the beginning of the twenty-first century it is now dwarfed by its competitors, the industry leaders. Woolworths' shares were renowned for being as safe as houses until they collapsed in the middle of the last decade. Marks & Spencer, until more recently, led their industry and others followed.

Their demise was even greater. This principle applies in all walks of life, just look at football teams, for example.

The property and construction industries behave similarly, where once great names whom others wanted to copy, have either ceased to exist or are just a shadow of their former self. In some cases they have been taken over by a competitor from abroad and their name has disappeared. It can be observed not just in the large building contractors or developers but real estate firms, surveyors and architects. It has been suggested that the construction industry of Great Britain will adopt the same pattern as the former British automobile manufacturers. There will be fewer firms and these will be foreign-owned.

Discontinuities change industry norms and create new industry orders. There are three roles in this process; the *shapers, adaptors* and *losers.* There are two final outcomes, winners and losers. The shapers are those who design their own futures by creating and exploiting discontinuities. These are the firms who set the new standards of practice. They determine the overall direction and are fashion leaders. The adaptors follow that lead by implementing changes to attempt to keep pace with the shapers or leaders. There are also victims and these are the firms that fail to respond or adapt to change.

The effectiveness of execution and the capability to adapt to ever changing circumstances and conditions ultimately decides who wins and who loses. Shapers can so easily become losers and adaptors winners. In order to deliver added value and become a winner it is necessary to go beyond customer led strategies and create added value led strategies.

Construction and property

The construction and property industries follow the same patterns as other industries. They face the same problems of securing sufficient workloads, providing a value led product and making a profit. Like other industries they have adapted the way in which they carry out their work. This is influenced by their own new ideas and innovation, by comparing and contrasting their work against other similar and dissimilar industries and through the impact of information and communication technologies.

Changes can be seen as either threats or opportunities. In many cases they are both, but the successful place a greater emphasis on the opportunities that are provided as a result of changing practices, markets, client needs and skills and competence.

- *Changes in practices*: The nature of pre-contract practices continues to change with a greater emphasis being placed on the role of the constructor in the design and development process. Comparisons, sometimes unfavourable, are made with other industries, with other countries abroad and through the competition being faced from non-construction professionals.
- *Changes in markets*: The growth of the European Union and the opening up of the market to foreign competition results in threats from incoming firms but also

opportunities for expansion and activity throughout the European Union. The shrinking of the world, largely through telecommunications, has also allowed work-loads to be expanded on a changing international scene.

- *Changes in client needs*: Clients require an increased emphasis on added value. They frequently require a single point responsibility in terms of design, procurement and management of construction.
- *Changes in skills and competence*: Increasingly more is being required with a greater emphasis on improved knowledge, skills and competence in attempting to find solutions for clients' problems.

The reasons for change

The construction industry is perceived as dirty, dangerous, exposed to the weather, unhealthy, insecure, underpaid, of low status and poor career prospects for educated people (Latham, 1994). It is widely agreed within the industry that it is too easy to set up in business as a general contractor. No qualifications are required, no experience and virtually no capital. Whilst market forces ultimately remove incompetent firms by depriving them of work, the existence of such unskilled producers is bad for clients and damages the wider reputation of the industry. Women are seriously under represented at all levels in the construction industry (Latham, 1994). The property industry has an equally poor image but for other reasons. It is believed that it often does little for the fees it charges and is easily confused with and misrepresented by estate agency in that occupation's worst guises.

The reasons why change is both necessary and desirable are not difficult to find and include, for example, the following:

- The apparent failure of the construction and property industries to satisfy the perceived needs of customers, particularly in the way that they organise and execute projects
- Government intervention in the construction industry through privatisation, PFI, CDM Regulations and compulsory competitive tendering
- European legislation and the challenges of the single market and the variability of practices that are employed
- Recent reports on the state of the industry
  - *Constructing the Team* (Latham, 1994)
  - *Improving Value for Money in Construction* (Atkin and Flanagan, 1995)
  - *Towards a 30% productivity improvement in construction* (Construction Industry Board, 1996)
  - *A Statement on the Construction Industry* (Royal Academy of Engineering, 1996)
  - *Value for Money: Helping the UK afford the buildings it likes* (Gray, 1996)
  - *Towards a New Construction Culture* (Powell, 1995)

- *Rethinking Construction* (Egan, 1998)
- *Rethinking Construction Innovation and Research* (Fairclough, 2002)[SBLX]

- Pressure groups formed to encourage change and improvement, such as the British Property Federation
- International comparisons, particularly with the USA, Japan and the European Union
- The influence of changing patterns of education and research and the implications for life long learning
- Trends within society for improved efficiency, effectiveness and economy
- Rapid changes expected from information and communications technologies in respect of design, management and manufacturing processes and practices
- The need to refocus these industries in high technology
- Varying attitudes amongst the wide number of different professions and professional bodies
- Developments occurring in other similar and different industries and the need to catch up
- A desire to reduce the adversarial nature associated with the development and construction of buildings
- The awareness of improved quality assurance mechanisms in other industries, especially in other product related industries
- A desire to establish best work practices
- The over-riding wish of clients for single point responsibility when developing building projects
- Changes in culture and work practices throughout society.

Private Finance Initiative

A Public Private Partnership (PPP) refers to any alliance between public bodies, local authorities or central government, and private companies to deliver a public project or service. The Private Finance Initiative (PFI) is a more formal approach to PPP that has been adopted in the UK.

In PFI, the public sector contracts to purchase quality services on a long-term basis so as to take advantage of private sector management skills incentivised by having private finance at risk. The private sector partner takes on responsibility for providing a public service, including maintaining, enhancing or constructing the necessary infrastructure and the public sector specifies a level of service in return for an annual payment, called a unitary charge. In choosing whether or not to use PFI the Government ensures:

- the choice of procurement route is based on an objective assessment of value for money;
- there is no bias between procurement options;
- value for money does not come at the expense of employee terms and conditions.

The PFI unitary charge is an annual payment made throughout the lifetime of the contract. This covers the cost of capital expenditure, private finance and the services needed to operate and maintain that asset. The total of all such PFI payments currently is less than 2% of the total annual resource budgets across Government departments. In 2005–6 this represented £6 billion from a budget of £304 billion. Its payment is conditional on the private sector reaching the required service levels set by the public sector. Payment only commences when the asset has been satisfactorily completed. Around 50% of PFI projects by capital value are reported on departmental balance sheets. The accounting treatment of a PFI project follows rules set and audited by a series of independent national and international organisations.

Partnerships UK published a report into operational PFI/PPP projects in 2006. The report, which commented on the largest survey of PFI projects ever undertaken, contains a comprehensive review of the performance of PFI projects during their operational phase. The findings show that public sector managers and users are content with their PFI/PPP projects. Specifically:

- 96% of projects are performing at least satisfactorily
- 66% of projects performing either to a very good or good standard
- 89% of projects are achieving the contract service levels either always or almost always
- 80% of all users of PFI projects are always or almost always satisfied with the service being provided

Public sector managers believe that they have developed an effective partnership with the private sector to deliver services. Over 97% believe that their relationship with their private sector partners is satisfactory or better and around 80% of public sector managers agree that the payment mechanism supports the effective contract management of the project.

PFI's record of delivery means that the Government is committed to using PFI as a procurement option wherever it is value for money to do so to deliver future infrastructure investment. In 2007, £26 billion of PFI investment across 200 projects is currently being proposed. This is one of the largest programmes worldwide. It includes the provision of over 60 new health facilities and 104 new schools. Figures 13.1 and 13.2 show the number of projects and the amount invested through PFI projects since 1992.

PFI is a small but important part of the UK Government's strategy for delivering high quality public services. In assessing where PFI is appropriate, the Government's approach is based on its commitment to efficiency, equity and accountability and on the principles of public sector reform. PFI is only used where it can meet these requirements and deliver clear value for money without sacrificing the terms and conditions of staff. Where these conditions are met, PFI delivers a number of important benefits. It requires the private sector to invest its own capital at risk and to deliver defined levels of service to the public over the long term. PFI also helps to deliver high quality public services and ensure that public assets are delivered on time and to budget.

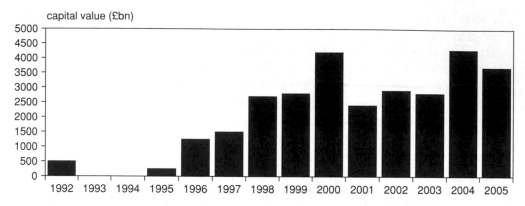

Figure 13.1   Value of PFI projects since 1992 (Source: PPP Forum)

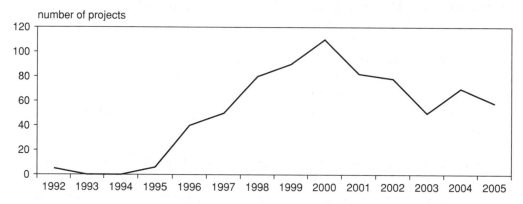

Figure 13.2   Number of PFI projects since 1992 (Source: PPP Forum)

PFI is used to fund major new public building projects, including hospitals, schools, prisons and roads. The Conservative Party first introduced the scheme in the early 1990s. Under New Labour, PFI has become a major, and sometimes controversial element of private sector involvement in Britain's public services. Since 1997, about £50bn worth of public–private deals have been signed, despite opposition from Labour Backbenchers and public sector unions. The Chancellor announced a £26bn expansion of PFI contracts across 20 public sector projects in the 2006 budget.

Private consortiums, usually involving large construction firms, raise the capital finance to design and build a public sector project. They are also contracted to maintain the buildings while a public authority, such as a council or NHS trust, uses them. This means the private sector is responsible for providing cleaning, catering and security services. Once construction is complete, the public authority begins to pay back the private consortium for the cost of the buildings and their maintenance, plus interest. The contracts typically last for 30 years, after which time the ownership of the buildings pass to the public authority.

It is likely that expanded capital building programmes to replace out-of-date schools, hospitals and housing could not be achieved over the same period of time if the funding was being provided only by the public purse. In the short term, PFI allows the government to keep its promise of improving public services without actually raising income tax to pay for it.

According to healthcare think-tank the Kings Fund, the physical condition of most hospitals is now vastly improved. By 2004, 30 major PFI schemes, with a capital value of around £3.36bn, had been opened or were being built in the NHS. Only four publicly funded schemes of a similar scale were completed or had begun construction during the same period.

Some critics suggest that PFI contracts are poor value for money and commit the public sector to expensive financial responsibilities for decades to come. Some NHS trusts have also found it too expensive to pay the annual charges to PFI contractors for building and servicing new hospitals. The Audit Commission warned that the deficit at Queen Elizabeth Hospital NHS trust in Woolwich, south-east London, would accumulate to £100m by 2008–09 unless the government restructured its PFI debt. The deal added about £9m a year to the costs met by an equivalent hospital built with money borrowed from the public purse. Unions, such as Unison, claim that PFI leads to poorer services because private companies maintain the buildings as cheaply as possible. The pay and conditions of cleaners, catering and security staff in PFI buildings are typically inferior than their counterparts in the public sector. A more recent Treasury report has suggested that PFI deals failed to provide value for money on the provision of many of these services. Future deals would be unlikely to automatically bundle cleaning, catering and security as part of new contracts. Some consortiums appear to have benefited considerably from PFI deals. Some have made excessive profits. Others have been plunged into financial crisis.

PFI has been further criticised as being too inflexible because it ties public services into 20- to 30-year deals despite the fact that it is difficult to plan how these services will be delivered even a few years ahead. For example, the government's latest health and social care White Paper has proposed moving 5% of the work of large general hospitals into the community. This shift in resources calls into question the plans for PFI investment into building large hospitals. The government's design watchdog has also attacked the cheap appearance of many PFI hospitals and schools.

Benchmarking

Benchmarking has been described as a method for organisational improvement that involves continuous, systematic evaluation of products, services and processes of organisations that are recognised as representing best practices. It is a system that uses objective comparisons of both processes and products. It may make internal comparisons within a single firm, by, for example, comparing the performances of different building sites. It can be an external system, where comparisons of performance are made against similar and dissimilar firms. The enterprises that are internationally recognised as being the world leaders are described as having the best practices and are the most efficient. The Construction Industry Board in the UK has published a list of key performance indicators that are updated and can be used by any firm to measure its performance (Ashworth and Hogg, 2007).

There is considerable interest in and practice of benchmarking techniques across a wide range of industries. It has been suggested that probably all of the top firms and companies use it in one way or another. It is also worth remembering that benchmarking is not just practised by those firms that are lagging behind and need to improve their performance, but by the international industry leaders, as a tool for maintaining their competitive edge. The focus of benchmarking is on the need for continuous improvement.

Origins

The technique of benchmarking originated in the USA and is now successfully applied in many industry sectors around the world. It owes much of its relatively recent development to the success achieved by the Xerox Corporation. In the late 1970s, Rank Xerox was severely challenged by Japanese competitors who were able to manufacture photocopiers of a better quality and at a lower cost. In response, Xerox successfully implemented a programme of benchmarking which incorporated the comprehensive examination of the organisation throughout. Better practice, in any relevant process used by another company, was identified and translated into new Xerox practice. This facilitated a significant improvement in company performance and an improvement in their competitive position (Pickrell *et al.*, 1997).

Practice

Benchmarking can be applied across the business sector and throughout the entire hierarchy of a company from strategic to operational levels. The aim of benchmarking is to improve the performance of an organisation by two methods:

- Identifying best-known practices relevant to the fulfilment of a company's mission. These may be found within or outside the company.
- Utilising the information obtained from an analysis of best practice to design and expedite a programme of changes to improve company performance. It is important to stress that benchmarking is not about blindly copying the processes of another company, since in many situations this would fail. It requires an analysis of why performance is better elsewhere, and the translation of the resultant information into an action suited to the company under consideration.

In theory, therefore, benchmarking is a very simple process involving a clear pattern of thought:

- Who does it better?
- How do they do it better?
- How can the company benefit from this knowledge?
  - Adapt/adopt the better practices
  - Improve performance.

Despite the acknowledged benefits of benchmarking in assessing company performance and prompting its improvement, its use in practice within the UK appears to be uncommon. Research by the CBI in the late 1990s concluded that benchmarking was weak overall in industry in general and that there was a need to raise awareness and practice of benchmarking across British industry. Where benchmarking practice occurs, it appears in the main to be restricted to large companies and in many instances is not carried out adequately. The level of use of benchmarking within the construction industry is consistent with this view. Some large construction organisations and their clients have begun starting to use the tool successfully; however, the extent of use is very limited and in many companies is in its infancy. The importance of benchmarking to the construction industry has recently been acclaimed at a very high level. The report *Rethinking Construction* of the Construction Task Force (Egan, 1998) to the Deputy Prime Minister, John Prescott, stated that 'Benchmarking is a management tool which can help construction firms to understand how their performance measures up to their competitors' and drive improvement up to world class standards.'

The Egan Report makes clear the underperformance of the construction industry and the need for fairly dramatic improvement, much of which is considered to depend upon successful benchmarking practice. There is a poor rate of implementation of new management practices, within both construction and property. Benchmarking practice in the UK is seen to fall into the category of fashionable but little used activities.

However, business survival may be dependent upon benchmarking and therefore, its wide acceptance may become a natural consequence. It is likely that the use of benchmarking practice will accelerate and as companies improve by it, the need for other organisations to follow will be apparent.

The need for benchmarking

The purpose of benchmarking is aimed at improving a company's performance with the aim of maintaining or achieving a competitive edge. The UK construction and property industries are witnessing rapid change in many areas, some of which are of great significance. These include, for example, methods of procurement, client expectations, client and contractor relationships, professional appointments, the Private Finance

Initiative, government focus, warranties and guarantees, payment provisions and safety regulations. Consideration of this incomplete list should otherwise convince anyone who believes that the construction industry stands still. In addition, there is an increasing level of competition from Europe and elsewhere, including the USA and Japan, where benchmarking practice is now part of the culture. There are good incentives for the use of benchmarking. The advantages that it may bring include:

- An opportunity to avoid complacency which may be nurtured by monopolistic situations resulting in the loss of market leadership
- An improvement in the efficiency of a company leading to an increase in profits
- An improvement in client satisfaction levels resulting in an improved reputation, leading to more work, better work and an increase in profits
- An improvement in employee satisfaction in working in a more efficient, profitable and rewarding environment.

Many organisations that have successfully used benchmarking in order to achieve a specific objective continue to use it thereafter as part of a total quality management programme. In this way it can contribute to the long-term objectives of the company which primarily may be to achieve a high level of customer satisfaction, recognised excellence in their sector and thereby a high return.

Best value

Best value (DETR, 2001) is a concept that has come out of the Local Government Act (1999) where it sets out the requirements that are expected. The key phrase in the Act is 'A Best Value authority must make arrangements to secure continuous improvement in the way in which its functions are exercised, having regard to a combination of economy, efficiency and effectiveness.' The so-called 3Es concept has been in existence for some time. The concept of best value applies equally as well in the private as well as the public sector. None of us wants to possess the cheaper things in life, but to possess the expensive things that cost less. Best value aims to achieve a cost-effective service, ensuring competitiveness and keeping up with the best that others have to offer. It embraces a cyclical review process with regular monitoring as an essential part of its ethos.

Best value extends the concepts of value money that have been identified for a long time within both construction and property. Egan (1998), for example, defines value in terms of zero defects, delivery on time, to budget and with a maximum elimination of waste. In order to show that best value and added value are being achieved, it becomes essential to benchmark performance including costs. It is also necessary to benchmark the overall cost of the scheme so that improved performance in the design can be assessed against its cost. The sharing of information underpins the whole best practice process. Even the leaders in an industry need to benchmark against their competitors in order to maintain that leading edge. Whilst the aspiration of best value is both admirable and essential, its demonstration in practice presents the challenge.

The best value concept for local authorities is being managed for the Department for Local Government, Transport and the Regions by the Audit Commission. It is a challenging performance framework placed on local authorities by central government and since April 2000, all local authorities in England and Wales have a duty to plan to provide their services under the principles of best value. Each service review must show that the local authorities have applied the four Cs of best value to the service, and show that it is:

- *Challenging*: why and how the service is provided
- *Comparing*: the performance with others, including non-local government providers
- *Competing*: the authority must show that it has embraced the principles of fair competition in deciding who should deliver the service
- *Consulting*: local service users and residents on their expectations about the service.

Local authorities are required to show that they are continuously improving the way in which their services are delivered.

The Audit Commission has set up a new inspection service to guide the work of best value. In common with other forms of inspection services and benchmarking, best value seeks out best practices and uses this to help all local authorities to improve their general levels of performance. Best practices today are unlikely to be best practices tomorrow, since the achievement of improvements in quality is always a journey and never a destination. The enhancement of quality remains the long-term goal. Best value inspectors use a simple framework of six questions to make sure that they collect the right information and evidence to support their judgements.

- Are the authority's aims clear and challenging?
- Does the service meet its aims?
- How does its performance compare?
- Does the best value review (BVR) drive improvement?
- How good is the improvement plan?
- Will the authority deliver the improvements?

Best value is a concept that is important not only to the public sector but also within private sector organisations, however it is achieved.

Partnering

Many of the problems that exist in building developments are attributed to the barriers that exist between individual firms and organisations, clients, contractors and subcontractors.

In essence, partnering is about breaking these barriers down by establishing more positive working relationships that are based on:

- Mutual objectives of teamwork
- Trust
- Sharing in risks and rewards.

In the UK construction industry, partnering activity is a relatively recent phenomenon
being given significant impetus by the report *Constructing the Team* (Latham, 1994) and
many subsequent publications and positive action. This highlights the benefits of partnering as one factor that is critical to the success of a construction project in providing
the basis for improving added value in construction (Ashworth and Hogg, 2007). The
report stated that 'an attitude of cooperation amongst the project team members and,
not least, with the client must be created. Partnering can help enormously without
running the risk of being uncompetitive or introducing complacency'. The relationships and implications of partnering can successfully be applied throughout the supply
chain of the different subcontractors, suppliers and manufacturers involved in a project.

Partnering also featured prominently in the report *Rethinking Construction* (Egan,
1998). This regarded partnering as central to improving the performance of the construction industry. The prominent recognition given to partnering towards the end of
the twentieth century is a clear indication that partnering can make major contributions
in improving or adding value within the construction industry.

What is partnering?

Partnering is reliant upon the principle that cooperation is a more efficient method of
working than confrontation. This latter approach is a product of traditional single stage
selective tendering in which each party is driven toward looking after their own independent aims and objectives. These principles are simply and well described by Bennett
and Jayes (1998) who stated that 'Firms are better off when they work to make the cake
bigger than when they fight to get a bigger share of the existing cake. This is true as long
as they make sure that everyone gets a fair share of the cake.' It is clear from the above
that the intentions of partnering are fairness and trust. Traditional methods of client
– contractor – subcontractor relationships expend a great deal of energy on retaining
as much of the cake as possible and little to prompt efficiency and improvement.
Partnering aims to remedy this.

There is no universally accepted version of partnering and the range of definitions
clearly demonstrates this. The following provides a relatively straightforward view:

Partnering is a management approach used by two or more organisations to achieve
specific business objectives by maximising the effectiveness of each participant's

resources. The approach is based on mutual objectives, an agreed method of problem resolution and an active search for continuous measurable improvements. (Bennett and Jayes, 1995)

From this definition three main components to a partnering arrangement can be identified.

(1) *Mutual objectives*: these should be accepted by all partners at commencement of the project and reviewed only if necessary during the project duration. It is common to incorporate these within a *charter* that may be publicly displayed to continually remind parties of their obligations.

(2) *An agreed method of problem resolution*: a better way of resolving disputes should be established based upon fairness and the willingness to find win – win solutions. These should be outside the usual contract methods.

(3) *An active search for continuous and measurable improvements*: this is particularly relevant to strategic partnering arrangements where continuous improvement is considered as a major benefit. Benchmarking is a method of monitoring this, using both internal and external sources for comparison. This principle fits easily within the aim of adding value in construction of projects.

The attributes and concerns of partnering

The adoption of partnering at a strategic level or for a specific project is considered to bring major improvements to the construction process resulting in significant benefits to each partner. Ashworth and Hogg (2007) have identified a number of attributes and concerns of partnering, as follows.

Attributes

- Reduction in disputes
- Reduction in time and expense in the settlement of disputes
- Reduction in costs
- Improvements in quality and safety
- Improvements in design and construction times and certainty of completion
- More stable workloads and income
- A better working environment.

Concerns

It is also necessary to acknowledge the existence of some important disadvantages to the use of partnering and other concerns regarding its use:

- Increased initial costs
- Complacency
- Single source employment
- Confidentiality
- The perceived need for competition
- Partnering through the supply chain
- Legal issues, which remain as yet unknown.

Conclusion

The benefits attributed to partnering and the level of promotion it has been given and continues to be given at high levels suggest it should have an important role to play in the future of any construction industry. However, it should also be recognised that significant negative opinion on the practice of partnering exists. Recent surveys indicate that nearly two-thirds of construction clients have never entered into a formal partnering arrangement with a contractor. Also, the prevailing culture that underlies the construction industry contains strong elements of mistrust, cynicism and a general resistance to change. These cannot be easily overturned. Market realities should not be ignored, but the industry appears to await new ideas that show only advantages and no disadvantages. This is not reality.

Value management

Value management is a style of management that is particularly dedicated to motivating people, developing skills and promoting synergies and innovation. Its aim is to maximise the overall performance of an organisation. It has evolved out of previous methods based on the concept of value and functional approach. These were pioneered by Lawrence D. Miles in the 1940s who developed the technique of *value analysis* as a method of improving value in existing products. Initially value analysis was used principally to identify and eliminate unnecessary costs. However, it is equally effective in increasing performance and addressing resources other than cost. Since its introduction, its application has widened beyond products into services, projects and administrative procedures. It was introduced into the construction industry a little over thirty years ago. The value management approach involves three root principles:

- a continuous awareness of value for the organisation, establishing measures or estimates of value, monitoring and controlling them;
- a focus on the objectives and targets before seeking solutions;
- a focus on function, providing the key to maximize innovative and practical outcomes.

The concept of value relies on the relationship between the satisfaction of many differing needs and the resources used. The fewer the resources used or the greater the satisfaction of needs, the greater the value. Stakeholders, internal and external customers may all hold differing views of what represents value. The aim of value management is to reconcile these differences and enable an organisation to achieve the greatest progress towards its stated goals with the use of minimum resources.

It is important to realise that value may be improved by increasing the satisfaction of need even if the resources used in doing so increase, provided that the satisfaction of need increases more than the increase in use of resources.

The key principles

Value management is distinct from other management approaches in that it simultaneously includes attributes which are not normally found together. It brings together within a single management system:

Management style

- an emphasis on teamwork and communication
- a focus on what things do, rather than what they are, i.e. the functional approach
- the encouragement of creativity and innovation
- a focus on the client's requirements
- a requirement to evaluate options qualitatively to enable the robust comparisons of options to be made.

Positive human dynamics

- teamwork that encourages people to work together towards a common solution
- personal satisfaction by recognising and giving credit
- communication through bringing people together by improving the lines of communication
- fostering a better common understanding and providing better group decision support
- encouraging change by challenging the status quo and introducing beneficial change
- ownership and outcomes of value management activities by those responsible for implementing them.

Consideration of external and internal environment

- external conditions by taking account of existing external conditions over which managers may have little influence

- internal conditions within an organisation which managers may or may not be able to influence.

The benefits of value management

These include:

- better business decisions by providing decision makers a sound basis for their choice
- improved products and services by clearly understanding and prioritising their needs
- enhanced competitiveness through technical and organisational innovation
- a common value culture within an organisation
- a common knowledge of the main success factors within an organisation
- the development of multidisciplinary and multitask teamwork
- decisions which can be supported by the stakeholders.

Lean construction

The lean construction process is a derivative of the lean manufacturing process. This is a concept that has been popularised since the early 1980s in the manufacturing sector. The original thinking was developed from Japan, although it is now being considered and introduced worldwide. It is concerned with the elimination of waste activities and processes that create no added value. It is about doing more for less.

Lean production is the generic version of the Toyota production system, which is recognised as the most efficient production system in the world today. Incidentally Toyota's activities in the construction industry are larger than those of its better known automobile business. It should be acknowledged at the outset that construction production is different from that of making automobiles, although it is possible to learn from and adapt successful methodologies from other industries. In the automobile manufacturing industry, spectacular advances in productivity, quality and cost reduction have been achieved since the late 1980s. Construction, by comparison, has not yet made these advances. It also remains the most fragmented of all industries, but this can be seen both as a strength and a weakness.

The application of lean production techniques in automobile manufacturing has been a huge success and is associated with three important factors:

- The simplification of manufacturing dies
- The development of long-term relationships with a small number of suppliers, to allow just-in-time management
- Changes in work practices, i.e. the culture and ethos of practices, most notably the introduction of team working and quality circles.

Lean thinking is aimed at delivering what clients want, on time and with zero defects. Lean construction has identified poor design information which results m a large amount of redesign work. Several organisations around the world have established themselves as centres for lean construction development. The aim, for example, of the Lean Construction Institute in the USA is a dedication towards eliminating waste and increasing value.

Few products or services are provided by a single organisation alone. The elimination of waste therefore has to be pursued throughout the whole value stream, including all who make any contribution to the process. Removing wasted effort represents the biggest opportunity for performance improvement.

Several companies around the world are attempting to introduce lean construction methods into their core businesses. The 1998 Egan Report provides two examples. One of the firms is based in Colorado and another in San Francisco. One of these has already reduced project times and costs by 30% through developments such as:

- Improving the flow of work on site
- Using dedicated design teams
- Innovation in design and assembly
- Supporting subcontractors in developing tools for improving processes.

This suggests that perhaps the most useful way of achieving cost reductions, while still maintaining value, is to consider profitable ways of reducing the time spent on construction work on site. Design readiness is the same principle. This suggests that to complete the design fully prior to starting work on site will save both construction time and the respective costs that are involved.

Lean construction is a philosophy that is about managing and improving the construction process to deliver profitably what the customer, the construction client, requires. Engineering, of its different kinds, has had to develop a wide range of strategies to remain at the competitive edge and to improve its products. Comparisons have been made on several occasions between the high technology engineering approach and the low technology that is adopted generally throughout the construction industry sector.

Because it is a philosophy, lean construction can be pursued through a number of different approaches. The lean principles have been identified as follows:

- The elimination of all kinds of waste. This includes not just the waste of materials on site, but all aspects, functions or activities that do not add value to the project.
- Specify value precisely from the perspective of the ultimate customer.
- Clearly identify the process that delivers what the customer values. This is sometimes referred to as the value stream.
- Eliminate all non-added value steps or stages in the process.
- Make the remaining added value steps flow without interruption, through managing the interfaces between the different steps.

- Let the customer pull: do not make anything until it is needed, then make it quickly. Adopt the philosophy of just-in-time management to reduce stockpiles and storage costs.
- Pursue perfection through continuous improvement.

The vision of lean construction stretches across many traditional boundaries. It challenges our current practices. It has required changes in work practices and in understanding new roles and different responsibilities. It has required a change in attitudes and culture. It recognises that technology has an important role to play today. It expects that those involved in a construction project will share a common purpose. It is setting new standards that can be measured to indicate improvement. It is a vision for the future.

The response to lean construction from the construction industry is not unanimous. Like many new ideas, when put into practice, some of them do not always live up to the theoretical expectations. But lean construction is not just a theory since its techniques have been successfully applied in other industries. There has also to be the desire for success and a belief system is therefore very important. Lean construction is not a phenomenon that is likely to disappear, since there is a groundswell of opinion and worldwide interest in its principles.

Facilities management

The development of building projects in the mid twentieth century was restricted to concepts that largely related to that of the initial building design alone. The role of the building surveyor and the systematic reuse of data and other information from practice were unknown. The procedures employed were applied to the actual building works and building contract alone, without any future consideration being given towards the long-term use of the project, its costs-in-use, repairs and future adaptations for which a client may be subsequently involved. The importance of total costs and how initial construction costs might be better considered within the principles of overall total cost appraisal were not considered. The consideration of future costs or costs-in-use was an idea that was not yet even being considered by the industry. During the high levels of construction activity during the 1960s, it was being recognised that existing procedures were inadequate in meeting the modern day needs of clients and their buildings.

Clients' expectations from their buildings, in line with expectations for other goods and services, were also increasing. Approximate estimating was frequently too approximate and some schemes had to be redesigned after tenders had been received. Hence the concept of cost planning was introduced. This is briefly considered in Chapter 9. Much later, life cycle costing, now referred to as whole life costing, which considered the total costs from inception through to demolition, helped to change and inform the philosophy of initial design solutions. During the design process it became apparent that to ignore future repair and maintenance problems was flawed. But even this philosophy was too restricted for the twenty-first century. The total concept of building design evaluation had to extend beyond those items that were traditionally thought of as building works.

As long ago as the beginning of the twentieth century it was recognised that spatial design was perhaps the most important single variable influencing the costs of buildings. This acknowledgement was the basis of the much used superficial floor area method for calculating the approximate initial costs of buildings. It was also recognised that buildings providing the same function and for the same numbers of occupants often resulted in different costs for both their construction and future maintenance. Procedures were therefore introduced to rationalise building design by setting parameters that identified just how much space should be allowed. The application of cost limits sought to reinforce these ideas.

However, the introduction of facilities management (FM) goes much further than the above important ideas alone. They were important in their time and still are as part of a fully integrated package. Facilities management considers not just the initial and future costs associated with the building, but the entire costs that are incidental in a client's business. Its aim is to manage the total facilities provided, in addition to those costs that are more traditionally related to buildings and building projects. The FM procedure includes examining the amenities of the business to see if they can be undertaken more effectively, efficiently and economically. This information is often set within a context of national patterns and trends, as well as in the particular organisation concerned. Benchmarking techniques can be applied in this respect, with consequent early days advantage.

Overview of facilities management

If you invite a group of individuals to describe the meaning of facilities management, you will result in as many definitions as there are members of the group. The many and varied functions involved in facilities management are not new but the way that these are managed or organised have become more integrated and extended. The role of a facilities manager, for example, is recognised as being more proactive and different to that of the traditional estates manager. Facilities management also differs from property management in that the range of activities that are carried out far exceeds those of just buildings and property.

Facilities management is concerned not just with the building structure, services and finishings but with the activities that go on within the building. It is much more than a combination of the traditional disciplines of estate management and building maintenance. It is also more than just the sum of the parts. It includes a range of activities that can be usefully summarised under the seven headings as shown in Figure 13.3 and Table 13.1. This table indicates the extent to which these services are carried out in-house or contracted out to external consultants.

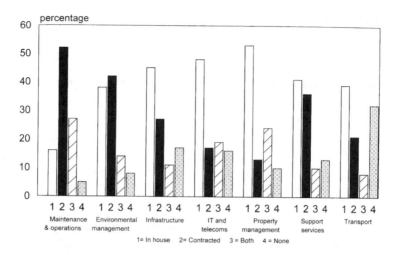

Figure 13.3   How FM services are procured by sector (percentage allocation) (Source: *Premises and Facilities Management Journal*)

This division of activities will vary depending upon the type, nature and size of organisation concerned. The headings can be further subdivided into the key facilities management services as shown in Table 13.1. These can also include pest control, porterage, laundry, environmental testing, furniture management, etc.

Table 13.1   Key facilities management services

Building maintenance and operations	Electrical
	Mechanical
	Fabric
Environmental management	Energy management
	Health and safety
	Waste management
Infrastructure	Utilities
IT & telecommunications	Telecommunications services
	Systems administration and management
	Customer response support
Property management	Asset management
	Space planning
Support services	Catering and vending
	Cleaning
	Office support services
	Security
Transport	Fleet management

(Source: *Premises and Facilities Management Journal*)

The professional role of facilities managers has been increasingly recognised since the 1990s. The emphasis has been on improving value for money in the provision of traditional facilities services. The use of external consultants for this purpose has largely been the result of their promise to reduce costs while at the same time helping to maintain or improving quality. This has been made possible through the wider experience gained by consultants in managing different clients and by having a clearer focus on their work, not being distracted by in-house politics. Consultants are better able to disseminate good practices because of their contacts with a range of different clients. External consultants also claim to remove some of the trouble and inconvenience from the client. The use of external consultants is partially threatened by the demands of accelerating organisational change. New work practices such as hot desking, hotelling and remote working call for changes in the way that the workplace is being used.

Facilities management has grown out of a range of different disciplines, most of which came from general management. About one-half of practising facilities managers are members of the British Institute of Facilities Management (BIFM). As many as 65% hold other professional qualifications but fewer than 10% have any academic qualifications in facilities management. Qualifications and skills development are therefore two important priorities for facilities managers. The BIFM's national professional qualification and training programme is being used by major public and private sector organisations.

Conclusion

The construction and property industries will continue to evolve and change for at least the rest of our lifetimes. New ideas will continue to emerge in seeking out improvements and efficiencies. The organisation and arrangement of these industries will continue, fuelled by ever more sophisticated improvements in information technology and the more extensive exploitation of the internet. The above headings are some of the more important examples that will continue to influence and shape the development of building projects for the first few decades of the twenty-first century.

Bibliography

Arrowsmith, S. (2000) *Public Private Partnerships and PFI*. Sweet and Maxwell, London.

Ashworth, A. (1980) The source, nature and comparison of published cost information. *Building Technology and Management*, March.

Ashworth, A. (1996) Estimating the life expectancies of building components in life cycle costing calculations. *Journal of Structural Survey*, **14** (2), 4–8.

Ashworth, A. (1997) *Obsolescence in Buildings: Data for Life Cycle Costing*. Construction Papers No. 74. Chartered Institute of Building, Ascot.

Ashworth, A. (2004) *Cost Studies of Buildings*. Pearson Prentice Hall, Harlow.

Ashworth, A. (2006) *Contractual Procedures in the Construction Industry*. Pearson Prentice Hall, Harlow.

Ashworth, A. and Hogg, K. I. (2000) *Added Value in Design and Construction*. Longman, Harlow.

Ashworth, A. and Hogg, K. I. (2007) *Willis's Practice and Procedure for the Quantity Surveyor*, 12th edn. Blackwell Publishing, Oxford.

Ashworth, A. and Skitmore, R. M. (1982) *Accuracy in Estimating*. Chartered Institute of Building, Ascot.

Atkin, B. and Flanagan, R. (1995) *Improving Value for Money in Construction*. RICS Books, London.

Badcoe, P. and Arrowsmith, S. (1999) *Public Private Partnerships*. Sweet and Maxwell, Andover.

Bailey, A. (1991) *How to be a Property Ddeveloper*. Mercury Business Books, London.

Balchin, P. N., Kieve, J. L. and Bull, G. H. (1995) *Urban Land Economics and Public Policy*. Macmillan, London.

Bennett, J. (1981) *Cost Planning and Computers*. University of Reading, Reading, and Property Services Agency.

Bennett, J. and Jayes, S. (1995) *Trusting the Team*. Centre for Strategic Studies in Construction, University of Reading, Reading.

Bennett, J. and Jayes, S. (1998) *The Seven Pillars of Partnering*. Thomas Telford Publishing, Reading.

Best, R. and De Valence, G. (1999) *Building in Value Pre-design Issues*. Arnold, Sydney.

Best, R. and De Valence, G. (2002) *Design and Construction: Building in Value*. Butterworth-Heinemann, Oxford.

Betts, M. (1999) *Strategic Management of IT in Construction*. Blackwell Science, Oxford.

354

BMI (2000) *Review of Occupancy Costs: BMI Special Report 290*. Building Maintenance Information, London.

Booker, A. (1994) *The European Property Industry*. College of Estate Management, Reading.

Bowcock, P. and Bayfield, N. (2000) *Excel for Surveyors*. Estates Gazette, London.

Brett, M. (1997) *Property and Money*. Estates Gazette, London.

Brook, M. (2004) *Estimating and Tendering for Construction Work*. Butterworth Heinemann, Oxford.

Buchan, R. D., Fleming, F. W. and Grant, F. E. K. (2003) *Estimating for Builders and Quantity Surveyors*. Butterworth Heinemann, Oxford.

Carroll, B. and Turpin, T. (2002) *Environmental Assessment Guide*. Thomas Telford Publishing, London.

Carsberg, B. *et al.* (2002) *Property Valuation: The Carsberg Report*. RICS, London.

CDM Consultants (2000) *Commercial Buildings*. CDM Consultants, London.

Chartered Institute of Building (2008) *CIOB Code of Estimating Practice*. Blackwell Publishing, Oxford.

Cole, R. (2001) *Property Finance*. Butterworth Tolley, London.

Confederation of British Industry (1997) *Benchmarking the Supply Chain*. Partnership Sourcing Ltd.

Construction Industry Board (1996) *Towards a 30% Productivity Improvement in Construction*. Thomas Telford Publishing, London.

Construction Industry Council (1999) *How to Rethink Construction: Practical Steps in Improving Quality and Efficiency*. Construction Industry Council, London.

Crosby, N., Murdoch, J., Hughes, C. and Mubanga, Y. (2004) *The Status of Property Valuations*. ESRC and the University of Reading.

Darlow, C. (1988) *Valuation and Development Appraisal*. Estates Gazette, London.

Davidson, A. W. (ed.) (2002) *Parry's Valuation and Conversion Tables*. Estates Gazette, London.

Davies, H. and Bourke, K. (1997) *Factors Affecting Service Life Predictions of Buildings: A Discussion Paper*. Building Research Establishment/CRC, Garston and London.

Department of the Environment (1989) *Environmental Assessment: A Guide to the Procedures*. HMSO, London.

Department of the Environment (1995) *The Green Belts*, Planning Policy Guidance Note 2. HMSO, London.

Department of the Environment, Transport and the Regions (1995) *Second Interim Evaluation of Enterprise Zones*. HMSO, London.

Department of the Environment, Transport and the Regions (2001) *Best Value and Procurement: The Handling of Workforce Matters in Contracting*, Circular 02/01. DETR, London.

DTZ (2006) *Money into Property*. DTZ Research, London.

Duffey, F. and Henney, E. (1989) *The Changing City, London*. The Bulstrode Press, London.

Eccles, T., Sayce, S. and Smith, J. (1999) *Property and Construction Economics*. International Thomson Business Press, London.

Edelman, D. B. (1986) *Statistics for Property People*. Estates Gazette, London.

Egan, J. (1998) *Rethinking Construction*. Department of the Environment, Transport and the Regions, London.

Emmitt, S. and Gorse, C. (2003) *Construction Communication*. Blackwell Publishing. Oxford.

Enever, N. and Isaac, D. (1995) *The Valuation of Property Investments*. Estates Gazette, London.

Erdman, E. (1992) *Property*. Mercury Business Books, London.

Esher, L. (1981) *A Broken Wave: The Rebuilding of England 1940–1980*. Allen Lane, London.

Fairclough, J. (2002) *Rethinking Construction Innovation and Research*. Department of Trade and Industry.

Felce, J. and Williamson, A. (2006) *Where to Find Value in the UK Property Market*. Schroders, London.

Flanagan, R. and Jewell, C. (2004) *Whole Life Appraisal*. Blackwell Publishing, Oxford.

Flanagan, R. and Norman, G. (1993) *Risk Management in Construction*. Blackwell Science, Oxford.

Fletcher, B. (1996) *A History of Architecture*. Architectural Press, Oxford.

Franks, J. (1998) *Building Procurement Systems*. Chartered Institute of Building, Ascot.

Fraser, W. D. (1993) *Principles of Property Investment and Pricing*. Macmillan, London.

Gann, D. (2000) *Building Innovation: Complex Constructs in a Changing World*. Thomas Telford Publishing, London.

Gilbert, B. and Yates, A. (1989) *Appraisal of Capital Investment in Property*. Surveyor's Publications, London.

Goldstone, A. (2000) *Freeman's Guide to the Property Industry*. Freeman Publishing, London.

Goobet, A. R. (1992) *Bricks and Mortals*. Century Business Books, London.

Gray, C. (1996) *Value for Money: Helping the UK Afford the Buildings It Likes*. Reading Construction Forum, London.

Green, S. (1986) *Who Owns London?* Weidenfeld and Nicolson, London.

Hackett, M. and Robinson, I. (2002) *Precontract Practice and Construction Administration for the Building Team*. Blackwell Science, Oxford.

Harvey, J. (1981) *The Economics of Real Property*. Macmillan, London.

Harvey, J. (2000) *Urban Land Economics*. Macmillan, Basingstoke.

Harvey, R. C. and Ashworth, A. (1997) *The Construction Industry of Great Britain*. Butterworth, Oxford.

Hillebrandt, P. (1984) *Analysis of the British Construction Industry*. Macmillan, London.

Hillebrandt, P. (2000) *Economic Theory and the Construction Industry*. Macmillan, London.

HMSO (various) *Housing and Construction Statistics*. The Stationery Office, London.

HMSO (various) *Annual Abstract of Statistics*. The Stationery Office, London.

Hudson, K. (1978) DHSS expenditure forecasting method. *Chartered Surveyor,* Building and Quantity Surveying Quarterly.

Investment Property Forum (2004) *Calculation of Worth*. RICS, Coventry.

Investment Property Forum (2007) *Understanding Commercial Property Investment: A Guide for Financial Advisers*. Investment Property Forum, London.

Johnson, T., Davies, K. and Shapiro, E. (2000) *Modern Methods of Valuation of Land, Houses and Buildings*. Estates Gazette, London.

Jones Lang LaSalle (2004) *The Glossary of Property Terms*. Estates Gazette, London.

Kelly, J., Macpherson, S. and Male, S. (1992) *The Briefing Process: A Review and Critique*. Royal Institution of Chartered Surveyors, London.

Kelly, J., Morledge, R. and Wilkinson, S. (2002) *Best Value in Construction*. Blackwell Science, Oxford.

Langford, D. and Male, S. (2001) *Strategic Management of Construction*. Blackwell Science, Oxford.

Latham, M. (1994) *Constructing the Team*. HMSO, London.

Littler, J., Cook, J. and Shariff, Y. (2001) *Sustainable Environments for the New Millenium*. Spon Press and Routledge, London.

Marriott, O. (1989) *The Property Boom*. Abingdon Publishing, London.

McGeorge, D. and Palmer, A. (2002) *Construction Management: New Directions*. Blackwell Publishing, Oxford.

MacGregor, B. (1996) *Understanding the Property Cycle*. RICS Books, London.

McIntosh, A. P. J. and Sykes, S. G. (1985) *A Guide to Institutional Property Investment*. Macmillan, London.

McWilliam, D. (1994) *Commercial Property and Company Borrowing*. RICS Books, London.

Merchant, K. A. and Van der Stede, W. A. (2003) *Management Control Systems: Performance Measurement, Evaluation and Incentives*. Financial Times, Prentice Hall, Harlow.

Millington, A. F. (2000) *Property Development*. Estates Gazette, London.

Millington, A. F. (2001) *An Introduction to Property Valuation*. Estates Gazette, London.

Ministry of Public Building and Works (1970) *The Decision to Build*. HMSO, London.

Modigliani, F. and Miller, M. (1958) The cost of capital, corporation finance and the theory of investment. *American Economic Review*, **48**, June, 261–297.

Moore, V. (2000) *A Practical Approach to Planning Law*. Blackstone Press, London.

Morton, R. (2002) *The Construction Industry*. Blackwell Science, Oxford.

Myers, D. (1994) *Economics and Property*. Estates Gazette, London.

Nutt, B. and McLennon, P. (2000) *Facility Management Risks and Opportunities*. Blackwell Science, Oxford.

Patel, K. (1999) *Change the Game, Change the Rules of the Game*. KPMG Books, London.

Philip, A. (2006) *Pension Funds and Their Advisers*. AP Information Services, Oxford.

Pickrell, S., Garnett, N. and Baldwin, J. (1997) *Measuring Up: A Practical Guide to Benchmarking in Construction*. Construction Research Communications, London.

Powell, J. (1995) *Towards a New Construction Culture*. Chartered Institute of Building, Ascot.

Property Services Agency (1990) *Building, Function and Location*. HMSO, London.

Raftery, J. (1991) *Principles of Building Economics*. Blackwell Science, Oxford.

Raftery, J. (1994) *Risk Analysis in Project Management*. Chapman and Hall, London.

Royal Academy of Engineering (1996) A Statement on the Construction Industry. RAE, London.

Royal Institution of Chartered Surveyors (1976) *An Introduction to Cost Planning*. RICS, London.

Royal Institution of Chartered Surveyors (1986) *A Guide to Life Cycle Costing for Construction*. RICS, London.

Royal Institution of Chartered Surveyors (1998) *The Surveyor's Construction Handbook*. RICS Books, London.

Royal Institution of Chartered Surveyors (2000) *Forecasting Office Supply and Demand*. RICS Books, London.

Royal Institution of Chartered Surveyors and the Building Research Establishment (1992) *Life Expectancies of Building Components*. RICS Research Paper, No. 11.

Ruddock, L. (1992) *Economics for Construction and Property*. Arnold, London.

Ryan, A. (1987) *Property*. Open University Press, Milton Keynes.

Salway, F. (1996) *Depreciation of Commercial Property*. College of Estate Management, Reading.

Scarrett, D. (1995) *Property Asset Management*. Spon Press, London.

Scott, H. (1992) *VAT and Property*. Butterworth, Oxford.

Skitmore, R. M. and Marsden, D. E. (1988) Which procurement system? Towards a universal procurement selection technique. *Construction Management and Economics*, **6** (2), 71–89.

Smith, N., Merna, T. and Jobling, P. (2006) *Managing Risk in Construction Projects*. Blackwell Publishing, Oxford.

Smith, T. (1992) *Accounting for Growth*. Business Books, London.

Stone, P. A. (1979) *Urban Development in Britain: Standards, Costs and Resources*. Cambridge University Press, Cambridge.

Stone, P. A. (1983) *Building Economy*. Pergamon Press, Oxford.

Then, D. and McGregor, W. (1999) *Facilities Management and the Business of Space*. Arnold, London.

Thomas, K. (1994) *An Introduction to Development Control*. Estates Gazette, London.

Upson, A. (1994) *Successful Property Development*. Blackwell Science, Oxford.

Van Gotzen, R. (2000) *The Property Finance Sourcebook*. Estates Gazette, London.

Walker, A. (1997) *Conserving Value: Making Effective Use of Listed Buildings*. Estates Gazette, London.

Warren, M. (1993) *Economic Analysis for Property and Business*. Butterworth-Heinemann, Oxford.

Westwood, F. (2000) *Achieving Best Practice: Shaping Professionals for Success*. McGraw-Hill, New York.

Wiggins, K. (2000) *Discounted Cash Flow*. College of Estate Management, Reading.

The World Markets Company plc (2006) *UK Pension Fund Service Annual Review*. The WM Service, London.

Index